Advances in Science, Technology & Innovation

IEREK Interdisciplinary Series for Sustainable Development

Advances in Science, Technology & Innovation (ASTI) is a series of peer-reviewed books based on important emerging research that redefines the current disciplinary boundaries in science, technology and innovation (STI) in order to develop integrated concepts for sustainable development. It not only discusses the progress made towards securing more resources, allocating smarter solutions, and rebalancing the relationship between nature and people, but also provides in-depth insights from comprehensive research that addresses the **17 sustainable development goals (SDGs)** as set out by the UN for 2030.

The series draws on the best research papers from various IEREK and other international conferences to promote the creation and development of viable solutions for a **sustainable future and a positive societal** transformation with the help of integrated and innovative science-based approaches. Including interdisciplinary contributions, it presents innovative approaches and highlights how they can best support both economic and sustainable development, through better use of data, more effective institutions, and global, local and individual action, for the welfare of all societies.

The series particularly features conceptual and empirical contributions from various interrelated fields of science, technology and innovation, with an emphasis on digital transformation, that focus on providing practical solutions to **ensure food, water and energy security to achieve the SDGs.** It also presents new case studies offering concrete examples of how to resolve sustainable urbanization and environmental issues in different regions of the world.

The series is intended for professionals in research and teaching, consultancies and industry, and government and international organizations. Published in collaboration with IEREK, the Springer ASTI series will acquaint readers with essential new studies in STI for sustainable development.

ASTI series has now been accepted for Scopus (September 2020). All content published in this series will start appearing on the Scopus site in early 2021.

M. Sumesh • João Manuel R. S. Tavares •
S. C. Vettivel • Mario Orlando Oliveira
Editors

2nd International Conference on Smart Sustainable Materials and Technologies (ICSSMT 2023)

Smart Sustainable Materials and Technologies (Volume 1)

Springer

Editors
M. Sumesh
Department of Civil Engineering
CARE College of Engineering
Trichy, India

S. C. Vettivel
Department of Mechanical Engineering
Chandigarh College of Engineering and Technology
Chandigarh, India

João Manuel R. S. Tavares
Faculdade de Engenharia da Universidade do Porto (FEUP)
Porto, Portugal

Mario Orlando Oliveira
Universidad Nacional de Misiones (UNaM)
Posadas, Argentina

ISSN 2522-8714 ISSN 2522-8722 (electronic)
Advances in Science, Technology & Innovation
IEREK Interdisciplinary Series for Sustainable Development
ISBN 978-3-031-49825-1 ISBN 978-3-031-49826-8 (eBook)
https://doi.org/10.1007/978-3-031-49826-8

This Springer imprint is published by the registered company Springer Nature Switzerland AG
The registered company address is: Gewerbestrasse 11, 6330 Cham, Switzerland

We are honored to dedicate the proceedings of ICSSMT 2023 to all the participants, organizers and editors of ICSSMT 2023.

Preface

The Second International Conference on Smart Sustainable Materials and Technologies (ICSSMT 2023) held between August 30 and 31, 2023 in Trichy, India, continued as a series of conference events targeting the implementation of smart materials for developing sustainable environments. The conference covered the fundamentals of material science and highlighted the new sustainable materials science research topics. Sustainable materials, structures, functional materials, process control and simulation, and analysis on different material characteristics were considered.

Smart sustainable materials have gained significant research traction over the past years related to their development and deployment. Its importance is unequivocally recognized by the United Nations Sustainable Development Goals (SDGs). To maintain a sustainable environment, governments are encouraging and actively supporting the adoption of sustainable materials for developing environment-friendly infrastructures. Keeping this as a goal, this international conference facilitates the dissemination and exchange of emerging technology and implementation-related information, to provide a forum for both academia and industry to share ideas and experiences in this field and that could accelerate the adoption of smart sustainable materials. The conference brought together both academia and industry experts and gave the audience a sound knowledge of the latest technological updates and was provided with examples of successful sustainable material deployments; an opportunity was offered to learn what all can be changed by implementing smart sustainable materials and how it can be done.

The conference welcomed technical contributions presenting a novel research idea and its practical results and some papers addressed the pros and cons of using sustainable materials, its industrial standards, key problems and solutions. A total of 108 submissions were received, and from that 26 papers were accepted by the technical committee members.

We take this opportunity to thank all the members of the 2nd ICSSMT 2023 Technical Program Committee as well as the reviewers. Conducting a high-quality conference program would not have been possible without their involvement. We also thank all the authors who dedicated much of their time and efforts to contribute to 2nd ICSSMT 2023. We are equally grateful to the members of the 2nd ICSSMT 2023 organizing committee for their help in handling all the event-related work to make this meeting a success.

Trichy, India — Dr. M. Sumesh
Porto, Portugal — Prof. João Manuel R. S. Tavares
Chandigarh, India — Dr. S. C. Vettivel
Posadas, Argentina — Dr. Mario Orlando Oliveira

Contents

A Review on Basalt Fibre-Reinforced Polymer Matrix Composite Material 1
P. C. Santhosh Kumar, M. Ravichandran, and R. Naveenkumar

Optimizing Tensile Strength and Hardness in Al-Based Hybrid Composites via Stir-Squeeze Technique 7
Srikanth Karumuri, M. Geetha, G. Raja Kumar, Suresh Seetharaman, D. R. Rajkumar, Bathula Srinivas, and P. Anusha

Evaluation of Hardness Properties of Al7475/B4C/Fly Ash Hybrid Composites Using Friction Stir Process 15
G. Mahesh, Vinay Kumar Domakonda, Shaik Farooq, Ram Subbiah, D. R. Rajkumar, N. Srinivasa Rao, and S. Vijayakumar

Multiscale Material Modelling for Evaluating Mechanical and Electrical Characteristics of Graphene/Glass Fibre Epoxy Hybrid Composites 21
M. Mahesh Kumar, S. Paul Vizhian, and M. Krishna

Fabrication and Tensile Impact and Hardness Test of Al6061/SiC/Coconut Shell Ash Composites 31
Thota Venkateswara Rao, S. Jeyakrishnan, D. R. Rajkumar, Ram Subbiah, S. Vijayakumar, M. Naga Swapna Sri, and P. Anusha

A Review on Rolling Process for Polymer Nanocomposites with Different Nanofillers to Enhance Properties 39
Aamir M. Shaikh and Pravin R. Kubade

Synthesis and Characterization of Co-precipitated Hematite-α-Fe_2O_3 Nanoparticles (AIO-NPs) 49
Dinesh Uthra, Krishna Patel, Jai Narayan Sahu, and M. Z. Khan

Influence of the Bio-lubricant Along with Nano Compounds as Additives on the Properties of the Lubricant 55
S. Nithya Poornima and V. Shantha

Experimental and Heat Transfer Analysis Using Nanofluid in Cylindrical Heat Pipe Heat Exchanger 65
P. Ramkumar, C. M. Vivek, P. Latha, and S. P. Manikandan

Two Au Atoms-Doped Silicene Nanoribbons in Unit Cell with an Electrical Field: A DFT Study 73
Hoang Van Ngoc

Enhancing Mechanical Characteristics and Cost-Efficiency of Composite Materials Through Hybridization and Nanoparticle Incorporation 83
G. Boopathy, G. Balaji, R. Bhoominathan, A. Bovas Herbert Bejaxhin, and N. Ramanan

Design and Development of Multipurpose Quadcopter Using Fused Deposition Modeling . . . 91
Yasin Sayyed, Nayan Patil, Nikhil Solanki, Bhargav Patil, Saurabh Kale, and Niranjan Padawale

Thermodynamic Characterization of Building Thermal Insulating Materials Using Cardboard Wastes and Zea Mays Everta . . . 103
Houda Ajabli, Amine Zoubir, Rabie Elotmani, Mohamed Louzazni, Khalid Kandoussi, and Abdelmajid Daya

Theoretical and Numerical Analysis of MHD Flows Through a Porous Material . . . 111
S. Krishnakumar, K. Lakshmi Narayanan, S. Thamizh Suganya, and L. Rajendran

Feasibility and Insights into the Optimization and Characterization of Friction Welded Aluminum–Steel Dissimilar Joints . . . 121
S. Hari Krishna Raj, G. Boopathy, S. Vijayananth, G. Balaji, and N. Ramanan

Friction Stir Welded Magnesium AZ31B Alloy Used to Evaluate Mechanical Properties at Various Rotational Speeds . . . 131
Bhiksha Gugulothu, P. S. Satheesh Kumar, N. Srinivasa Rao, S. Vijayakumar, D. R. Rajkumar, T. V. Janardhana Rao, and M. Naga Swapna Sri

Evaluation of Mechanical and Thermomechanical Properties of Shape Memory Polymers Reinforced with Graphite, Graphene, and CNT . . . 139
M. K. Ranganatha Swamy, Nagaraj Patil, U. S. Mallik, and V. Udayakumar

Conductivity Tensor in an Infinite Potential Cylindrical Quantum Wire with Electron-Acoustic Phonon Interaction . . . 145
Hoang Van Ngoc

Manufacturing of Eco-Friendly Brick Using Wasted Textile Apparel . . . 153
S. Ispara Xavier, R. Divahar, P. S. Aravind Raj, P. J. Adarsh, Akshay Manoj, and Badusha Jalal

Performance Analysis of Savonius VAWT with and Without Dimples and Fins by Using CFD . . . 165
Shivaji Suryabhan Aher and Netra Pal Singh

Effect of Different Rib Shapes on the Performance of Solar Thermal Collector . . . 179
Rehan Fazal, Ravindra Mohan, Geetesh Goga, Santosh Kumar, Anil Singh Yadav, and Subhendu Chakroborty

Exploring the Complexities and Challenges of Plastic Recycling: A Comprehensive Research Review . . . 189
Sunil Sable, Mitesh Ikar, and Priyanka Dudheinamdar

Pyrolysis-Based Fuel Extraction from Household Waste Plastics for Diesel Generators . . . 203
K. R. Deepa, Nayana shetty, N Shamala, and P. Ashwini Kumari, R. Likitha, and K. A. Manohar

Experimental and Performance Study of Diesel Engine Using Inlet Preheated Air by Heat Pipe Heat Exchanger . . . 211
P. Ramkumar, M. Sivasubramanian, S. Ramasamy, C. M. Vivek, and P. Latha

Metal Fused Filament Fabrication Process—A Review 219
S. Jasper and M. Ravichandran

Mineralogical Identification of Clays Using K-Mean Clustering 227
Siva Kumar Prasad Chebiyyam and Mallikarjuna Rao Kattamuri

A Review on Basalt Fibre-Reinforced Polymer Matrix Composite Material

P. C. Santhosh Kumar, M. Ravichandran, and R. Naveenkumar

Abstract

Polymer matrix composite materials are most frequently used in engineering applications due to its biodegradable, environmentally friendly and better mechanical properties. Basalt fibre composites are widely used due to their high performance in automobile applications. When compared with traditional fibres like glass fibre and carbon fibre, basalt fibre has superior mechanical properties. This review paper presents the mechanical, physical and chemical properties of basalt fibre composites along with various polymer matrices and also the various combination of basalt fibre with other polymer matrix composite materials. From the literatures, it is observed that, the plasma treatment increases the thermal stability, alkali treatment increases the corrosion resistance and additive treatment increases the mechanical properties of the basalt fibre based composites.

Keywords

Basalt fibre • Polymer matrix • Mechanical properties • Low manufacturing cost

1 Introduction

Basalt fibre material utilization is recently increased due to its better properties such as eco-friendly, economical, lightweight and biodegradable property, which has the promising chemical, mechanical and physical properties. Dhand et al. (2014) conducted an analysis of basalt fibre and found that large portions are covered with thermal and mechanical properties. Fiore et al. (2015) analysed the structural, physical and heat transfer properties of basalt fibre and conducted a study on the effect of addition of different materials. Balaji et al. (2020) found that the surface treatment on basalt fibre attains good improvement in mechanical properties and its properties are in between glass fibre-reinforced composite and carbon fibre-reinforced composites.

Jain et al. (2018) indicated that to overcome certain challenges of basalt fibre such as low thermal stability and high moisture absorption properties, the fibre is subjected to chemical, surface and additive treatment. Lebedev et al. (2019) signified basalt fibre as the best alternative for glass fibre due to its high resistance to climate variation. David Plappert et al. (2020) found basalt fibre as the best alternative for engineering applications due to its better recycling behaviour and economical manufacturing cost based on the results obtained in tension, compression and shear specimen analysis. Kalmeshwar Ullegaddi et al. (2019) conducted a review on tribological properties of basalt fibre-reinforced composite materials and found that compared with carbon and glass fibre, basalt fibre has good abrasive, erosion and sliding properties. Li Yan et al. (2021) conducted another study of analysis on physicochemical and mechanical abilities of basalt fibre and concluded that the shorter fibre increases the flexural behaviour and tensile features of the material than the glass fibre-reinforced composite material. Zhensheng Guo et al. (2018) found that pre-treated basalt fibre has good alkali resistance, and consequently, there is no change in compression strength. Hafsa Jamshaid and Rajesh Mishra (2015) conducted a detailed study on manufacturing, properties and behaviour of basalt fibre and found that basalt fibre is 100% inert material and can be used in compressed natural gas cylinder, muffler and headliners. Figure 1 represents the various compositions present in the basalt fibre.

P. C. Santhosh Kumar · M. Ravichandran (✉) · R. Naveenkumar
Department of Mechanical Engineering, K. Ramakrishnan College of Engineering, Tiruchirappalli, Tamil Nadu 621112, India
e-mail: smravichandran@hotmail.com

M. Sumesh et al. (eds.), *2nd International Conference on Smart Sustainable Materials and Technologies (ICSSMT 2023)*, Advances in Science, Technology & Innovation, https://doi.org/10.1007/978-3-031-49826-8_1

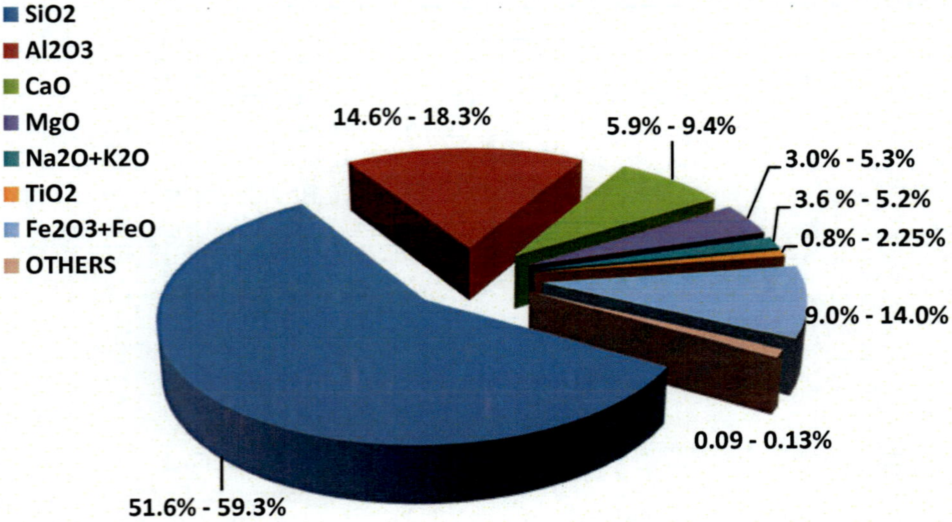

Fig. 1 Composition of basalt fibre (Plappert et al., 2020)

2 Literature Survey

2.1 Basalt Fibre Polymer Matrix Composite

Basalt fibre is a material obtained from basalt rock, which is composed of plagioclase, pyroxene and olivine. Basalt fibre has good mechanical strength, thermal resistance, chemical resistance and ecological factors. Vikas Sharma et al. (2019) conducted an experimental comparison of basalt fibre and glass fibre with epoxy resin composite plate with a varying weight percentage of 10, 20, 30 and 40% fibre. Composite are prepared by using hand lay-up technique. Evaluate the various mechanical qualities such as tensile values, shock resistance, bend resilience and abrasive wear rate also calculate the density and void content. Concluded that an increase in fibre content increases the mechanical properties and Epoxy with 40% basalt fibre having good mechanical properties than glass fibre composite. Xing Zhao et al. (2019) conducted an investigation on mechanical properties of basalt fibre within temperature. Concluded that an increase in temperature decreases the interfacial bonding between fibre and resin which reduces the tensile and flexural strength. Jingjing He et al. (2018) conducted an experiment on basalt fibre with epoxy resin by varying the fibre volume fraction and fibre orientation angle. Concluded that varying the orientation angle decreases the mechanical properties and increasing the fibre content up to 1.5% increases the mechanical properties. Dhand et al. (2015) conducted experimentation on basalt fibre and concluded that basalt fibre has high specific mechanical, physical, chemical biodegradability and non-abrasive qualities. Basalt fibres are used in anti-ballistic applications and automobile and aerospace applications and it has superior properties than glass fibres and conventional asbestos. Zhiming Yang et al. (2019) investigated the fibre hybridization, mechanical performance and less-than-average velocity impact behaviour of woven basalt fibre, woven ultrahigh molecular weight polyethylene fibre with epoxy. 60BF/40PEF composite material exhibits better energy absorption capacity and low-velocity impact behaviour when compared with individual basalt and PEF fibre epoxy resin composite material.

2.2 Hybrid Polymer Matrix Composites

To gain more advantages, the basalt fibre is mixed with another fibre to make a hybrid composite plate, which signifies the improvement in mechanical properties. Tirillo et al. (2017) conducted a high-speed shock resistance behaviour of hybrid material of basalt and carbon with epoxy composite and found that basalt content increases the impact strength when compared with carbon epoxy composite plate. Hafsa Jamshaid and Rajesh Mishra (2016) conducted an experimental comparison of basalt woven with jute and polypropylene and conducted various thermal experiments such as thermal resistance, diffusivity and thermal conductivity. Omer Yavuz Bozkurt (2008) conducted an experimental study on mechanical properties such as tensile and bending behaviour of basalt fibre with epoxy and basalt, aramid fibre with epoxy resin. Based on experimental results, the presence of basalt fibre improves the mechanical properties when compared with aramid epoxy plate. Omer Yavuz Bozkurt and Muhammet Enver Gokdemir (2018) conducted an experimental investigation of accumulation of basalt fibre with the corresponding carbon fibre epoxy composites. The result shows that the incorporation of basalt fibre increases the damping properties. Lopresto et al. (2011) made an experimental investigation on basalt fibre with epoxy resin and glass fibre with epoxy resin and concluded that bending properties and crush behaviour of basalt fibre is better than glass fibre in the range of 35 to 42 percentage higher elastic behaviour is obtained in basalt fibre. Whereas better elongation behaviour was found in glass fibre. Colombo et al. (2012) conducted a comparative study on basalt fibre with epoxy resin and vinyl ester resin, and also compared with carbon fibre and glass fibre. Concluded that epoxy with basalt fibre shows better mechanical properties than vinyl ester; also basalt fibre is a replacement for carbon fibre because of its affordable cost when compared to glass fibre for its toughness. David Plappert et al. (2020) conducted a comparative study on basalt fibre with epoxy, carbon fibre with epoxy, E glass with epoxy and S glass with epoxy. Concluded that basalt fibre has high tensile strength than E glass but not S glass fibre. Maximum compressive strength is obtained at carbon fibre and maximum shear strength is obtained at E glass epoxy resin. Hao Zhou (2020) conducted a comparison of basalt fibre concrete with conventional concrete and concluded that reinforcement of basalt fibre has a huge increase in the tensile and flexural strength than compressive strength; also if there is an increase in the basalt beyond the optimized value, the mechanical properties get decreased. The breakdown mode also transformed from brittle to non-brittle failure.

2.3 Silane-Treated Polymer Matrix Composites

In order to enlarge the interfacial bond between fibres with resin, the fibre was treated with silane under various composition which in turn increased the tensile, compression, wear and impulsive force of the material. Nahid Jamali et al. (2019) completed an investigational basalt fibre, epoxy with varying silanized graphene oxide nanoparticles by varying the weight percentage of 0, 0.1, 0.2, 0.3, 0.4 and 0.5% and study various mechanical properties and concluded that the presence of 0.4% weight ratio having significant improvement in mechanical properties. Russell J. Varley et al. (2015) conducted an experimental investigation on surface treatment on basalt fibre to improve the chemical bonding which in turn further recover the mechanical behaviour of basalt fibre. Epoxy silane method shows the largest improvement of tensile strength up to 66%. Nahid Jamali et al. (2019) concluded with an experiment to improve the mechanical properties of basalt fibre with an epoxy plate by adding the Silanized Graphene Oxide nanoplatelets. 0.4 wt% addition of nano fillet increase the tensile strength 16%, flexural strength 47% and compressive strengths as 51% (Tirillò et al., 2017). Espana et al. (2013) researched the effect of totalling different types of silane coupling agent with basalt fibre and found that the best interaction of fibre and epoxy was obtained when we use Silane B ([3-(2-aminoethylamino)propyl]-trimethoxysilane).

2.4 Nanoparticles Based Polymer Matrix Composites

Addition of nanoparticles to any fibre increases the strength drastically. Basalt fibre shows better improvement in physical properties when it is combined with nanoparticles. Arezoo Abdi et al. (2018) experimentally investigated the basalt fibre and epoxy resin by varying the silanized $CaCO_3$ with the weight percentage of 0, 1, 3 and 5%. Evaluated the various mechanical attributes such as bendability, elongation behaviour and compare with conventional one and found that 3% weight fraction having 28, 35, 20 and 30 % enhanced properties than conventional composite plate. Ary Subagia et al. (2013) conducted the mechanical aspects of tensile behaviour, flexural modulus and its corresponding strength of epoxy resin composites in the presence of basalt fibre as a reinforcement with tourmaline microparticles. Four specimens are prepared by varying the weight percentage of tourmaline microparticles by using vacuum assisted resin infusion moulding process. That is, 0, 1, 2 and 1% with surfactant and found that 1% with surfactant had good mechanical improvement of 16% improvement in expansion and curving strength and 27.4 and 153.3% increase in expansion and curving modulus. Majka et al. (2020) conducted a comparison of epoxy and vinyl ester resin by using basalt fibre, basalt powder and basalt textile. The presence of epoxy resin having good mechanical properties and basalt textile will make it more difficult to prepare the composite plate than the remaining two; among the various compositions, basalt powder with epoxy resin exhibits better mechanical properties and heat absorption capacity. Shicheng Yan et al. (2019) concluded with a study on thermal, mechanical and tribological effects of Polyethersulfone (PES) with sodium–montmorillonite (Na–MMT) nanoparticles in the presence of polytetrafluoroethylene (PTFE). By varying the weight percentage of Na–MMT of 1, 5, 10 and 20%, it was found that 10% Na–MMT exhibits better optimal wear and friction performance which in turn protects the polymer from damage. Somaiah Chowdary et al. (2020) performed an experimental study on Sisal fibre, Kevlar Fibre in mat form with varying nanosilica powder with epoxy resin stirring was done both mechanical and ultrasonic form and over and done with that inclusion of nanoparticles improved the mechanical aspects and better results is obtained at 4% nanosilica powder. Yingwei Hou et al. (2017) conducted a study on the addition of Tungsten

Powder to basalt fibre epoxy composite. The presence of Tungsten Powder increases the Gamma Ray Shielding Property and advanced mechanical behaviour than steel. Suresha et al. (2020) performed an experimental comparison of the physical and inherent properties of nano-SiO_2, epoxy resin with the presence and without the presence of carbon fibre. Plates are manufactured by hand lay-up and found that the addition of 3% weight ratio shows better improvement of mechanical properties for nano-SiO_2. Mehmet Bulut et al. (2019) deport an experiment for mechanical and flowing characteristics of basalt fibre and epoxy resin with varying nano clay particles. The presence of nano clay improves the various mechanical properties, and 2% of nano clay has better dynamic and natural frequency. Ahmet Erklig and Nurettin Furkan Dogan (2020) manufactured a basalt/glass fibre, epoxy with varying Nanographene particles. Investigate the tensile, flexural and low-velocity impact tests. SEM images are taken to ensure the uniform distribution of fibre and nanoparticles. Concluded that 0.1% weight fraction has a significant improvement of mechanical properties than conventional one. Abd El-baky and Attia (2020) experimentally concluded that the addition of halloysite clay nanotubes to the glass fibre with aluminium alloy improves the mechanical properties. The addition of nanoparticle improves the bonding strength. Results found that 1% weight ratio of nanoparticles shows better mechanical properties. Zike Wang et al. (2016) conducted an elevated temperature exposure of basalt fibre with epoxy resin under 0.5, 1 and 1.5 h with a temperature of 150 and 250 °C. And found that elevated temperature treatment of basalt fibre and epoxy will fire debonding and decomposition of the resin. Logesh and Bupesh Raja (2018) conducted a study on E glass with hybrid magnesium and aluminium layered double hydroxides in varying weight percentage and concluded that 5% weight fraction shows good mechanical properties. Reddy and Reddy (2020) conducted an experiment with used jute fibre, epoxy resin and sic powder; sic is varied under the weight ratio of 0, 3, 6, 9 and 12%. And concluded that the augmentation of sic improves several mechanical properties.

3 Extraction of Basalt Fibre

Basalt fibre is obtained from raw materials such as basalt rock which is naturally available. The first step is to crush the basalt rock into small pieces with the help of machinery as shown in Fig. 2. Then it will be completely cleaned with water to take away the unwanted particles and then it is transferred to the furnace to melt the basalt rock the temperature required for melting the basalt rock is more than 1450 °C. At that temperature, it becomes a molten stage then it is extracted with the help of centrifugal blowing, centrifugal multirole or melt blowing techniques. The properties of basalt fibre are Density in the range of 2.63 to 3.05 g/cm^3, Elongation strength in the range of 3000 to 4840 MPa, Elastic modulus in the range of 79.3–110 GPa and a Maximum service temperature is 650 °C.

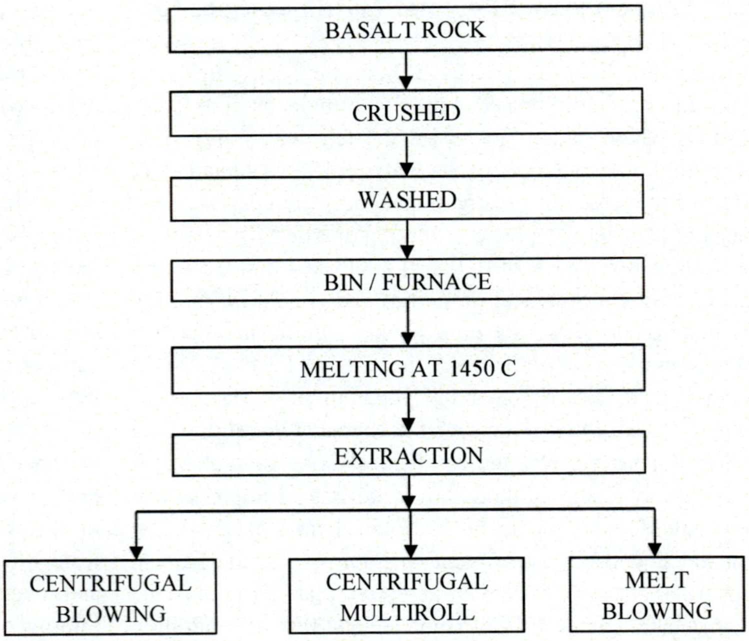

Fig. 2 Extraction of basalt fibre

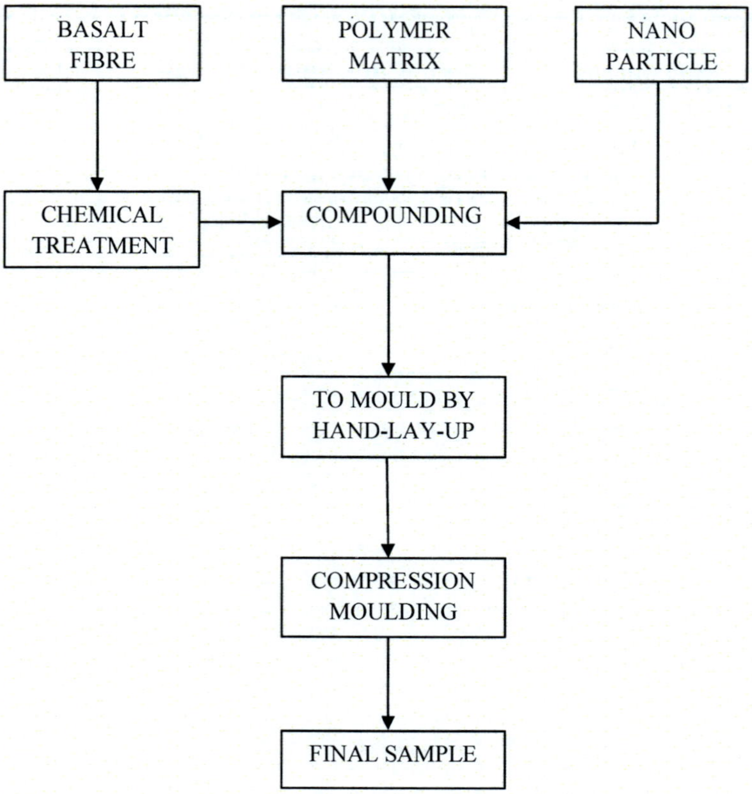

Fig. 3 Composite plate manufacturing process

4 Fabrication of Polymer Matrix Composite Material

After the complete extraction of fibre, the composite plate is manufactured with the polymer matrix, usually epoxy resin is mostly used as shown in Fig. 3. In addition to that polymer matrix, the incorporation of nanoparticles also increases the bonding strength which further increases the mechanical stamina of the material.

5 Conclusion

This review paper focused the comprehensive features of basalt fibre reinforced composite materials. Basalt fibre composites have slightly higher mechanical properties than E Glass fibre composite; also basalt fibre with other fibres also increases the impact strength and thermal properties of the material. The presence of basalt fibre improves the mechanical properties when compared with aramid epoxy plate, which also increases the damping properties. Further, the inclusion of nanoparticles also increases the bend resistance and modulus of the materials. Also, the preparation process of basalt fibre-reinforced composite material and extraction of basalt fibre are also discussed.

References

Abdi, A., Eslami-farsani, R., & Khosravi, H. (2018). Evaluating the mechanical behavior of basalt fibers/epoxy composites containing surface-modified $CaCO_3$ Nanoparticles. *19*(3), 635–640. https://doi.org/10.1007/s12221-018-7755-x.

Balaji, K. V., Shirvanimoghaddam, K., Sankar, G., Ellis, A. V., & Naebe, M. (2020). Surface treatment of Basalt fi ber for use in automotive composites. *Materials Today Chemistry, 17*, 100334. https://doi.org/10.1016/j.mtchem.2020.100334.

Barkoula, N. M., Alcock, B., Cabrera, N.O., & Peijs, T. (2008). Flame-retardancy properties of intumescent ammonium poly(Phosphate) and mineral filler magnesium hydroxide in combination with graphene. *Polymers and Polymer Composites*, *16*(2), 101−113. https://doi.org/10.3390/polym6112875.

Bozkurt, Ö. Y., & Gökdemir, M. E. (2018). Effect of basalt fiber hybridization on the vibration-damping behavior of carbon fiber/epoxy composites. *Polymer Composites, 39*, E2274–E2282. https://doi.org/10.1002/pc.24606.

Bulut, M., Yavuz, Ö., Ahmet, B., Hakan, E., & Özkan, Y. (2019). Mechanical and dynamic properties of basalt fiber reinforced composites with nanoclay particles. *Arabian Journal for Science and Engineering*, (0123456789). https://doi.org/10.1007/s13369-019-04226-6.

Chowdary, M. S., & Raghavendra, G. (2020). Influence of nano-silica on enhancing the mechanical properties of sisal/kevlar fiber reinforced polyester hybrid composites.

Colombo, C., Vergani, L., & Burman, M. (2012). Static and fatigue characterisation of new basalt fibre reinforced composites. *Composite Structures, 94*(3), 1165–1174. https://doi.org/10.1016/j.compstruct.2011.10.007.

Dhand, V., Mittal, G., Rhee, K. Y., Park, S. J., & Hui, D. (2015). A short review on basalt fiber reinforced polymer composites. *Composites Part b, Engineering, 73*, 166–180. https://doi.org/10.1016/j.compositesb.2014.12.011.

Dhand, V., Mittal, G., Rhee, K. Y., & Hui, D. (2014). A short review on basalt fiber reinforced polymer composites. *Composites Part B*, (December). https://doi.org/10.1016/j.compositesb.2014.12.011.

El-baky, M. A. A. (2020). Experimental study on the improvement of mechanical properties of GLARE using nanofillers. (June), 1–14. https://doi.org/10.1002/pc.25699.

Erkliğ, A., & Furkan, N. (2020). Nanographene inclusion effect on the mechanical and low velocity impact response of glass/basalt reinforced epoxy hybrid nanocomposites. *The Journal of the Brazilian Society of Mechanical Sciences and Engineering*. https://doi.org/10.1007/s40430-019-2168-y.

Espan, J. M., Samper, M. D., Fages, E., Sa, L., Balart, R., & Polite, U. (2013). Investigation of the effect of different silane coupling agents on mechanical performance of basalt fiber composite laminates with biobased epoxy matrices. https://doi.org/10.1002/pc.22421.

Fiore, V., Scalici, T., Di Bella, G., & Valenza, A. (2015). SC. *Composites Part B*. https://doi.org/10.1016/j.compositesb.2014.12.034.

Guo, Z., Wan, C., Xu, M., & Chen, J. (2018). Review of basalt fiber-reinforced concrete in China: Alkali resistance of fibers and static mechanical properties of composites. *2018*.

He, J., Shi, J., Cao, X., & Hu, Y. (2018). Tensile mechanical properties and failure modes of a basalt fiber/epoxy resin composite material. *2018*.

Hou, Y., Li, M., Gu, Y., Yang, Z., Li, R., & Zhang, Z. (2017). Gamma ray shielding property of tungsten powder modified continuous basalt fiber reinforced epoxy matrix composites. 1–10. https://doi.org/10.1002/pc.24469C.

Jain, N., Singh, V. K., & Chauhan, S. (2018). Review on effect of chemical, thermal, additive treatment on mechanical properties of basalt fiber and their composites. 1–7.

Jamali, N., Khosravi, H., Rezvani, A., & Tohidlou, E. (2019). Mechanical properties of multiscale graphene oxide/basalt fiber/epoxy composites. *20*(1), 138–146. https://doi.org/10.1007/s12221-019-8794-2.

Jamshaid, H., & Mishra, R. (2016). Thermomechanical characteristics of basalt hybrid and nonhybrid woven fabric–reinforced epoxy composites. *Polymer Composites, 37*(10), 2982–2994. https://doi.org/10.1002/pc.23495.

Lebedev, M. P., Startsev, O. V., & Kychkin, A. K. (2020). Heliyon the effects of aggressive environments on the mechanical properties of basalt plastics. *Heliyon, 6*(November), e03481. https://doi.org/10.1016/j.heliyon.2020.e03481.

Logesh, K., & Raja, V. K. B. (2018). Evaluation of mechanical properties of Mg-Al layered double hydroxide as a filler in epoxy-based FML composites.

Lopresto, V., Leone, C., & De Iorio, I. (2011). Mechanical characterisation of basalt fibre reinforced plastic. *Composites Part b, Engineering, 42*(4), 717–723. https://doi.org/10.1016/j.compositesb.2011.01.030.

Majka, T. M., Byrdy, A., & Pielichowski, K. (2020). Applied sciences basalt textile-reinforced vinylester and epoxy resins for anchors used to fasten ventilated building facades.

Plappert, D., Ganzenmüller, G. C., & May, M. (2020). Mechanical properties of a unidirectional basalt-fiber/epoxy composite. 1–12. https://doi.org/10.3390/jcs4030101.

Reddy, K. M., & Reddy, B. C. M. (2020). Materials today: Proceedings mechanical characterization of chemically treated used jute fiber reinforced epoxy composite with SIC fillers. *Materials Today: Proceedings*, 917–921. https://doi.org/10.1016/j.matpr.2020.06.055.

Sharma, V. (2020). Mechanical and three-body abrasive wear behavior analysis of glass and basalt fiber-reinforced epoxy composites. 1–15. https://doi.org/10.1002/pc.25670.

Subagia, I. D. G. A., Tijing, L. D., Kim, Y., Kim, C. S., Iv, F. P. V., & Shon, H. K. (2013). Mechanical performance of multiscale basalt fiber-epoxy laminates containing tourmaline micro/nano particles. *Composites, Part B: Engineering*. https://doi.org/10.1016/j.compositesb.2013.10.034.

Suresha, B. (2020). Physico-mechanical properties of nano silica-filled epoxy-based mono and hybrid composites for structural applications.

Taylor, P., & Jamshaid, H. (2015). A green material from rock: Basalt fiber–a review. *The Journal of The Textile Institute*. https://doi.org/10.1080/00405000.2015.1071940.

Tirillò, J., et al. (2017). High velocity impact behaviour of hybrid basalt-. *Composite Structures*. https://doi.org/10.1016/j.compstruct.2017.02.039.

Ullegaddi, K., & Mahesha, C. R. (2019). Tribological properties of basalt fibers-a review. *969*, 335–342. https://doi.org/10.4028/www.scientific.net/MSF.969.335.

Wang, Z., Yang, Z., Yang, Y., & Xian, G. (2016). Flexural fatigue behavior of a pultruded basalt fiber reinforced epoxy plate subjected to elevated temperatures exposure. 1–11. https://doi.org/10.1002/pc.24124.

Yan, S., Xue, Y., Hu, Y., & Wang, S. (2019). Thermal, mechanical, and tribological properties of sodium–montmorillonite-nanoparticle-reinforced polyethersulfone and poly-tetrafluoroethylene ternary composites.

Yan, L., Chu, F., Tuo, W., Zhao, X., & Wang, Y. (2021). Review of research on basalt fibers and basalt fiber-reinforced composites in China (I): Physicochemical and mechanical properties. (I). https://doi.org/10.1177/0967391120977396.

Yang, Z., Liu, J., Wang, F., Li, S. & Feng, X. (2019). Effect of fiber hybridization on mechanical performances and impact behaviors of basalt fiber/UHMWPE fiber reinforced epoxy composites. *Composite Structures. 229*(September), 111434. https://doi.org/10.1016/j.compstruct.2019.111434.

Zhao, X., Wang, X., Wu, Z., Keller, T., & Vassilopoulos, A. P. (2019). Temperature effect on fatigue behavior of basalt fiber-reinforced polymer composites. *Polymer Composites, 40*(6), 2273–2283. https://doi.org/10.1002/pc.25035.

Zhou, H., Jia, B., Huang, H., & Mou, Y. (2020). Experimental study on basic mechanical properties of basalt fiber reinforced concrete.

Optimizing Tensile Strength and Hardness in Al-Based Hybrid Composites via Stir-Squeeze Technique

Srikanth Karumuri, M. Geetha, G. Raja Kumar, Suresh Seetharaman, D. R. Rajkumar, Bathula Srinivas, and P. Anusha

Abstract

This research study focused on the hybrid composites using AA 7178 as a base alloy reinforced with two materials ZrO_2/SiC. Nine specimens were prepared using stir-squeeze processing techniques by varying their speed (rpm), time(s) and pressure (N/m^2). To analyze the mechanical strength and resistance, a Taguchi design was utilized on hybrid composites. The experimental results indicated that the ultimate tensile strength (UTS), (the maximum value 265.5 MPa) was achieved in the S9 experiment, while the minimum value (175 MPa) was observed in the S1 experiment. To perform the analysis of variance (ANOVA) for the parameters and to analyze SN ratio, a state-of-art software Minitab was utilized. The ANOVA results revealed that stirring speed had the most significant impact on improving tensile strength which contributed approximately to 43.2%. The next substantial factor was the Stirring time which contributed around 42.5% and followed by Squeeze pressure at 12.1%. From the hardness ANOVA result, Stirring time was the first factor contributing to 46.7% followed by a Squeeze pressure of 35.2% and a Stirring speed of 7.6% respectively.

Keywords

Metal matrix composite • Squeeze casting • Optimization • ANOVA and minitab

1 Introduction

Metal Matrix composites have many benefits, including increased tensile and impact strength, increased abrasion resistance and superior physical qualities. Out of the various methods available for producing Advanced Metal Matrix Composite (MMC), the Stir Casting (SC) process stands out as the most cost-effective approach compared to other methods such as powder metallurgy, plasma sintering (PS), etc. This cost advantage makes it a popular choice for materials gathering and processing under a wide range of conditions. Transferring molten metal into common molds with stir casting is a successful technique for creating high-quality pieces. Weight ratio, variables and the bonding of the base metals and additives all affect the mechanical properties. Conventional casting processes often result in various defects, such as shrinkage cavity, porosity, and misrun. However, the Squeeze Casting process, which involves pressurized extrusion molding of materials, offers a high degree of manufacturing flexibility (Arvind & Kaushik, 2022; Lee et al., 2012; Shorowordi et al., 2003; Zhao et al., 2020).

Using electronic waste plastic transport properties of concrete at different ratios, Senthil Kumar et al. 2018 carried

S. Karumuri
Department of Mechanical Engineering, College of Engineering & Technology, Mizan Tepi University, Tepi Campus, Ethiopia
e-mail: srikanth.karumuri@mtu.edu.et

M. Geetha
Department of Electronics and Communication Engineering, Sri Eshwar College of Engineering, Coimbatore, India
e-mail: geetha.m@sece.ac.in

G. R. Kumar
Department of Mechanical Engineering, Swarna Bharati Institute of Science and Technology, Khammam, TS, India

S. Seetharaman (✉)
Department of Electrical and Electronics Engineering, Sri Eshwar College of Engineering, Coimbatore, India
e-mail: falseengrsuresh@gmail.com

D. R. Rajkumar
Department of Mechanical Engineering, CARE College of Engineering and Technology, Tiruchirappalli, 620 009, India
e-mail: rajkumardevapiriam@gmail.com

B. Srinivas
Department of Mechanical Engineering, BVC Engineering College (Autonomous), Odalarevu, Andhra Pradesh, India

P. Anusha
Department of Mechanical Engineering, P V P Siddhartha Institute of Technology, Vijayawada, India

M. Sumesh et al. (eds.), *2nd International Conference on Smart Sustainable Materials and Technologies (ICSSMT 2023)*,
Advances in Science, Technology & Innovation, https://doi.org/10.1007/978-3-031-49826-8_2

out experimentation. Generally, the 7XXX series of alloys are widely utilized in various businesses such as aviation, navigation etc. SiC/ZrO_2 reinforced AA7178 (Sekar et al., 2018) composites that were created using a stir-and-squeeze casting technique. In comparison to base metal, the findings for UTS and HN were much greater. The inclusion of silicon carbide and zirconium dioxide in the composites is responsible for this improvement (Park et al., 2008). The widespread use of reinforcing MMCs across various industries has been impeded by the higher production costs associated with producing multipart components with contours. Umanath et al. 2020 conducted a study on the improvement of Al6063 alloys with SiC-MoS_2 reinforcements, mainly focusing on the particle sizes of additives. They found that corrosion resistance decreased as the temperature increased.

Prabhu et al. (2006) reported in their present paper that a high-silicon aluminum alloy SiC metal grid composite with 10% SiC content was effectively controlled using different stirring rates and stirring times. Optical filtering glass and Brinell hardness tests were conducted to analyze the composite made by bottom-up casting. The microstructure and hardness of the composite were affected by mixing speed and time. Dao et al., (2012) investigated the microstructure and mechanical properties of connecting rods fabricated squeeze casting method. From the result, it was observed that 100 MPa, 250 °C and 575 °C are the best parameters for improving the mechanical performance of connecting rods. Gobalakrishnan et al. (2018a, 2018b) evaluated the mechanical properties of Al 6061 metal matrix composites as per ASTM Standards and concluded that increasing the wt % of TiB_2 particles was the most important factor in improving the strength properties. Manjunath et al. (xxxx) ANOVA analysis of the AL7075/SiC/Gr MMC revealed that SiC weight percentage, followed by Al7075 particle size, had the greatest impact on hardness, tensile, and compressive strength. Vengatesh and Rajkumar (2019) did an experimental study to calculate the influence of filler material on the tensile and hardness properties of the laminated composite plate. Shanmughasundaram and Subramanian (2012) identified optimum parameters by using Taguchi and ANOVA methods to improve the tensile and hardness properties of aluminum matrix composites. Emin et al. (2019) used the hot press method to fabricate MMC which is composed of spheroidal graphite cast iron as reinforcement with tin bronze matrix. To predict the production parameters, he used ANOVA to decide the surface roughness and thrust force which helps in drilling MMC.

This research aims to fabricate MMC using AA7178/ZrO_2/SiC and determine the optimal parameters to enhance their mechanical properties, thereby promoting the utilization of Al7178 in various applications. The study employs the Taguchi technique to evaluate four-level parameters, seeking to identify a specific set of input parameters that can enhance the desired outputs of tensile strength and hardness of the prepared hybrid composites.

2 Materials and Methodology

AA 7178 was mixed with ZrO_2 and SiC to produce a hybrid composite through stir casting, as shown in Fig. 1. An electric furnace was used to melt the aluminum alloy, which was put inside of a crucible and heated to about 800 °C. The compound of AA 7178 is provided in Table 1.

In a furnace, the reinforcement powders were warmed to 400 °C before being weighed on a digital weighing scale (Vijayakumar et al., 2022a). At nearly 780 °C, Alloys and reinforcement were then joined. The melted grains were continually disturbed about 4–8 min by a motor before being poured into the necessary molds to create Al-based composites. Several samples were created by adjusting parameters such as stirrer speeds, mixing duration and %wt of reinforcement ingredients.

Table 2 considers the three constraints used in this research work to analyze the mechanical properties. Table 3 summarized the obtained experiment results. The molten metal was transported into the appropriate mold having a

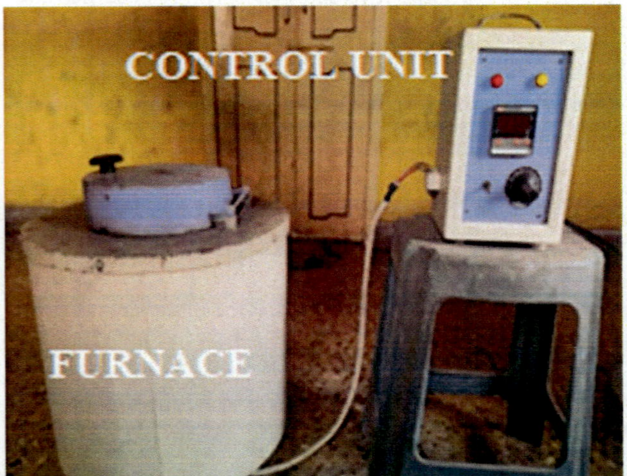

Fig. 1 Stir casting setup

Table 1 Composition of AA 7178 elements

Si	Fe	Cu	Mn	Mg	Zn	Ti	Cr	Al
0.4–0.7	0.4–0.8	0.2–0.8	0.9–1.5	1.2–1.8	0.1–0.5	0.2–0.8	0.3–0.6	Balance

Table 2 Casting constraints and stages

Constraints	Stages		
	1	2	3
Stir speed (rpm)	700	800	900
Stir time (min)	20	30	40
Squeeze pressure (N/m^2)	40	60	80

Table 3 Experimental observations

Exp. No	Process constraints			Experimental results	
	Stir speed	Stir time	Squeeze pressure	Tensile strength	Hardness
S1	700	20	40	175.0	83.60
S2	700	30	60	182.7	74.60
S3	700	40	80	187.8	77.60
S4	800	20	60	186.0	70.60
S5	800	30	80	200.8	81.60
S6	800	40	40	242.7	77.80
S7	900	20	80	188.6	73.10
S8	900	30	40	242.5	75.96
S9	900	40	60	265.5	82.76

size of 60 × 80 × 90 mm. During the squeeze, a mechanical plunger was used to provide pressure (40 MPa) for a certain amount of time before the solidified component was removed.

3 Mix Design

3.1 Taguchi Method of Design of Experiment

The primary objective of the Taguchi method is to enhance the quality of manufacturing processes by minimizing variability. By doing so, it aims to bolster product reliability and performance while concurrently curtailing costs. This approach holds significant applicability across diverse sectors, including engineering, manufacturing, and quality management. Performing confirmation experiments is a crucial step in the Taguchi method's implementation. These experiments serve to validate that the optimized parameters genuinely result in improved performance and decreased variability.

In this statistical investigation, Taguchi L9 OA was selected. Taguchi calculated S/N ratio for each experimental setting. The higher S/N ratio determines the level of process parameters that should be used. The input parameters are stir speed (700–900 rpm), stir speed (20–40 min), and squeeze pressure (40–80 N/m^2) have been selected. The optimum parameter setting was designed by ANOVA with the help of Minitab software.

3.2 Analysis of Variance (ANOVA)

The ANOVA (Analysis of Variance) table is a fundamental component of statistical analysis used to assess the significance of differences among group means in an experiment. It breaks down the total variability observed in the data into different sources of variation and quantifies how much of the total variation can be attributed to each source. It consists of five important factors which are the Sum of Squares, Degrees of Freedom, Mean Square, F-statistic and P-value. Based on the P-value, researchers can decide whether to reject the null hypothesis (indicating that there are significant differences among groups) or fail to reject it.

4 Result and Discussion

The fabricated specimens as per ASTM B557M shown in Fig. 2 underwent tensile tests. All specimens were polished and cleaned with abrasive paper to eliminate any debris on

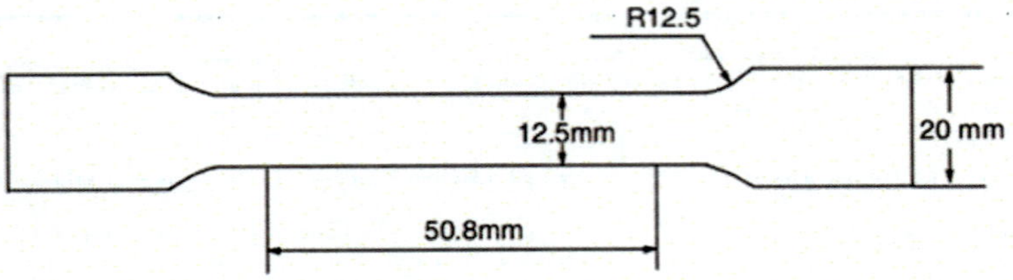

Fig. 2 ASTM B557M standard-tensile test

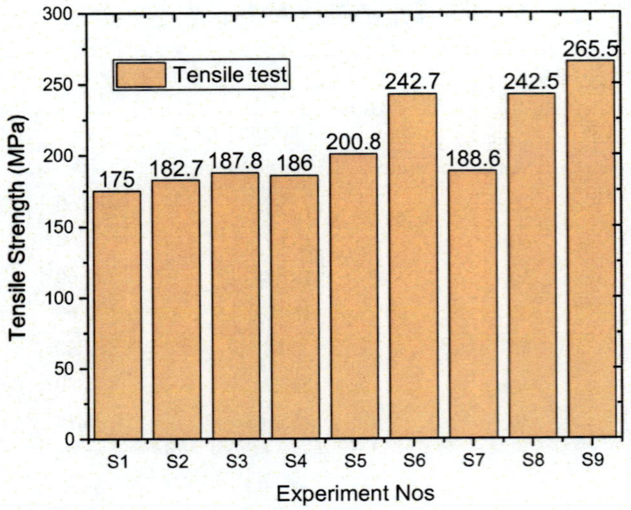

Fig. 3 Tensile test result

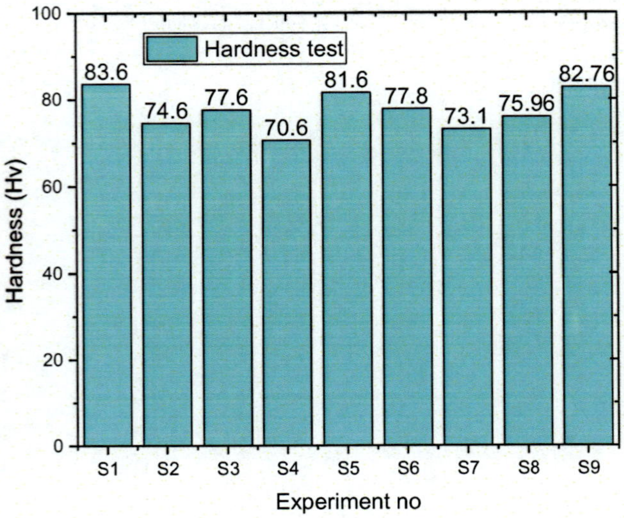

Fig. 4 Hardness observation results

the surface before the load was applied (Vijayakumar et al., 2022b). The tensile strength was calculated based on L9 orthogonal array and its results were presented in Fig. 3. The findings indicated that the maximum Tensile strength (265.5 MPa) was achieved in the S9 experiment, while the minimum tensile strength (175 MPa) was observed in the S1 experiment. This variation in tensile strength can be subjected to applied force, reinforcement strength and grain size of reinforcements. Squeeze pressure and grain refining both contribute to the decrease in the porosity gap. However, the produced composites might have low ductility and brittleness. The discrepancy between the Al matrix and nanoparticles at room temperature is along with the hard reinforcements in the matrix.

Three trials were performed at multiple locations on every specimen, and the average value was determined (Gugulothu et al., 2023). The Hardness varied depending on the %wt of the reinforcement mixed with Al7178. The highest hardness value (83.7 Hv) was obtained in sample S1, with parameters of 700 rpm of stir speed, 20 min of stir time and 40 MPa of squeeze pressure. From the experiments, sample S4 recorded the lowest hardness value (70.6 Hv), with parameters of 800 rpm of stir speed, 20 min of stir time, and 80 MPa of squeeze pressure.

Figure 4 displays the overall hardness readings for all experiments. The design of experiments was performed using Minitab-18 software, considering various input factors. The SN ratio responses are illustrated for both tensile strength and hardness in Figs. 5 and 6 respectively. After analyzing Table 4, it was observed that the optimum UTS was achieved at the 3rd level of Stir speed, the 3rd level of Stir time (2nd rank), and the 1st level of Squeeze pressure (3rd rank).

From Table 5, an approximately 43.2% contribution due to stir speed to improve tensile property and Stirring time was 42.5% and followed by Squeeze pressure 12.1%. It was noticed that the optimum hardness was achieved at the 1st, 3rd and 2nd stages of speed, time and pressure. From Tables 6 and 7, Stirring time was the first factor which contributed 46.7% followed by Squeeze pressure 35.2% and Stirring speed 7.6% respectively.

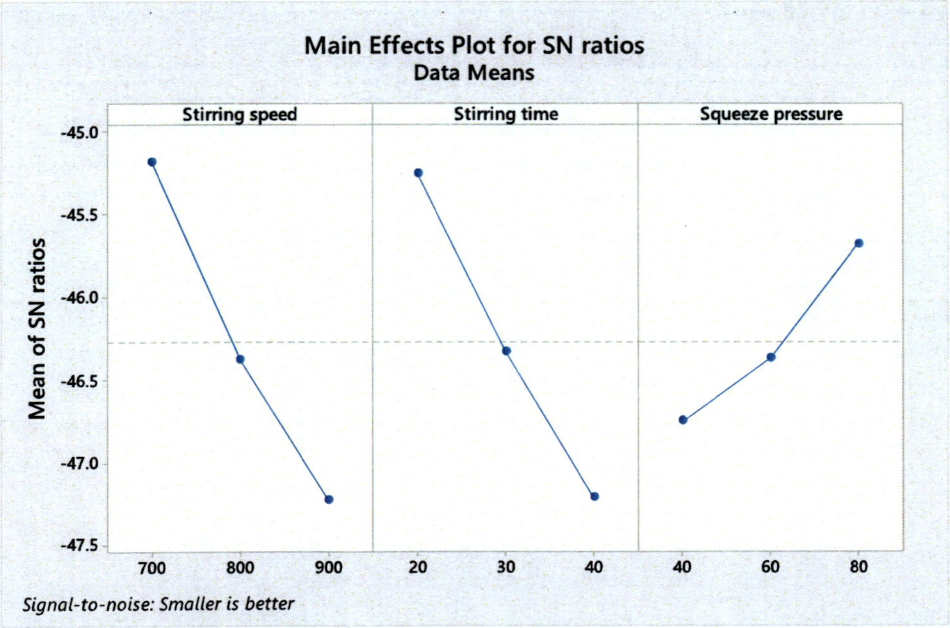

Fig. 5 SN ratio for tensile strength

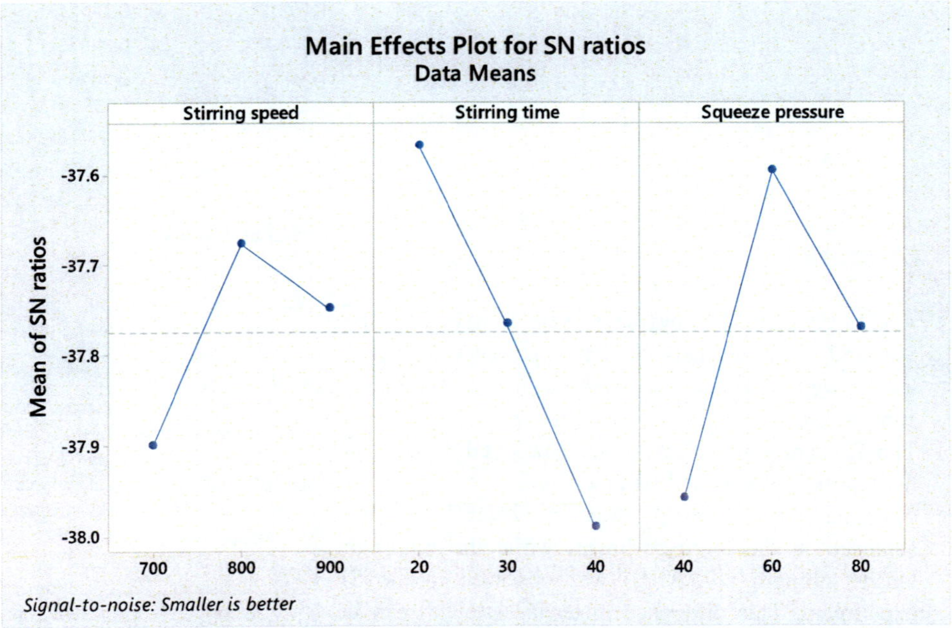

Fig. 6 SN ratio for hardness

Table 4 SN ratio—tensile strength

Stages	Stirring speed	Stirring time	Squeeze pressure
1	−45.19	−45.25	−46.75
2	−46.38	−46.33	−46.37
3	−47.23	−47.22	−45.68
Delta	2.04	1.96	1.07
Merit	1	2	3

Table 5 Tensile strength—ANOVA result

Parameters	DoF	Sum of squares	Mean square	F ratio	P ratio	%
Stir speed	2	6.2960	3.14799	34.99	0.028	43.2
Stir time	2	5.8081	2.90407	32.28	0.030	42.5
Squeeze pressure	2	1.7708	0.88541	9.84	0.092	12.1
Residual error	2	0.1799	0.08996			2.2
Total	8	14.0549				

Table 6 SN ratio–hardness

Stages	Stirring speed	Stirring time	Squeeze pressure
1	−37.90	−37.57	−37.96
2	−37.68	−37.77	−37.60
3	−37.75	−37.99	−37.77
Delta	0.22	0.42	0.36
Rank	3	1	2

Table 7 ANOVA result-hardness

Parameters	DoF	Sum-of-squares	Mean square	F ratio	P ratio	%
Stir speed	2	0.05708	0.03854	0.05	0.055	7.6
Stir time	2	0.27108	0.13554	0.18	0.044	46.7
Squeeze pressure	2	0.19745	0.09872	0.13	0.048	35.2
Residual error	2	0.16727	0.73363			10.5
Total	8	0.71287				

5 Conclusion

(1) AA7178/SiC/ZrO_2 composites were prepared by the Stir and Squeeze method. Three parameters with 3 levels were selected to analyzes the metal matrix composites.
(2) The Taguchi approach (L9) was employed to calculate the number of tests and parameters.
(3) The maximum Tensile strength (265.5 MPa) was achieved in the S9 experiment, while the minimum tensile strength (175 MPa) was observed in the S1 experiment. This difference in tensile strength can be subjected to the applied stress, the strengthening effect and the grain size of reinforcements.
(4) An approximately 43.2% contribution due to stir speed to improve tensile property and Stirring time was 42.5% followed by Squeeze pressure 12.1%.
(5) From the Hardness ANOVA result in Table 7, Stirring time was the first factor contributing 46.7% followed by Squeeze pressure at 35.2% and Stirring speed at 7.6% respectively.

References

Arvind, S., & Kaushik, M. P. (2022). Metal matrix composites fabricated by stir casting process–a review. *Advanced Materials and Processes, 8*, 1270–1291.

Dao, V., Zhao, S., Lin, W., & Zhang, C. (2012). Effect of process parameters on microstructure and mechanical properties in AlSi9Mg connecting-rod fabricated by semi-solid squeeze casting. *Materials Science and Engineering A, 558*, 95–102.

Gobalakrishnan, B., Lakshminarayanan, P. R., & Varahamoorthi, R. (2018a). Mechanical properties of Al 6061/TiB2 in-situ formed metal matrix composites. *Journal of Advanced Microscopy Research, 13*(1), 125–130.

Gobalakrishnan, B., Lakshminarayanan, P. R., & Varahamoorthi, R. (2018b). Effect of TiB2 particle addition on the mechanical properties of Al/TiB2 in situ formed metal matrix composites. *Materials Testing, 60*(12), 1221–1224.

Gugulothu, B., Bharadwaja, K., Vijayakumar, S., Rao, T. V. J., Sri, M. N. S., Anusha, P., & Agrawal, M. K. (2023). Modeling and parametric optimization of electrical discharge machining on casted composite using central composite design. *IJIDeM*.

Lee, et al. (2012). Microstructures and thermal properties of A356/SiCp composites fabricated by liquid pressing method. *Materials and Design, 37*, 313–316.

Manjunath, T. S., Seenappa, Kumar, G. B., Veeresh, Kumar, M. V. P., & Rudresh, M. Study of mechanical properties of SiC and Gr reinforced Al7075 powder metallurgy composites and analysis of results using ANOVA. *Journal of The Institution of Engineers (India): Series D, 103*, 341–348.

Park, B. G., Crosky, A. G., & Hellier, A. K. (2008). Fracture toughness of microsphere Al_2O_3-Al particulate metal matrix composites. *Composites Part B Engineering, 39*, 1270–1279.

Prabu, S. B., Karunamoorthy, L., Kathiresan, S., & Mohan, B. (2006). Influence of stirring speed and stirring time on distribution of particles in cast metal matrix composite. *Journal of materials processing technology*, *171*(2), 268–273.

Salur, E., Aslan, A., Kuntoglu, M., Gunes, A., & Sahin, O. S. (2019). Experimental study and analysis of machinability characteristics of metal matrix composites during drilling. *Composites Part B: Engineering*, *166*, 401−413.

Sekar, K., Jayachandra, G., & Aravindan, S. (2018). Mechanical and welding properties of A7178-SiC-ZrO2 hybrid composite fabricated by stir and squeeze casting. *Materials Today: Proceedings, 5*(9), 20268–20277.

Kumar, K. S., Premalatha, P., & Baskar, K. (2018). Evaluation of transport properties of concrete made with E-Waste plastic. *Journal of Testing and Evaluation*, *45.*

Shanmughasundaram, P., & Subramanian, R. (2012). Influence of magnesium and stirrer model in production of Al–fly ash composites–a taguchi approach. *Journal of Applied Sciences Research, 8* (3), 1646–1653.

Shorowordi, K. M., Laoui, T., Haseeb, A. S. M. A., Celis, J. P., & Froyen, L. (2003). Microstructure and interface characteris SiCs of B4C, SiC and Al2O3 reinforced Al matrix composites: A comparative study. *Journal of Materials Processing Technology, 142*(3), 738–743.

Umanath, K., Palanikumar, K., Sankaradass, V., & Uma, K. (2020). Optimizations of friction stir welding process parameters of AA6063 aluminium alloy by taguchi technique. *Journal of Materials Processing Technology, 46*, 4008–4013.

Vengatesh, S., & Rajkumar, D. R. (2019). Influence of natural fiber powders with carboxy methyl cellulose on mechanical and physical properties of laminated plate. *IRJET, 06*, 4221–4225.

Vijayakumar, S., Anitha, S., Arivazhagan, R., Hailu, A. D., Rao, T. V. J., & Pydi, H. P. (2022). Wear investigation of aluminum alloy surface layers fabricated through friction stir welding method. *Advances in Materials Science and Engineering*, 1–8.

Vijayakumar, S., Satheesh Kumar, P. S., Sampath kumar, P., Manickam, S., Ramaiah, G. B., & Pydi, H. P. (2022). The effect of stir-squeeze casting process parameters on mechanical property and density of aluminum matrix composite. *Advances in Materials Science and Engineering*, 1–10.

Zhao, G., Yuan, Y., Zhang, P., Zhou, T., & Zhou, H. (2020). Influence of orientations biomimeSiC units processed by laser on wear resistance of 7178 aluminium alloy. *Optics & Laser Technology, 127*, 106196.

Evaluation of Hardness Properties of Al7475/B4C/Fly Ash Hybrid Composites Using Friction Stir Process

G. Mahesh, Vinay Kumar Domakonda, Shaik Farooq, Ram Subbiah, D. R. Rajkumar, N. Srinivasa Rao, and S. Vijayakumar

Abstract

Nowadays, Advanced composite materials along with good properties are playing a significant role in many industries. Researchers and scientists use many methods for making composite material. Friction stir process (FSP) is one of the unconventional processes for producing surface composites. In this study, Al7475/B4C/fly ash hybrid composites are formed by FSP by changing the level of the processing parameters such as Tool rotation speed (750, 1000 and 1250 rpm), Tool travel speed (20, 30 and 40 mm/min) and percentage of reinforcements (B4C -4, 6, 8 wt% and constant Fly ash 5%). Boron carbide and Fly ash are used as primary and secondary supporting elements on Al 7475. A Hardness test has been conducted for 15 prepared samples. Design expert software is employed to perform the optimization process to identify the best parameters to improve the hardness value of prepared composites. From the ANOVA result, the tool rotational speed is the best parameter to increase the hardness (88HV). The prediction of Hardness values can also be found by varying the input process parameter values through the ramp diagram. The optimized value of Hardness 70.9 is obtained for input process parameters of TRS-1232.72 rpm, TTS-33.96 mm/min and RE 12.4 wt%.

Keywords

RSM • Aluminum matrix composites • Fly ash • Hardness • Optimization • ANOVA

1 Introduction

Metal matrix composites have been formed with many methods by converting solid/liquid materials into desired shapes. The assembly of different elements or alloys generally welding operation is used (Paramasivam & Vijayakumar, 2021). Dwivedi et al. (2021) used a friction stir process to merge different dissimilar materials such as copper, aluminum, iron and their alloys. With the aid of the rotation part which makes contact with workpiece surface to join various zones can be identified at the metal layer during FSP which depends on the parameters section at many levels like tool speed, tilt angle, depth and metal behavior and lubrication. Sathish et al. found the hardness of the composite prepared by FSW improved due to an increase in Tool speed (TS). Rana et al. (2016) used AA 7075 reinforced with boron-carbide composites and observed the wear rate using SEM analysis. Akshay et al. (2018) got improved values on UTS and yield strength by adding extra passes of reinforced BN on AA6061. Karpasand et al. (2020) explained Al7075 alloy with Boron carbide/Titanium boride composite was made through FSP and showed the effect of material behavior which was more than base metal. Mohanasundaram et al. (2020) prepared different AMC(Al-6061/B4C) on FSP

G. Mahesh
Department of Mechanical Engineering, Saranathan College of Engineering, Tiruchirapalli, Tamil Nadu, India
e-mail: maheshmech1@saranathan.ac.in

V. K. Domakonda · S. Farooq
Department of Mechanical Engineering, Vignan's Foundation for Science, Technology and Research, Vadlamudi, Andhra Pradesh, India

R. Subbiah
Department of Mechanical Engineering, Gokaraju Rangaraju Institute of Engineering and Technology, Hyderabad, India

D. R. Rajkumar (✉)
Department of Mechanical Engineering, CARE College of Engineering and Technology, Tiruchirappalli, 620 009, India
e-mail: rajkumardevapiriam@gmail.com

N. S. Rao
Department of Mechanical Engineering, Shri Vishnu Engineering, College for Women, Bhimavaram, India
e-mail: srinivasaraon@svecw.edu.in

S. Vijayakumar
Department of Mechanical Engineering, BVC Engineering College (Autonomous), Odalarevu-533210, Andhra Pradesh, India

M. Sumesh et al. (eds.), *2nd International Conference on Smart Sustainable Materials and Technologies (ICSSMT 2023)*,
Advances in Science, Technology & Innovation, https://doi.org/10.1007/978-3-031-49826-8_3

by making changes to input consideration, in which optimal coating reading of thickness and depth are noted as around 3 mm and 20 mm respectively. Pitchipoo et al., (2021) and Kumar et al., (2021) revealed tensile and impact properties of the weld process were raised for AA6082/T6, and maximum strain (23%) has been observed at TRS (1400 RPM, 50 mm/Min and 1° tool tilt angle) and better UTS and HV.

The effectiveness of invented joint underneath tool pin shape of welding-tool was examined and thermal stuff also improved by 14%. After going through many articles and literature evaluations (Christy et al., 2020; Prakash Pasupulla et al., 2022), FSW of Al 5456 was examined by Chen et al. (Chen et al., 2006) under various conditions. During welding, the plastic material was unable to flow correctly toward the end of the pin due to the tool's modest tilt angle of 1.5°, and the nugget zone did not get enough plastic material as a result of the tool's tilt angle rising to more than 4.5°. According to Bahemmat et al. (2012), altering welding and rotational speeds causes a decrease in the quantity of heat generated and dissipated, which in turn causes the hardness value in the AA7075/AA2024-T4 Stir zone to increase. The pin's shoulder diameter has a significant impact on the heat generated during the FSW process, and various research have examined the link between shoulder and pin diameter to give better mechanical properties (Kumar et al., 2012; Rajakumar et al., 2011; Zhang et al., 2009).

The objective of this investigation is to make Al7475-B4C-Fly ash composites fabricated via Friction stir processing and to analyze the best hardness contribution of selected constraints through RSM technique. Many researchers have used Taguchi design but in this paper, we have discussed RSM technique. The significance of Al7475-B4C-Fly ash hybrid composite would depend on its intended application. Aluminum alloys are known for their lightweight, high strength-to-weight ratio and corrosion resistance. Boron carbide is a ceramic material with exceptional hardness and high-temperature stability. Fly ash, when properly processed, can be used as a filler material to enhance the mechanical properties of composites including compressive strength (Alnahhal et al., 2022; Velumani et al., 2016). The future scope of Al7475/B4C/fly ash hybrid composites can be explored in various areas, including aerospace, automotive, construction and other industries that require materials with enhanced mechanical properties.

2 Experimental Procedure

The elements of base alloy (Al7475) are listed in Table 1. Milling machine helps to cut the base plate 7475 into standard dimensions 100 × 50 × 4 mm. Groove is made near to center of the plate to fill the chosen reinforcement mixture in the gap the fabrication of composites on FSP was carried out as per the parameter values in Table 2. Due to heat generated by inserting tool on the workpiece. Afterward, the creation sample and micro-hardness were observed by using a Rockwell tester; response surface methodology has been used to investigate mathematical and automation values of measured HV. HSS-13 tool is employed as a tool to join the plates of Al alloy which is shown in Fig. 1.

Table 1 Compositions of Al 7475

Material-elements	Zn	Mg	Si	Cr	Al
Contribution (%)	5.5	2.2	1.4	0.15	Remaining

Table 2 FSW process factors

Parameters	Level-I	Level-II	Level-III
A-tool rotational speed (rpm)	750	1000	1250
B-tool travel speed (mm/min)	20	30	40
C-RE % w.t (B4C/FA)	8	11	14

Fig. 1 HSS tool for friction stir welding

2.1 Central Composite Design

Popular experimental design methods within response surface methodology (RSM) include Central Composite Design (CCD). It involves a combination of factorial and axial points to explore the response surface of a system and optimize the desired response. As per CCD approach with block 2 category 15 experiments are conducted.

2.2 Hardness Test

In most hardness tests, the material is identified while being loaded or forced with a predetermined force using equipment (Gugulothu et al., 2022; Vijayakumar et al., 2022a). Making various impressions on the surface of the composites of welded areas with a 1/16-inch indenter and a 40-kg load will reveal the hardness of the welded composites (Vijayakumar et al., 2022b).

3 Result and Discussions

ANOVA (Analysis of Variance) is a statistical method utilized to assess and determine the significance of differences between groups. Its primary application lies in comparing the means of two or more groups to identify statistically significant variations among them. By evaluating both the between-group variance (the variability between groups) and the within-group variance (the variability within each group), ANOVA provides insights into potential meaningful differences among the compared groups. If the between-group variance significantly outweighs the within-group variance, it indicates the presence of substantial distinctions among the groups under examination (Bag et al., 2023; Chandra et al., 2020; Ramadoss et al., 2020).

Based on central composite design of RSM technique, fifteen samplings were shaped under Friction stir processing (FSP). Every investigational output clarifications are registered in Table 3. In the Optimize procedure on RSM, P and F-values are the most important aspects that influence predicting output responses in ANOVA model. F and P values play a crucial role in determining how well a chosen quadratic regression model can be predicted. A detailed investigation of the model's significance using ANOVA was conducted. The chosen factors should be regarded as insignificant if P is greater than 0.05, and the variables should be regarded as significant if P is less than 0.05. The highest BHN (88 HV) is found at run no 4; however, a minimum (59HV) is seen for run 6 among all samples (Table 3). HBN is increased due to increasing TRS and RE but reduced on TTS. This has happened because of heat generated or friction on the work surface of samples and the rate of hardness depending on the rotational speed of the tool and the addition of reinforcements on Al alloy. ANOVA–Hardness for models is shown in Table 4.

Actual and predicted values of HB are shown in Fig. 3 and also the interaction of particular constraints vs BHN is also indicated (Fig. 2). The line that fits the probability plots points only shows minimal variations. It appears that these data are best modeled by the normal distribution. The actual and predicted values are very close to each other in Fig. 3. A ramp diagram, commonly used in Design Expert software and other experimental design software, is a graphical representation that illustrates the experimental conditions or factor levels in a designed experiment (Anusha et al., 2023; Gugulothu et al., 2023; Nathan et al., 2019). The horizontal axis represents each factor, while the vertical axis shows the

Table 3 Experimental result-BHN

Run	A: TRS (rpm)	B: TTS (mm/min)	C: RE (W.t %)	Hardness (HV)
1	750	30	6	68
2	1250	20	4	75
3	1000	20	8	72
4	1250	30	8	88
5	1000	40	4	74
6	1000	30	4	59
7	750	30	8	62
8	750	20	4	82
9	1000	40	6	67
10	1250	30	6	78
11	1250	40	4	70
12	750	40	4	81
13	1000	20	6	77
14	1000	30	4	76
15	1000	30	8	63

Table 4 ANOVA–hardness

Source	SS	Df	MS	F-value	p-value
Model	388.90	9	43.21	0.4127	0.08824
A-TRS	40.50	1	40.50	0.3868	0.04121
B-TTS	24.50	1	24.50	0.2340	0.08902
C-RE	4.50	1	4.50	0.0430	0.04439
A.B	4.00	1	4.00	0.0382	0.08527
A.C	64.00	1	64.00	0.6113	0.04697
B.C	36.00	1	36.00	0.3438	0.05831
A^2	144.23	1	144.23	1.38	0.02934
B^2	83.31	1	83.31	0.7957	0.04133
C^2	11.31	1	11.31	0.1080	0.07557
Residual	523.50	5	104.70		
Lack of fit	365.50	3	121.83	1.54	0.04166
Pure error	158.00	2	79.00		
Total	912.40	14			

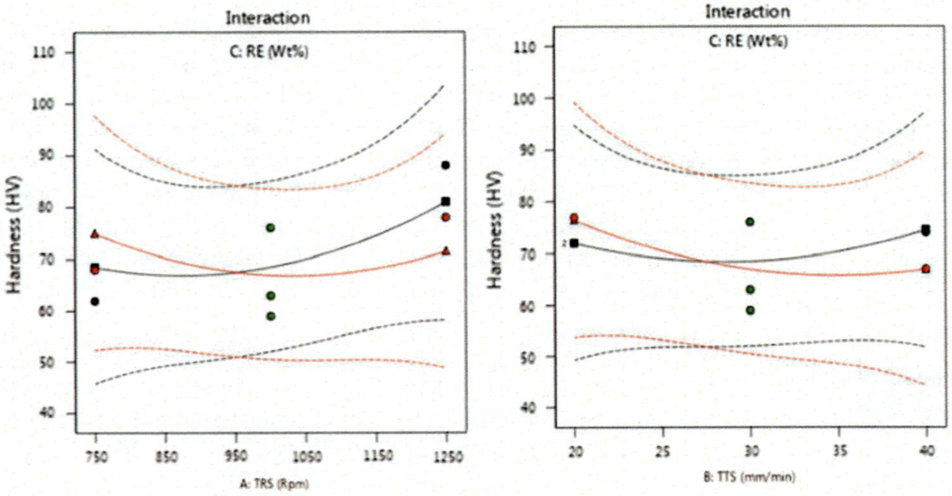

Fig. 2 Interaction among parameters

levels or settings of factors. It displays the progression of factor levels throughout the experiment and understands the arrangement and sequencing of factor levels. Furthermore, It facilitates the examination of the relationship between factor settings and the response variables being investigated. The prediction of Hardness values also can be found by varying the input process parameter values through the ramp diagram. Ramp diagram response surface process found that optimum hardness value was 70.9 HV for optimal parameters of TRS (1233 RPM), TTS (33.6 mm/min) and RE (12.45w.t %). Optimized rate of BHN from input responses is shown in Fig. 4.

4 Conclusion

The supports of boron-carbide along with Fly ash are mixed on AA7475 item through a Friction stirring process. Fifteen specimens are made by changes in stirring process factors such as TRS (750–1250 rpm), TTS (20–40 mm/min) and reinforcement level (8–14%). Response surface method (RSM) in design software is used as the main source to examine optimized values of experimental BHN evaluations. The following results are to be below.

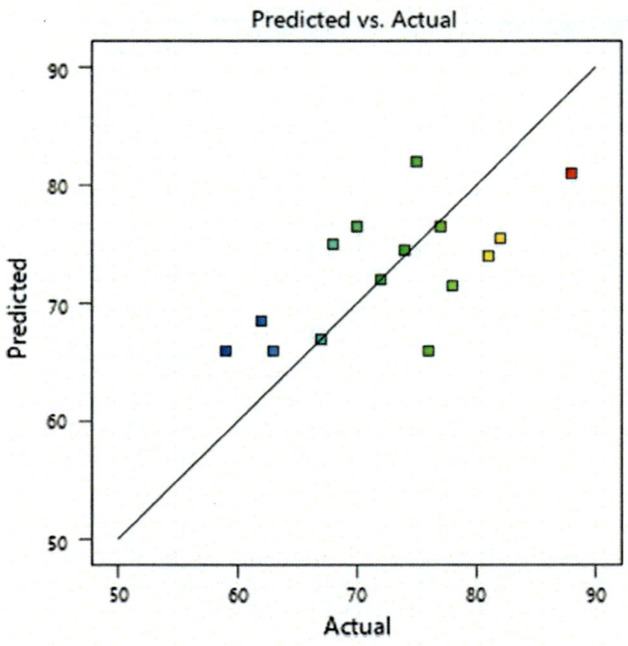

Fig. 3 Actual versus predicted-BHN

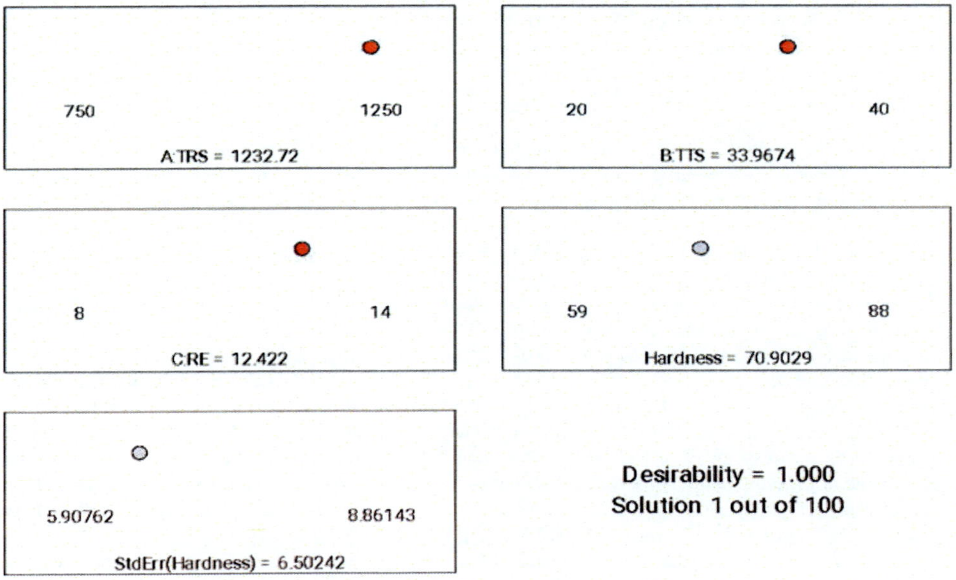

Fig. 4 Optimized rate of BHN from input responses

- The highest hardness (88 HV) is found at specimen no 4 with process parameters of tool rotational speed 1250 rpm, tool transverse speed 30 mm/min and reinforcement percentage 8 wt%, whereas minimum hardness (59HV) is obtained at specimen no 6 by maintaining the tool rotational speed at 1000 rpm, tool transverse speed at 30 mm/min and reinforcement percentage at 4 wt%.
- Hardness value is increased by increasing tool rotational speed and reinforcement percentage. This has happened because of the heat generated or fiction on the work surface of samples.
- Hardness prediction values can be found by varying the input process parameter values through the ramp diagram. From the ramp result, the optimum hardness value 70.9 HV for optimal parameters of TRS (1233 RPM), TTS (33.6 mm/min) and RE (12.45w.t %).

References

Akshay, B. R., Keshavamurthy, R., Kuppahalli, P., & Sudhan, J. M. (2018). Mechanical properties of friction stir processed Al6061-BN surface composite. *Materials Today: Proceedings, 5*(11), 24568–24577. https://doi.org/10.1016/j.matpr.2018.10.254.

Alnahhal, A. M., Alengaram, U. J., Yusoff, S., Darvish, P., Srinivas, K., & Sumesh, M. (2022). Engineering performance of sustainable geopolymer foamed and non-foamed concretes. *Construction and Building Materials, 316*(125601), 125601. https://doi.org/10.1016/j.conbuildmat.2021.125601.

Anusha, P., Sri, M. N. S., Vijayakumar, S., Rao, T. V. J., Paramasivam, P., Jeyakrishnan, S., & Saxena, K. K. (2023). Design and optimization the wear characteristics for Al7178/TiO2/B4C/FA central hybrid composite. *International Journal on Interactive Design and Manufacturing*. https://doi.org/10.1007/s12008-023-01341-5.

Bag, R., Panda, A., Sahoo, A. K., & Kumar, R. (2023). Sustainable high-speed hard machining of AISI 4340 steel under dry environment. *Arabian Journal for Science and Engineering, 48*(3), 3073–3096. https://doi.org/10.1007/s13369-022-07094-9.

Bahemmat, P., Haghpanahi, M., Besharati Givi, M. K., & Reshad Seighalani, K. (2012). Study on dissimilar friction stir butt welding of AA7075-O and AA2024-T4 considering the manufacturing limitation. *The International Journal of Advanced Manufacturing Technology, 59*(9–12), 939–953. https://doi.org/10.1007/s00170-011-3547-4.

Chandra, D., Chauhan, N. R., & Rajesha, S. (2020). Hardness and toughness evaluation of developed Al metal matrix composite using stir casting method. *Materials Today: Proceedings, 25*, 872–876. https://doi.org/10.1016/j.matpr.2019.12.026.

Chen, H.-B., Yan, K., Lin, T., Chen, S.-B., Jiang, C.-Y., & Zhao, Y. (2006). The investigation of typical welding defects for 5456 aluminum alloy friction stir welds. *Materials Science & Engineering. A, Structural Materials: Properties, Microstructure and Processing*, *433*(1–2), 64–69. https://doi.org/10.1016/j.msea.2006.06.056.

Christy, J. V., Arunachalam, R., Mourad, A.-H.I., Krishnan, P. K., Piya, S., & Al-Maharbi, M. (2020). Processing, properties, and microstructure of recycled aluminum alloy composites produced through an optimized stir and squeeze casting processes. *Journal of Manufacturing Processes, 59*, 287–301. https://doi.org/10.1016/j.jmapro.2020.09.067.

Dwivedi, R., Kumar Singh, R., Kumar, S., & Kumar Srivastava, A. (2021). Parametric optimization of process parameters during the friction stir processing of Al7075/SiC/waste eggshell surface composite. *Materials Today: Proceedings, 47*, 3884–3890. https://doi.org/10.1016/j.matpr.2021.03.545.

Gugulothu, B., Anusha, P., Swapna Sri, M. N., Vijayakumar, S., Periyasamy, R., & Seetharaman, S. (2022). Optimization of stir-squeeze casting parameters to analyze the mechanical properties of Al7475/B4C/Al2O3/TiB2 hybrid composites by the Taguchi method. *Advances in Materials Science and Engineering, 2022*, 1–9. https://doi.org/10.1155/2022/3180442.

Gugulothu, B., Bharadwaja, K., Vijayakumar, S., Rao, T. V. J., Sri, M. N. S., Anusha, P., & Agrawal, M. K. (2023). Modeling and parametric optimization of electrical discharge machining on casted composite using central composite design. *International Journal on Interactive Design and Manufacturing*. https://doi.org/10.1007/s12008-023-01323-7.

Karpasand, F., Abbasi, A., & Ardestani, M. (2020). Effect of amount of TiB2 and B4C particles on tribological behavior of Al7075/B4C/TiB2 mono and hybrid surface composites produced by friction stir processing. *Surface & Coatings Technology, 390* (125680), 125680. https://doi.org/10.1016/j.surfcoat.2020.125680.

Kumar, R., Singh, K., & Pandey, S. (2012). Process forces and heat input as function of process parameters in AA5083 friction stir welds. *Transactions of the Nonferrous Metals Society of China, 22* (2), 288–298. https://doi.org/10.1016/s1003-6326(11)61173-4.

Kumar, R., Singh Dhami, S., & Mishra, R. S. (2021). Optimization of friction stir welding process parameters during joining of aluminum alloys of AA6061 and AA6082. *Materials Today: Proceedings, 45*, 5368–5376. https://doi.org/10.1016/j.matpr.2021.01.958.

Mohanasundaram, S., Vijay, S. J., Vasanth, X. A., Rai, R. S., & Kantharaj, I. (2020). Optimization of coating thickness and coating width for friction surfaced Al6061-B4C over Al6061. *Materials Today: Proceedings, 33*, 939–945. https://doi.org/10.1016/j.matpr.2020.06.492.

Nathan, V.B., Soundararajan, R., Abraham, C. B., & Rahman, A. F. (2019). Evaluation of mechanical and metallurgical properties on aluminium hybrid metal matrix composites. *Materials Today: Proceedings, 18*, 2520–2529. https://doi.org/10.1016/j.matpr.2019.07.109.

Paramasivam, P., & Vijayakumar, S. (2021). Mechanical characterization of aluminium alloy 6063 using destructive and non-destructive testing. *Materials Today: Proceedings*. https://doi.org/10.1016/j.matpr.2021.04.312.

Pitchipoo, P., Muthiah, A., Jeyakumar, K., & Manikandan, A. (2021). Friction stir welding parameter optimization using novel multi objective dragonfly algorithm. *The International Journal of Lightweight Materials and Manufacture, 4*(4), 460–467. https://doi.org/10.1016/j.ijlmm.2021.06.006.

Prakash Pasupulla, A., Abebe Agisho, H., Seetharaman, S., & Vijayakumar, S. (2022). Characterization and analysis of TIG welded stainless steel 304 alloy plates using radiography and destructive testing techniques. *Materials Today: Proceedings, 51*, 935–938. https://doi.org/10.1016/j.matpr.2021.06.305.

Rajakumar, S., Muralidharan, C., & Balasubramanian, V. (2011). Influence of friction stir welding process and tool parameters on strength properties of AA7075-T6 aluminium alloy joints. *Materials in Engineering, 32*(2), 535–549. https://doi.org/10.1016/j.matdes.2010.08.025.

Ramadoss, N., Pazhanivel, K., & Anbuchezhiyan, G. (2020). Synthesis of B4C and BN reinforced Al7075 hybrid composites using stir casting method. *The Journal of Materials Research and Technology, 9*(3), 6297–6304. https://doi.org/10.1016/j.jmrt.2020.03.043.

Rana, H. G., Badheka, V. J., & Kumar, A. (2016). Fabrication of Al7075 / B4C surface composite by novel friction stir processing (FSP) and investigation on wear properties. *Procedia Technology, 23*, 519–528. https://doi.org/10.1016/j.protcy.2016.03.058.

Velumani, P., SenthilKumar, S., & Premalatha, P. V. (2016). An innovative approach to evaluate the performance of sludge-incorporated fly ash bricks. *Journal of Testing and Evaluation, 44*, 2155–2163.

Vijayakumar, S., Anitha, S., Arivazhagan, R., Hailu, A. D., Rao, T. V. J., & Pydi, H. P. (2022a). Wear investigation of aluminum alloy surface layers fabricated through friction stir welding method. *Advances in Materials Science and Engineering, 2022*, 1–8. https://doi.org/10.1155/2022/4120145.

Vijayakumar, S., Satheesh Kumar, P.S., Sampath kumar, P., Manickam, S., Ramaiah, G. B., & Pydi, H. P. (2022b). The effect of stir-squeeze casting process parameters on mechanical property and density of aluminum matrix composite. *Advances in Materials Science and Engineering*, 1–10. https://doi.org/10.1155/2022/3741718.

Zhang, Z., Liu, Y. L., & Chen, J. T. (2009). Effect of shoulder size on the temperature rise and the material deformation in friction stir welding. *The International Journal of Advanced Manufacturing Technology, 45*(9–10), 889–895. https://doi.org/10.1007/s00170-009-2034-7.

Multiscale Material Modelling for Evaluating Mechanical and Electrical Characteristics of Graphene/Glass Fibre Epoxy Hybrid Composites

M. Mahesh Kumar, S. Paul Vizhian, and M. Krishna

Abstract

Investigating the impact of graphene incorporation in epoxy–glass fibre hybrid composites on mechanical and electrical properties through Multiscale Material Modelling (MMM). In this paper, it is examined how the inclusion of graphene in epoxy and glass fibre hybrid composites affected their mechanical and electrical characteristics. This analysis is carried out by employing a MMM and Finite Element Modelling (FEM) to evaluate elastic constants, Poisson's ratio, and electrical properties. The representative volume elements were employed to model the graphene and glass fibre phases. By employing the Mori–Tanaka and Fei Deng models, multiscale analysis was conducted to predict the elastic constants and electrical conductivity of the hybrid composites. The outcomes revealed that the addition of 5 wt% graphene enhanced the longitudinal and transverse Young's modulus, in-plane and out-plane Poisson ratio, and in-plane and out-plane shear modulus of glass fibre hybrid composites by 0.95%, 7.12%, 6.44%, 2.66%, 2.47%, and 3.65% respectively. Furthermore, the electrical conductivity of these composites increased by 0.69% with the addition of 5 wt% graphene, compared to 1 wt% addition of graphene, and the orientation does not significantly affect the overall electrical conductivity because of the establishment of conductive pathways.

Keywords

Glass fibre/graphene hybrid composites • Multiscale material modelling • Finite element analysis

M. Mahesh Kumar (✉) · S. Paul Vizhian
Department of Mechanical Engineering, UVCE, Bangalore, India
e-mail: maheshkumarm0026@gmail.com

M. Krishna
Department of Mechanical Engineering, RVCE, Bangalore, India
e-mail: krishnam@rvce.edu.in

1 Introduction

Glass/Epoxy composites have gained widespread utility because of exceptional specific stiffness and strength. Conversely, glass/epoxy composites infused with nanofillers exhibit heightened structural and functional attributes, including sensing capabilities, electromagnetic interference shielding, and electrical properties (Aguilar et al., 2010; Pal & Kumar, 2016; Sharma & Shukla, 2014). Incorporating graphene into polymer composites augments electrical conductivity—a pivotal non-structural parameter. Mathematical models of elastic constants in a nano-enhanced composites are integral for optimizing designs and industrial applications. Consequently, the non-structural facets of nanofiller-infused glass/epoxy composites are explored using molecular dynamics and micromechanics models (Chwał et al., 2013). Some researchers (Tsai et al., 2010) have engaged a multistage micromechanics model to forecast the electrical resistance of nano-FRP structures.

The Mori–Tanaka (Chatzigeorgiou et al., 2012) micromodel has been widely adopted by researchers to prognosticate effective elastic constants in CNT-reinforced polymer structures using composites and self-consistent methods. Likewise, this model has been applied to predict both along-axis and perpendicular-axis elastic moduli of carbon nanotube-based composites (Tsai et al., 2010). The micro-mechanical model employing a unit cell approach was used to forecast elastic constants for diverse orientations of carbon nanotube (CNT)-based composites (Ansari & Hassanzadeh-Aghdam, 2016). The Mori–Tanaka model (Feli et al., 2017) has been employed to compute the effective elastic behaviour of CNT-enhanced polymer-based composites. Other researchers (Pan et al., 2016) have devised novel micromodels to forecast effective mechanical properties in nanopolymer composites.

The Multiscale Modelling (MM) of combined electrical conductivity in nanofiller-based polymer composites significantly impacts electronic application design. Various

M. Sumesh et al. (eds.), *2nd International Conference on Smart Sustainable Materials and Technologies (ICSSMT 2023)*, Advances in Science, Technology & Innovation, https://doi.org/10.1007/978-3-031-49826-8_4

theoretical, experimental, and analytical methods have computed effective electrical conductivity in nanofiller-based composites. Deng et al. (2008) formulated a simple micro-mechanical model accounting for percolation, anisotropy, and the conductive network to predict effective electrical conductivity. Takeda et al. (2011) considered electron hopping mechanisms among CNTs and adapted the Deng and Zheng model by treating CNTs as effective nanofibers to evaluate overall conductivity. Electron hopping and conductive network formation effects on effective electrical conductivity have been investigated by other researchers (Li et al., 2010), employing the Mori–Tanaka micromechanics approach (Seidel & Lagoudas, 2009). Feng et al. (2013) introduced an MM model.

Although considerable research has delved into modelling and simulating nanofiller-based polymer composites for mechanical and electrical properties, these investigations are mainly confined to two-phase systems with randomly oriented fillers. Conversely, there is a dearth of literature concerning three-phase nanofiller-modified GFRP composites. This study aims to scrutinize the impact of graphene as a nanofiller addition to GFRP, evaluating. Consequently, parametric analyses of these properties are conducted, encompassing varying weight percentages (wt%) of graphene and its orientations within the GFRP matrix.

2 Multiscale Micromechanics Modelling for Mechanical Properties

Figure 1 illustrates the methodology of Multistage Material Modelling (MMM) employed to calculate the effective elastic modulus of stages in a GFRP reinforced with graphene. The orientation of the graphene is aligned with the z-axis. The computation of the elastic tensor Cr involves transforming the elastic stiffness of graphene.

Given Expression:

$$C_r = [T]^T C_{Gr} [T]^{-1} \tag{1}$$

Where T is transformation matrix

$$T = \begin{bmatrix} 1 & 0 & 0 & 0 & 0 & 0 \\ 0 & cos^2\alpha & sin^2\alpha & \cos\alpha\sin\alpha & 0 & 0 \\ 0 & sin^2\alpha & cos^2\alpha & -\cos\alpha\sin\alpha & 0 & 0 \\ 0 & -2\cos\alpha\sin\alpha & 2\cos\alpha\sin\alpha & cos^2\alpha - sin^2\alpha & 0 & 0 \\ 0 & 0 & 0 & 0 & \cos\alpha & -\sin\alpha \\ 0 & 0 & 0 & 0 & \sin\alpha & \cos\alpha \end{bmatrix} \tag{2}$$

The elastic stiffness of epoxy composites with graphene flakes is evaluated using the Mori Tanka model (Chatzigeorgiou et al., 2012) given in

$$C_{eq} = C_M + V_{gra} \times O \tag{3}$$

$$\text{Where } O = (C_{gra} - C_M)\left[I + S_{ESH} : C_{gra} - C_M\right]^{-1} \tag{4}$$

$$S_{ESH} = 1/(1 - V_{gra}) * \left[V_{gra}\left((CM)^{-1} : C_\alpha - I\right)^{-1} - (C_{gra} - C_M)^{-1} : C_M\right) \tag{5}$$

$$C_M = \begin{pmatrix} 1-\mu & \mu & \mu & 0 & 0 & 0 \\ \mu & 1-\mu & \mu & 0 & 0 & 0 \\ \mu & \mu & 1-\mu & 0 & 0 & 0 \\ 0 & 0 & 0 & \frac{(1-2\mu)}{2} & 0 & 0 \\ 0 & 0 & 0 & 0 & \frac{(1-2\mu)}{2} & 0 \\ 0 & 0 & 0 & 0 & 0 & \frac{(1-2\mu)}{2} \end{pmatrix}$$

$$C_M = \frac{E}{(1+\mu)(1-2\mu)} \tag{6}$$

Here, C_{eq}, C_m, and C_{gra} stand for the elastic stiffness tensors corresponding to the hybrid composites, epoxy, and graphene respectively. Additionally, V_{gra}, O, and I, represent the graphene volume fraction, the orientation of graphene, and the IV order identity tensor. S_{ESH} refers to the Eshelby tensor of the hybrid composites (Hassanzadeh-Aghdam et al., 2018). The volume fraction of graphene within the epoxy matrix can be determined by relating it to the graphene's weight percentage and the densities of graphene and epoxy, as follows:

$$V_{gra} = \frac{w_{gra}}{w_{gra} + \frac{\rho_{gra}}{\rho_m} - \frac{\rho_{gra}}{\rho_m} w_{gra}} \tag{7}$$

$$C_{glass} = \begin{pmatrix} \frac{1}{E_1} & \frac{\mu_{21}}{E_2} & \frac{\mu_{31}}{E_3} & 0 & 0 & 0 \\ \frac{\mu_{12}}{E_1} & E_2 & \frac{\mu_{32}}{E_3} & 0 & 0 & 0 \\ \frac{\mu_{13}}{E_1} & \frac{\mu_{23}}{E_2} & E_3 & 0 & 0 & 0 \\ 0 & 0 & 0 & \frac{1}{G_{23}} & 0 & 0 \\ 0 & 0 & 0 & 0 & \frac{1}{G_{13}} & 0 \\ 0 & 0 & 0 & 0 & 0 & \frac{1}{G_{12}} \end{pmatrix} \tag{8}$$

$$E_1 = E_{equl} \times (1 - V_f) + V_f \times E_f \tag{9}$$

$$E_2 = \left[\frac{1}{E_1 X (1 - V_f)} + \frac{1}{E_f X V_f}\right]^{-1} \tag{10}$$

$$\mu_{12} = \mu_{equl12} x (1 - V_f) + V_f x\ \mu_{12f} \tag{11}$$

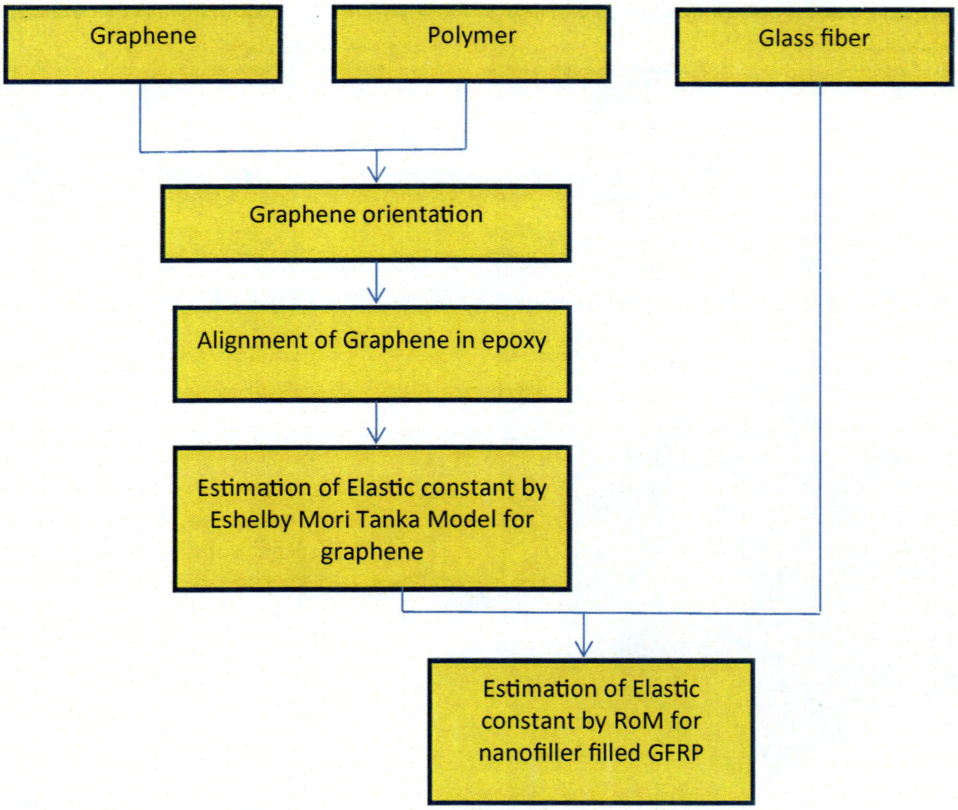

Fig. 1 Presents a multi-scale micro-mechanics method for estimating the effective elastic constants of glass/graphene reinforced hybrid composites

$$G_2 = \left[\frac{1}{G_{equ12}X(1 - V_f)} + \frac{1}{G_{12f}XV_f}\right]^{-1} \quad (12)$$

$$\mu_{23} = \left[\frac{1}{\mu_{equ23}X(1-V_f)} + \frac{1}{\mu_{23f}XV_f}\right]^{-1} \quad (13)$$

Where,

v_m and v_f are the volume fraction of glass fibre and graphene respectively. With $v_m + v_f = 1$.

3 Finite Element Analysis for Mechanical Properties

The multiscale approach using FEM for predicting average elastic constants of graphene/GFRP composites is given in three stages. In the initial stage, the nanoscale graphene model, and in the second stage, the microscale model of glass fibre and epoxy are modelled using FE software, as shown in Figs. 2 and 3.

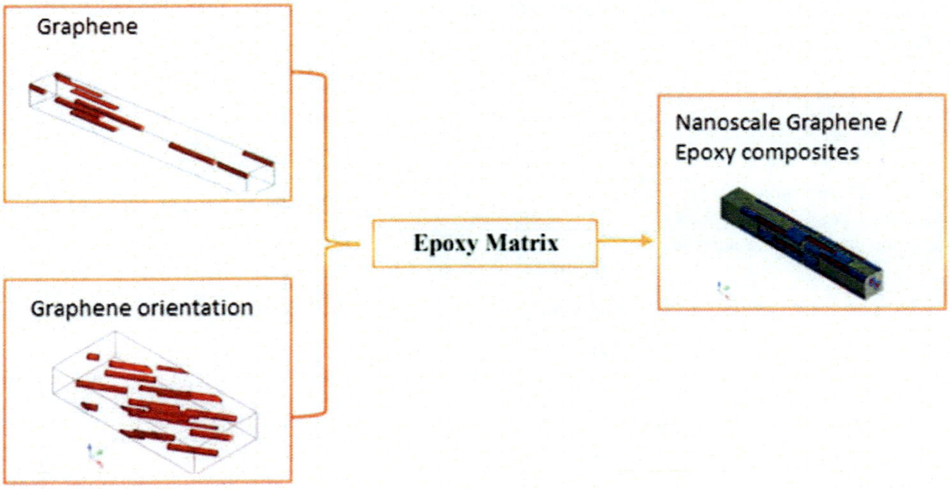

Fig. 2 Schematic representation of nanoscale RVE of MWCNT/epoxy composites

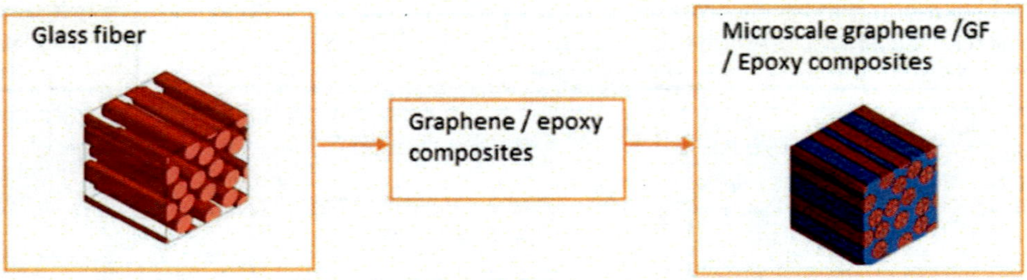

Fig. 3 Schematic representation of microscale glass fibre/graphene/epoxy composite

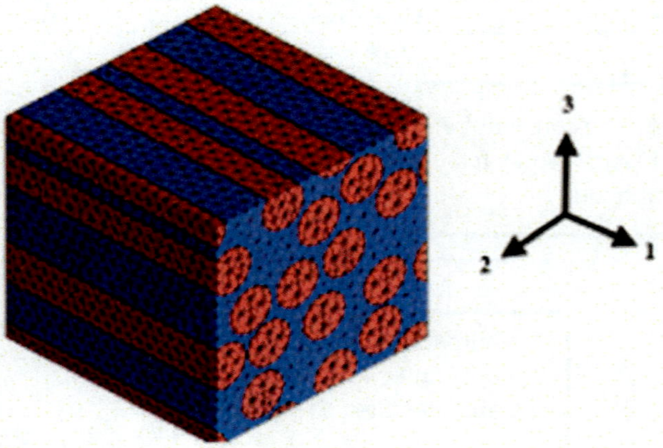

Fig. 4 Graphene-reinforced glass/epoxy composites meshed with tetrahedron mesh

The third stage involves the generation of mesh and post-processing. The nanographene was meshed with linear tetrahedron elements (Fig. 4).

Post-processing (Xia et al., 2003) entails solving the weak formulation of the equilibrium equation for six distinct loading scenarios in isolation. Subsequently, the fluctuating strain tensor is projected at each integration point and utilized to establish the microstructural strain tensor, M_{ijkl}. The determination of the effective stiffness, C_{ijkl}, involves integrating the characteristics of the composites divided by the overall composite volume.

$$\breve{C} = \frac{1}{V_{Com}} \sum_{i=1}^{NE} \sum_{j=1}^{NI} |C||M||J| w_j \tag{14}$$

In this context, C signifies the stiffness tensor of the constituents, M represents the microstructural strain tensor, J denotes the Jacobian determinant, NE stands for the count of elements in the composites, wj signifies the weights, and NI represents the quantity of integration points utilized in numerical integration. The attributes of glass fibre, graphene, and epoxy utilized in both models are detailed given in Table 1.

4 Multiscale Micromechanics Modelling to Predict Electrical Conductivity

The strategy for estimating the electrical conductivity of hybrid composites reinforced with graphene is a multiscale approach that unfolds in three stages. The initial stage involves aligning the graphene in a straight orientation at an angle θ within the z-direction. The electrical conductivity tensor S is derived through a specific transformation of the graphene's electrical conductivity tensor Sgra and can be formulated as per (Seidel & Lagoudas, 2009).

$$S = [Q]^T S_{Gr} [Q]^{-1} \tag{15}$$

$$Q = \begin{bmatrix} (\cos a)(\sin \varnothing) & (\sin \alpha)(\sin a) & \cos\phi \\ -\sin a & \cos a & 0 \\ -(\cos a)(\sin \varnothing) & -(\sin \alpha)(\sin a) & \sin\phi \end{bmatrix} \tag{16}$$

Table 1 Elastic properties of glass fibre, graphene, and epoxy are inputs to modelling analysis

Materials	Density	Diameter	E_1	E_2	G_{12}	G_{23}	μ_{12}	μ_{23}
	gm/cc	μm	GPa					
Glass fibre	2.58	7	72.4	9.2	3.4	2.2	0.2	0.35
Graphene	2.26	$4–20 \times 10^{-3}$	130	22.5	277	5.98	0.139	0.91
Epoxy	1.14		3				0.35	

Here, Q represents the transformation matrix. In the subsequent stage, the electrical conductivity of the epoxy nanocomposite featuring oriented straight-type graphene 's' is assessed utilizing the Fei Deng model (Subramanian et al., 2006), as indicated by:

$$\frac{S_{m1}}{S_m} = 1 + \frac{V_{gra}}{3}\left[\frac{1}{\left(\frac{S_a(1,1)}{S_m} - 1\right)^{-1} + H} + \frac{2}{\left(\frac{S_a(2,2)}{S_m} - 1\right)^{-1} + \frac{(1-H)}{2}}\right] \quad (17)$$

where Sa, Sm, and $Sm1$ are electrical conductivities of graphene, epoxy resin, and graphene/epoxy matrix. V_{gra} is graphene volume fraction. Which can be obtained by correlating it to the weight percentage, and densities of graphene and epoxy expressed as

$$V_{gra} = \frac{W_{gra}}{W_{gra} + \frac{\rho_{gra}}{\rho_m} - \frac{\rho_{gra}}{\rho_m} W_{gra}} \quad (18)$$

where W_{gra}, ρ_{gra}, and ρ_m are weight fractions of graphene, the density of graphene and density of matrix, respectively, and H is a non-dimensional number (Subramanian et al., 2006), based on the aspect ratio (Ar = l/d) of graphene given as.

$$H = \frac{1}{A_r^2 - 1} X \left[\frac{A_r}{\sqrt{A_r^2 - 1}} \ln(A_r + \sqrt{A_r^2 - 1}) - 1\right] \quad (19)$$

In the third stage, the electrical conductivity of a three-phases such as glass fibre, graphene, and epoxy nanocomposite is obtained using the rule of mixtures and can be expressed as (Haider et al., 2016).

$$S_1 = V_f S_f + V_m S_{m1} \quad (20)$$

5 Material Properties

Before performing any analysis, the adequate material properties of all the constituent phases must be fully defined in Table 2.

6 Results and Discussion

6.1 Effect of Graphene on Elastic Constants

MM models, as shown in Figs. 5, 6, and 7 for Young's modulus, Poisson's ratio, and shear modulus, respectively.

The effectiveness of the graphene reinforcement with glass fibre composites was predicted by both FEM and MM models, as shown in Figs. 5, 6, and 7 for elastic, Poisson, and shear modulus, respectively.

Figure 5a shows that the longitudinal Young's modulus of hybrid composites increased by 0.23 to 0.95% for the addition of 5% graphene, and the MM model shows that the increase in longitudinal Young's modulus of composites is 1.35 to 6.025% by the addition of graphene to the composites.

In Fig. 5b, FE result showed that the Transverse modulus of graphene-reinforced epoxy–glass fibre composites increases by 1.68 to 7.12% for the addition of 5% graphene, and the MM model showed that the increase in transverse modulus of composites were found to be 2.94 to 9.41% by the addition of 5% graphene into the epoxy/glass composites.

From Fig. 6a, it would be inferred that the FE analysis of the in-plane Poisson's ratio of hybrid composites increased by 3.36 to 6.44% by addition of graphene, and the MM also showed that the same trend for the glass-reinforced epoxy hybrid composite increased by 3.6 to 8.6% by addition of 5% graphene.

For the out-of-plane Poisson ratio, the FEM showed that it increased with the addition of graphene by 0.44 to 2.66% for a 5% wt addition. The MM model results showed an increase in the out-plane Poisson ratio of glass fibre-reinforced composites of 1.03 to 3.07%, which is depicted in Fig. 6b.

From Fig. 7a, it could be inferred that from FE analysis, the in-plane shear modulus of graphene-based composites increases by 0.84% at 1 wt% to 2.47% at 5 wt% by the addition of graphene, and through MM results, the increase in in-plane shear modulus of composites was found to be 1.01% at 1 wt% to 3.70% at 5 wt% by the addition of graphene.

Table 2 Electrical conductivity of constituents materials (Martone et al., 2011)

Materials	Diameter, μm	Electrical conductivity (S/m)	Aspect ratio
Glass fibre	7	10^{-11}	NA
Graphene	4–20 × 10^{-3}	80 × 10^{6}	20–1000
Epoxy	NA	9 × 10^{-9}	NA

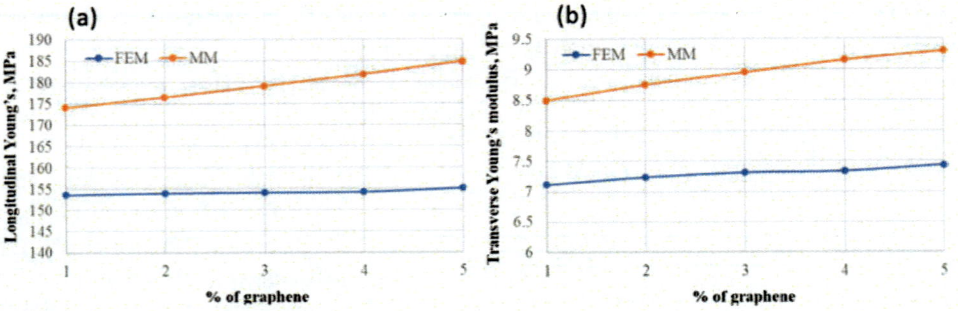

Fig. 5 Effect of graphene on elastic modulus in **a** Longitudinal-direction and **b** Transverse direction hybrid composites at zero-degree orientation

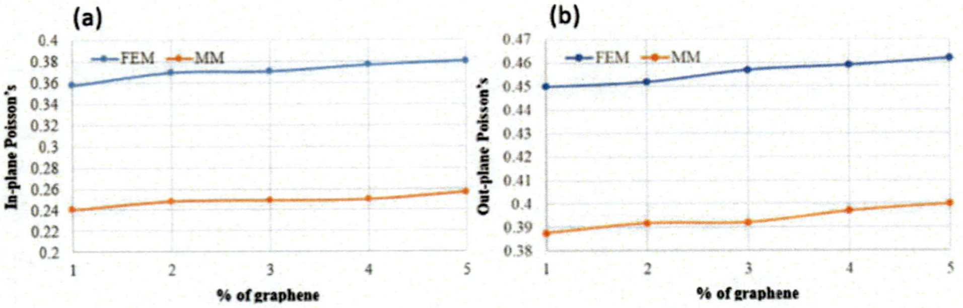

Fig. 6 Effect of graphene on **a** In-plane and **b** out-plane poisson's ratio of graphene reinforced glass fiber composites at zero-degree orientation

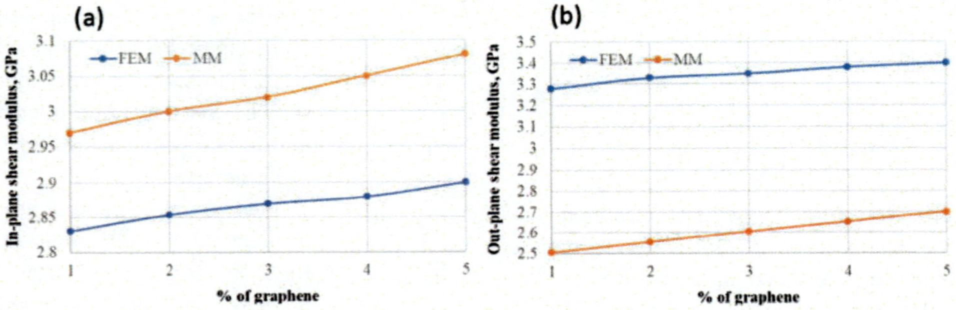

Fig. 7 Effect of graphene on **a** In-plane and **b** out of plane shear modulus of graphene reinforced glass fiber composites at zero-degree orientation

Similarly, for out-plane shear modulus results, FE analysis showed that composite increases by 1.52% at 1 wt% to 3.65% at 5 wt% by the addition of graphene, and through MM results, the increase in out-of-plane shear modulus of composite was found to be 1.99% at 1 wt% to 7.82% at 5 wt % by the addition of graphene.

6.2 Effect of Graphene Orientation on Elastic Constants

The FEM and MM results for longitudinal Young's modulus, Transverse elastic, in-plane shear, out-plane shear, in-plane Poisson's ratio and out-plane Poisson's ratio are shown in Fig. 8a-f, respectively, for 5% of graphene reinforcement glass fibre epoxy composites.

Figure 8a shows that for FE and MM results, Young's modulus is highest at zero orientation, but the modulus decreases with the angle of orientation, with a smaller enhancement at 90 degrees.

Figure 8b, which could be inferred from FE results that the transverse elastic modulus of hybrid composites increases from 0° to 90° orientation of graphene in glass fibre epoxy composites, and the same trend is also obtained through the MM approach.

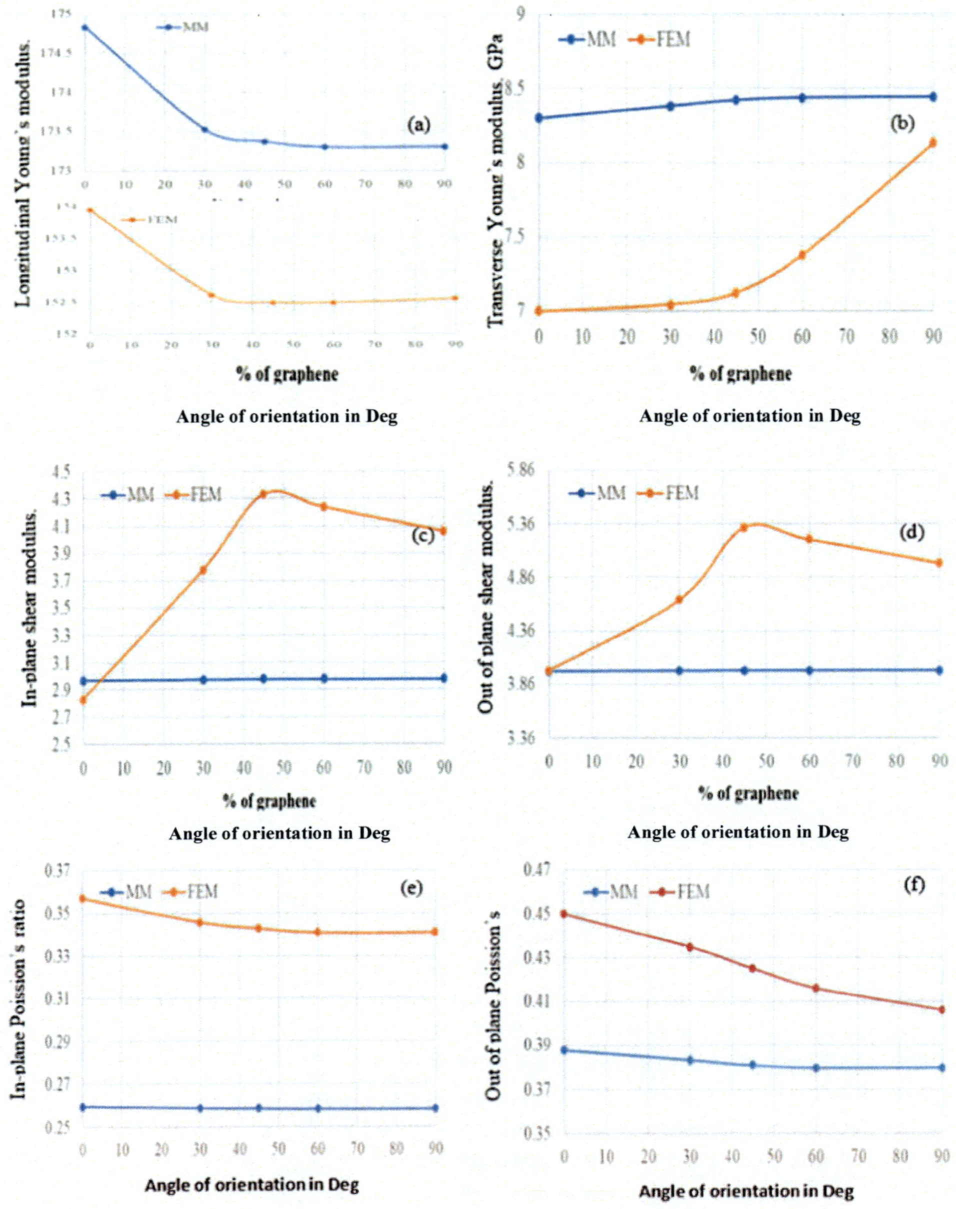

Fig. 8 Effect of graphene angle orientation on **a** E1, **b** E2, **c** G12, **d** G23, **e** μ12 and **f** μ23 of graphene-reinforced glass fibre composites for different graphene orientation degree orientation

Figures 8c and d showed that the FE results of the in-plane shear and out-plane modulus of the graphene hybrid composite increase from 0° to 45° and remain constant for the 45° to 90° graphene orientation hybrid composite, and the similar result is obtained for MM approach.

Figure 8e and f showed that the FE results of the in-plane and out-plane Poisson ratio, respectively, of glass fibre-reinforced composite decreases from 0° to 90° graphene orientation in hybrid composite, and the same is observed in MM approach.

6.3 Electrical Conductivity of Graphene-Reinforced Glass Fibre Epoxy Composites

The influence of the addition of graphene and its orientation on the electrical conductivity of graphene hybrid composites by FEM and MM models is shown in Fig. 9. It is inferred that the electrical conductivity of graphene or hybrid composite is almost constant until 2 wt% addition of graphene, there is a sudden increase in electrical conductivity between 2 wt and 3

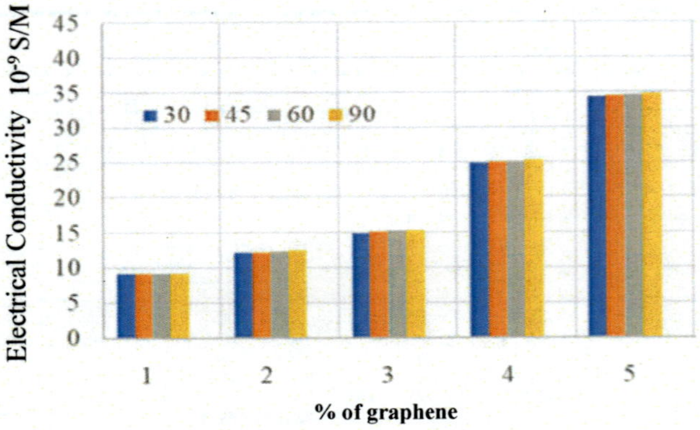

Fig. 9 Effect of % of graphene and graphene orientation on electrical conductivity graphene hybrid composite

wt% and it remains constant with further addition. Composite, because of the conductive paths formed at different orientations, has no effect on its overall electrical properties.

6.4 Conclusion

In conclusion, the impact of graphene incorporation and their angle of orientation influencing on the elastic constants and electrical conductivity of epoxy/glass hybrid composites has been examined. The hybrid composite demonstrated enhancements in longitudinal (E_1), transverse (E_2), in-plane shear modulus (G_{12}), and out-plane shear modulus (G_{23}), in-plane (μ_{12}), and out-of-plane Poisson's ratio (μ_{23}) by ranges of 0.23 to 0.95%, 1.68 to 7.12%, 0.84 to 2.47%, 1.52 to 3.65%, 3.36 to 6.44%, and 0.44 to 2.66% with graphene additions ranging from 1 to 5 wt%.

This improvement in elastic constants of graphene composites has been attributed to the introduction of high-specific stiffness graphene. Furthermore, a noteworthy observation was that, except for the transverse elastic as decided as the orientations increased. Notably, at a 90-degree orientation, the transverse Young's modulus (E_2) witnessed a remarkable 16.57% increase compared to the 0-degree orientation, particularly at 1 wt% graphene addition. This orientation adjustment to 90 degrees exhibited the potential to enhance transverse stiffness without compromising longitudinal stiffness.

Regarding electrical conductivity, the graphene-reinforced hybrid composites displayed consistent electrical conductivity up to a 2 wt% graphene addition, followed by a sudden surge between 2 and 3 wt%. Beyond this point, the electrical conductivity remained constant with further increases in graphene content. This abrupt conductivity increase can be attributed to the establishment of conductive networks within the composite structure. Notably, the overall electrical conductivity of these composites is augmented by 0.69% with the incorporation of 5 wt% graphene. Importantly, the orientations exhibited no discernible impact on the overall electrical conductivity, primarily due to the formation of efficient conductive pathways.

References

Aguilar, J. O., Bautista-Quijano, J. R., & Avilés, F. (2010). Influence of carbon nanotube clustering on the electrical conductivity of polymer composite films. *Express Polymer Letters, 4*(5), 292–299.

Ansari, R., & Hassanzadeh-Aghdam, M. K. (2016). Micromechanical investigation of creep recovery behavior of carbon nanotubereinforced polymer nanocomposites. *International Journal of Mechanical Sciences, 115*, 45–55.

Chatzigeorgiou, G., Seidel, G. D., & Lagoudas, D. C. (2012). Effective mechanical properties of "fuzzy fibre" composites. *Composites Part b: Engineering, 43*(6), 2577–2593.

Chwał, M., Kędziora, P., Barski, M. (2013). Carbon nanotube/polymer nanocomposites: A brief modeling overview. *Key Engineering Materials, 542*, 29–42. Trans Tech Publications.

Deng, F., & Zheng, Q. S. (2008). An analytical model of effective electrical conductivity of carbon nanotube composites. *Applied Physics Letters, 92*, 071902.

Feli, S., Karami, L., & Jafari, S. S. (2017). Analytical modeling of low velocity impact on carbon nanotube-reinforced composite (CNTRC) plates. *Mechanics of advanced materials and structures*, 1–13.

Feng, C., & Jiang, L. (2013). Micromechanics modelling of the electrical conductivity of carbon nanotube (CNT)–polymer nanocomposites. *Composites Part a: Applied Science and Manufacturing, 47*, 143–149.

Haider, M. F., Haider, M. M., & Yasmeen, F. (2016). Micromechanics model for predicting anisotropic electrical conductivity of carbon fiber composite materials. In *AIP Conference Proceedings* (vol. 1754, no. 1, p. 030011). AIP Publishing.

Hassanzadeh-Aghdam, M. K., Ansari, R., & Darvizeh, A. (2018). Multi-stage micromechanical modeling of effective elastic properties of carbon fibre/carbon nanotube-reinforced polymer hybrid composites. *Mechanics of Advanced Materials and Structures*, 1–15.

Li, X., Rong, J. P., & Wei, B. Q. (2010). Electrochemical behaviour of single-walled carbon nanotube supercapacitors under compressive stress. *ACS Nano, 4*(10), 6039–6049.

Martone, A., Faiella, G., Antonucci, V., Giordano, M., & Zarrelli, M. (2011). The effect of the aspect ratio of carbon nanotubes on their effective reinforcement modulus in an epoxy matrix. *Composites Science and Technology, 71*(8), 1117–2112.

Pal, G., & Kumar, S. (2016). Multiscale modelling of effective electrical conductivity of short carbon fibre-carbon nanotube-polymer matrix hybrid composites. *Materials & Design, 89*, 129–136.

Pan, J., Bian, L., Zhao, H., & Zhao, Y. (2016). A new micromechanics model and effective elastic modulus of nanotube reinforced composites. *Computational Materials Science, 113*, 21–26.

Seidel, G. D., & Lagoudas, D. C. (2009). A micromechanics model for the electrical conductivity of nanotube–polymer nanocomposites. *Journal of Composite Materials, 43*(9), 917–941.

Sharma, K., & Shukla, M. (2014). Three-phase carbon fibre amine functionalized carbon nanotubes epoxy composite: processing, characterisation, and multiscale modelling. *Journal of Nanomaterials*, 2.

Subramanian, V., Zhu, H., & Wei, B. (2006). Synthesis and electrochemical characterizations of amorphous manganese oxide and single walled carbon nanotube composites as supercapacitor electrode materials. *ElectrochemCommun, 8*(5), 827–832.

Takeda, T., Shindo, Y., Kuronuma, Y., & Narita, F. (2011). Modelling and characterization of the electrical conductivity of carbon nanotube-based polymer composites. *Polymer, 52*(17), 3852–3856.

Tsai, J. L., Tzeng, S. H., & Chiu, Y. T. (2010). Characterizing elastic properties of carbon nanotubes/polyimide nanocomposites using multiscale simulation. *Composites Part b, Engineering, 41*(1), 106–115.

Xia, Z., Zhang, Y., & Ellyin, F. (2003). A unified periodical boundary conditions for representative volume elements of composites and applications. *International Journal of Solids and Structures, 40*(8), 1907–1921.

Fabrication and Tensile Impact and Hardness Test of Al6061/SiC/Coconut Shell Ash Composites

Thota Venkateswara Rao, S. Jeyakrishnan, D. R. Rajkumar, Ram Subbiah, S. Vijayakumar, M. Naga Swapna Sri, and P. Anusha

Abstract

A research study was made to examine the mechanical properties of metal matrix composite which consisting of Al alloy reinforced with Coconut Shell Ash (CSA) and SiC. The experiment involved three distinct mix ratios of CSA and SiC as follows: Sample A (Al6061 96%, SiC 2%, CSA 2%), Sample B (Al60601 92%, SiC 4%, CSA 4%), and Sample C (Al6061 88%, SiC 6%, CSA 6%). Al/SiC/CSE Composites were fabricated by the Stir casting method. For examining the mechanical properties of the composites, Tensile, Charpy, and Hardness tests were conducted as per ASTM D638, ASTM E23, and ASTM E10 Standards respectively. The results revealed that ultimate tensile strength increased due to CSA addition, but hardness decreased. However, there was only a marginal improvement in the percentage of elongation, and the stiffness decreased as well. The experimental study concluded that tensile properties were increased while increasing the amount of Sic and CSA with aluminum also CSA had a favorable impact on the mechanical properties 6061 of hybrid composites.

Keywords

Metal matrix composites • UTM • Impact test • Hardness test • Al-Sic-CSA composite

1 Introduction

The potential of metal matrix composites (MMCs) to enhance the characteristics of alloys has led to their rising popularity. MMCs have a better specific strength and wear resistance than unreinforced alloys. The utilization of low-cost, low-density reinforcements to make MMCs has recently attracted increased attention, making these composites even more available and practical (Eswara Kumar & Seshu Kumar, 2018).

A research study by Siva Prasad (Alo et al., 2017; Siva Prasad & Krishna, 2011). On rice husk, ash-based composites prepared utilizing the vortex process showed a reduction in density of the composite but hardness of the composite is elevated. Arun Kumar and Swamy (Arun Kumar & Swamy, 2011) in their detailed examination of the microstructure and mechanical behavior of hybrid composites concluded that the compressive strength of graphite-reinforced composites significantly rose. Rajan et al. (2007) examined hybrid composites with fine fly as particles made by using three distinct stir casting techniques. Coconut fly ash is widely available, reasonably priced, and produced as a solid waste in daily life. Therefore, it is anticipated that using coconut fly ash to make composites will help small engine and automobile applications overcome cost-related problems. Ejiofor and Reddy (1997) compared unreinforced alloys, cast by using the stir casting

T. V. Rao
Department of Mechanical Engineering, Visakha Institute of Engineering and Technology, Narava, Visakhapatnam, India

S. Jeyakrishnan
Department of Mechanical Engineering, ARM College of Engineering and Technology, Maraimalainagar, Chennai, 603209, India

D. R. Rajkumar (✉)
Department of Mechanical Engineering, CARE College of Engineering and Technology, Tiruchirappalli, 620 009, India
e-mail: rajkumardevapiriam@gmail.com

R. Subbiah
Department of Mechanical Engineering, Gokaraju Rangaraju Institute of Engineering and Technology, Hyderabad, India

S. Vijayakumar
Department of Mechanical Engineering, BVC Engineering College (Autonomous), Odalarevu-533210, Andhra Pradesh, India

M. N. S. Sri · P. Anusha
Department of Mechanical Engineering, P V P Siddhartha Institute of Technology, Vijayawada, India
e-mail: mswapnasri@pvpsiddhartha.ac.in

M. Sumesh et al. (eds.), *2nd International Conference on Smart Sustainable Materials and Technologies (ICSSMT 2023)*, Advances in Science, Technology & Innovation, https://doi.org/10.1007/978-3-031-49826-8_5

technique. Results revealed that a higher weight % of SiC reinforcements resulted in a higher surface roughness.

Alo et al.'s (2012) investigation of Al-Si-SiC composites revealed that the thermal and hardness characteristics of the material depend on the silicon and silicon carbide concentrations. When mixed with silicon carbide and aluminum alloy composites, groundnut shell ash (Alaneme et al., 2018) provided influence to a variety of mechanical properties. According to the findings of Surappa, Madakson, Raju, Adeosun, Aigbodion, Manohar, and Venkataramana (Surappa & Ramesh 2004, Adeosun & Aigbodion, 2017; Aigbodion, 2010; Gobalakrishnan et al., 2021; Madakson et al., 2012; Manohar & Mukunda, 2014; Raju et al., 2018; Venkataramana et al., 2017), the composite material used in automotive applications showed a noticeable improvement in strength, wear resistance, hardness, and sudden load application. Vijayakumar, Gugulothu, and Gobalakrishnan utilized a simple and inexpensive way for producing liquid state is stir casting (Devapiriam et al., 2022; Gobalakrishnan et al., 2023; Gugulothu et al., 2022; Vijayakumar et al., 2022a, 2022b). The unequal distribution of particles, which is brought about by poor wettability and gravity-related segregation, poses a substantial problem for this procedure. Stir casting is still a widely utilized technique in spite of these drawbacks since it is inexpensive and simple to perform.

This experimental work investigates the incorporation of Silicon Carbide (SiC) and Coconut shell ash (CSA) into Al alloy 6061-based MMC using the vortex method which was shown in Fig. 1. The MMCs were prepared by adding SiC and CSA composites at varying concentrations of 2%, 4%, and 6% to the molten aluminum 6061 alloy, which was changed its liquid at 500 °C. The addition of composites was carried out by generating a vortex in the melt with a mechanical graphite stirrer rotating at 500 rpm. The particles were preheated to 500 °C and introduced to the melt through the vortex at a rate of 0.1 kg/min.

2 Materials and Methods

2.1 Materials

In this work, Al 6061, SiC, and CSA were selected for fabrication of hybrid composites. The materials selected for the experimental study The Al6061 and SiC were purchased from Bangalore Fine Chemicals, Bengaluru, Karnataka, and CSA was from Wellon official store in Surat, Gujarat. The selected materials and their proposition to fabricate composites are shown in Fig. 1 and Table 1.

2.2 Stir Casting Method

Liquid metallurgy technique called Stir Casting used molten matrix after permitted to harden. It is essential to ensure the proper mixing of reinforcements and molten aluminum to achieve optimal performance. The success of the Stir Casting process (SCP) largely depends on the appropriate mixing of matrix with reinforcement particles to achieve uniform distribution and improve the tensile, impact, and hardness properties of the end product. To reach this, the most commonly used and the vortex or stir casting technique is a simple and result-oriented method. The stir casting setup abstract is displayed in Fig. 2 and the actual image is shown in Fig. 3.

The vortex method was considered to reduce gravitational segregation in the crucible by introducing pre-treated particles of ceramics into molten alloy created by the impeller. When manufacturing MMCs using the melt stirring technique; higher reinforcement volume ratios (RVRs) and smaller particle sizes can make the production process more challenging and increase the likelihood of porosity and particle agglomeration. Despite these challenges, this technique remains attractive due to its simplicity, flexibility, and cost-effectiveness for the fabrication of larger components (Pal et al., 2022).

Fig. 1 Selected materials for preparation of composite

Table 1 Composition of materials in percentage (%)

Samples	Aluminum Al 6061 (%)	Silicon carbide (Sic) (%)	Coconut shell ash (CSA) (%)
Sample A	96	2	2
Sample B	92	4	4
Sample C	88	6	6

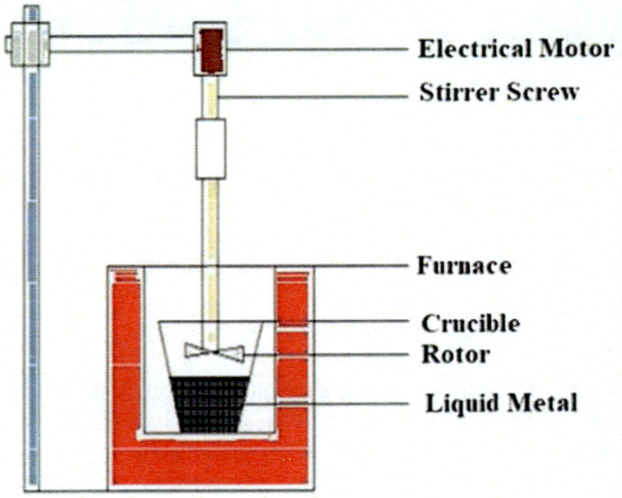

Fig. 2 Stir casting setup

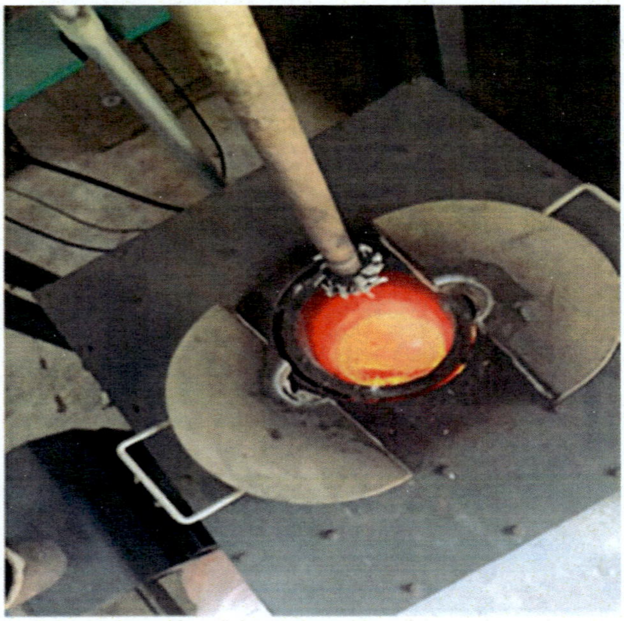

Fig. 3 Mixing of Al 6061-Sic-CSA

The stir casting machine is utilized to transfer the aluminum alloy, in billet form, to the furnace. The alloy is heated to 850 °C and kept for a few hours until it melts into a liquid state. Once it is confirmed that the alloy is liquid, the mixture is allowed to cool in the furnace until it reaches a semi-solid state at approximately 600 °C. Next, preheated particles of SiC and coconut shell ash are manually introduced to the mixture and stirred for 5–10 min. The resulting slurry of composites is then heated to a temperature of 800 ± 50 °C and mechanically stirred at a speed of 400 rpm for 10 min. Finally, the mixture is poured into sand molds that are equipped with metallic chills. The final stir-casted specimens are shown in Fig. 4.

3 Mechanical Properties

3.1 Tensile Test

Hybrid composites are made according to ASTM D638 (Devapiriam et al., 2022) dimensions for tensile inspection, and a test specimen was prepared as illustrated in Fig. 5, and was put under tensile load to evaluate tensile properties. The tensile strength of pure aluminum can be measured by observing the material failure under a maximum tensile load of 17.46 KN, which resulted in the obtained fracture.

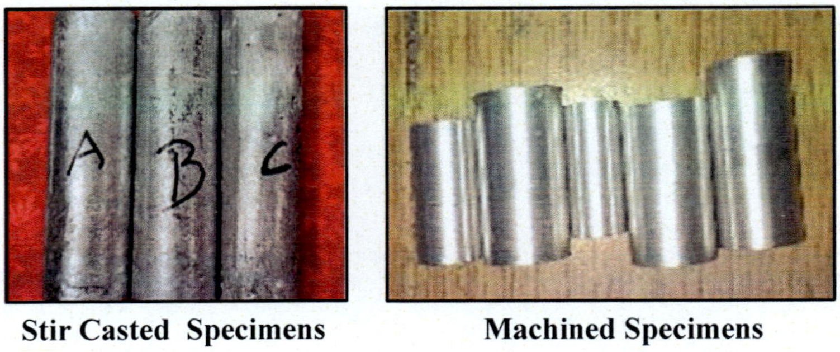

Fig. 4 Stir-casted specimens (before and after machined)

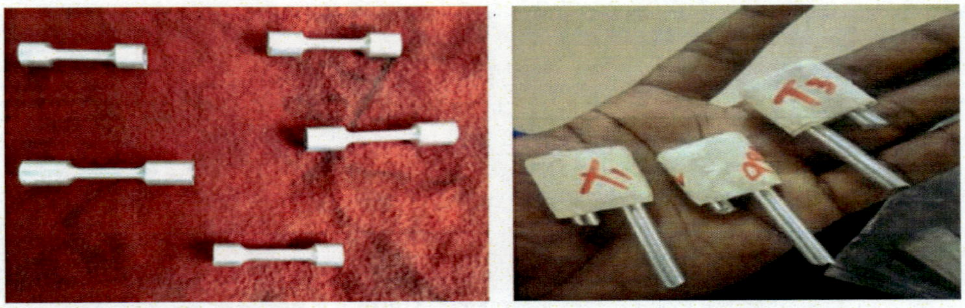

(a) (b)

Fig. 5 Tensile test specimen **a** before testing and **b** after testing

Table 2 Tensile properties of samples

Specimen	Ultimate tensile strength (MPa)	% of elongation in 25 mm GL	Yield strength (MPa)
Sample A	141	10.50	84
Sample B	145	9.00	92
Sample C	149	11.50	92

Furthermore, during testing, it was able to withstand a tensile force of up to 20.57 KN. Table 2 displays the tensile characteristics of three samples.

3.2 Charpy Impact Test

The Charpy and Izod tests differ in the orientation of the test specimen as shown in Fig. 8a, with the former being held horizontally between two vertical bars and the latter being positioned vertically. A specimen Fig. 8b with a notch is subsequently impacted by a heavy pendulum or hammer, which falls through a predetermined distance at a fixed velocity. The amount of energy required to break the specimen in a single blow is typically measured, as is the case in both the Charpy and Izod impact tests and the absorbed values tabulated in Table 3.

3.3 Hardness Test

Table 4 below displays the hardness values of the three samples at varying compositions. The Brinell hardness number can be determined using equation (Devapiriam et al., 2022).

Table 3 Charpy test obtained values

Specimen size (mm)	Notch type	Test temperature	Absorbed energy (J)			Average
			ID–1	ID–2	ID–3	
10 X 10 X 55	V	+ 24 °C	24	62	34	40.00

Table 4 Hardness test obtained values

Sample ID	Sample A	Sample B	Sample C
BHN (5 mm ball/250 kg load)	70.2	68.8	66.2
	69.5	68.2	65.5
	71.0	67.5	65.5

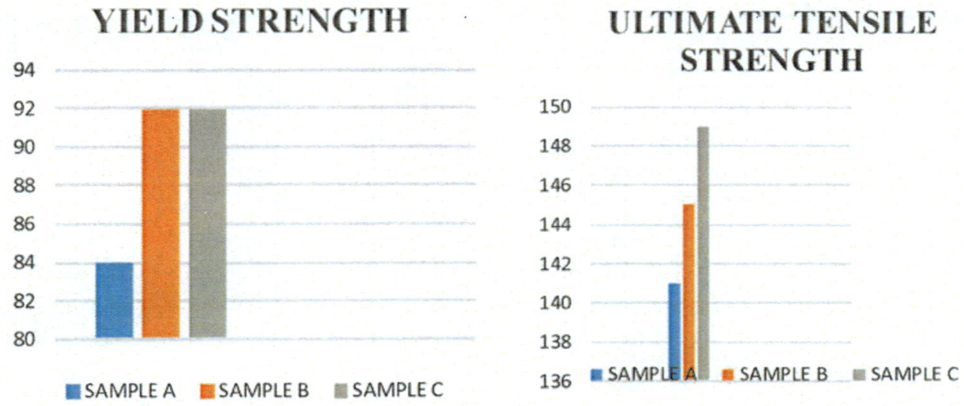

Fig. 6 Yields strength and ultimate tensile strength comparison of samples

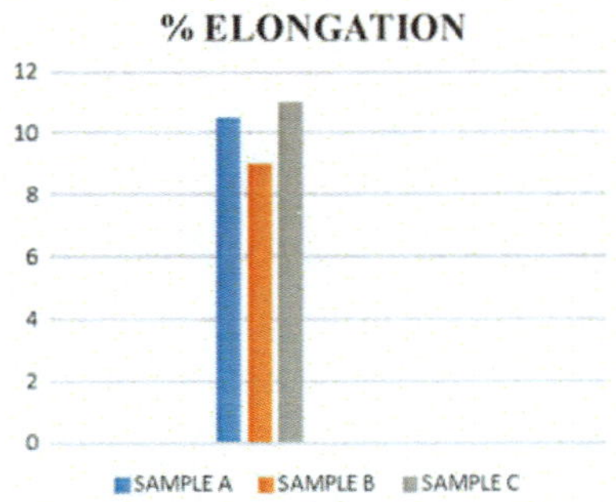

Fig. 7 % elongation comparison of samples

$$BHN = Load \Big/ \left(\pi D\left(D - (D2 - d2)^{1/2}\right)\right) \quad (1)$$

where,

D = diameter of steel ball intender, d = diameter of intender; Load = 250 kg.

4 Results and Discussion

4.1 Tensile Test

From Figs. 6 and 7, it was observed that the Sample C has a good Ultimate tensile strength of 149 MPa and 11.50% of elongation in 25 mm GL and 92 MPa of Yield strength when compared to Sample A and Sample B.

4.2 Charpy Test

From Fig. 8c, it showed that Sample B (92% Al, 4%SiC, and 4% CSA) absorbed more energy when compared with Sample A and Sample C.

4.3 Hardness Test

From Fig. 9b, it showed that Sample A (96% Al, 2%SiC, and 2% CSA) has more hardness than Sample B and Sample C as more amount of aluminum in the composites.

5 Conclusion

This research work concludes that.

(1) Aluminum 6061 with silicon carbide and coconut shell at three different (96%Al, 2%SiC, 2%CSA; 92%Al, 4% SiC, 4%CSA; and 88%Al, 6%SiC, 6%CSA) weight percent metal matrix composites were produced using stir casting method.
(2) Mechanical properties like Tensile Strengh, Ultimate Tensile Strengh, % of elongation, resistance to penetration, and hardness of Al-SiC-CSA composites were found using the tensile test, Charpy test, and hardness test at room temperature.

Fig. 8 Charpy test: **a** Specimen dimension (ASTM International 2016; Cubides-Herrera et al., 2019), **b** Test specimens, **c** Result

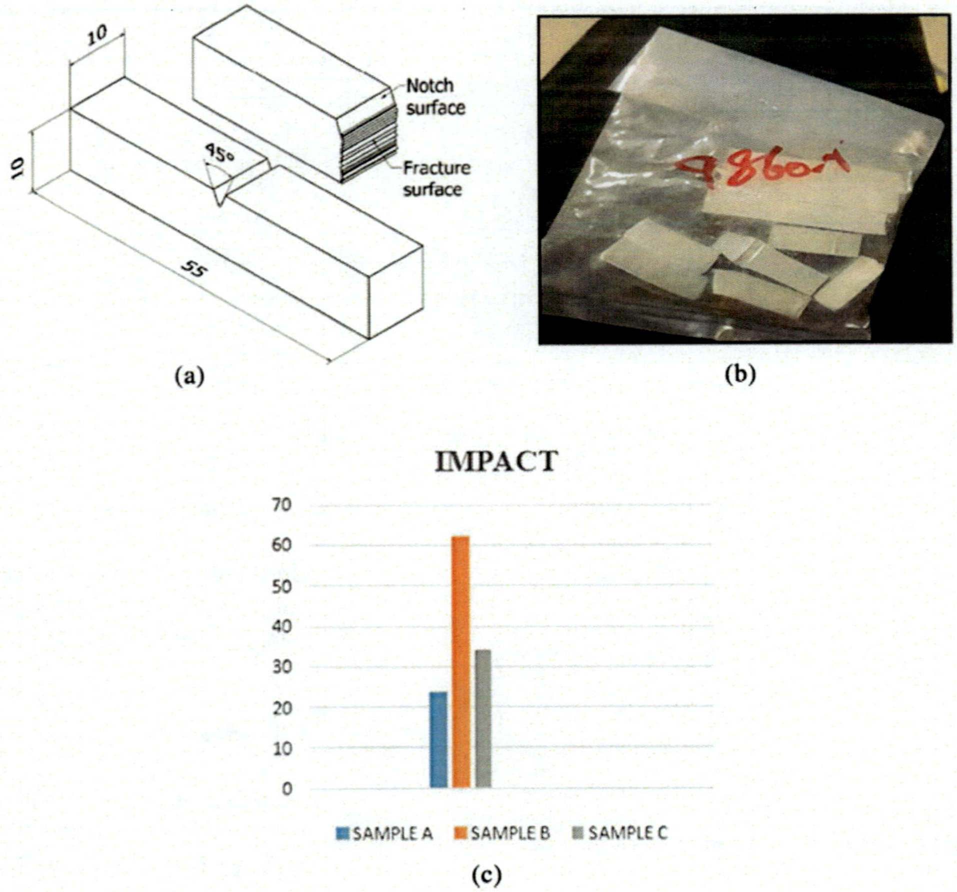

Fig. 9 Hardness test **a** Equipment and **b** samples result

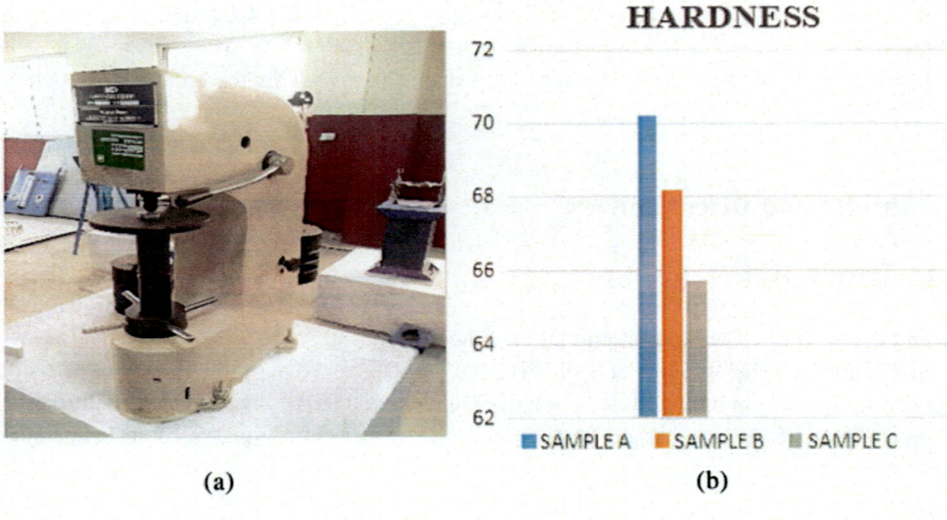

(3) 88%Al, 6%SiC, and 6%CSA composites produced good Tensile properties due to 6% of SiC and 6% of CSA.

(4) 92% Al, 4%SiC, and 4% CSA compositesabsorbed 62 J of energy due to SiC and CSA.

(5) 96%Al, 2%SiC, and 2%CSA composites showed good resistance to penetration.

This suggests that MMCs have the potential to be widely employed.

References

Adeosun, S. O., & Aigbodion, V. S. (2017). Comparative studies on the mechanical properties of aluminum metal matrix composites reinforced with coconut shell ash and silicon carbide. *IJMPERD, 7*, 57–64.

Aigbodion, V. S. (2010). Potential of using coconut shell ash particle in metal matrix composite. *Journal of Metals, Materials and Minerals, 9*(7), 637–643.

Alaneme, K. K., Bodunrin, M. O., & Awe, A. A. (2018). Microstructure, mechanical and fracture properties of groundnut shell ash and silicon carbide dispersion strengthened aluminium matrix composites. *Journal of King Saud University-Engineering Sciences, 30*, 96–103.

Alo, O. A., Umoru, L. E., Ajao, J. A., & Oluwasegun, K. M. (2017). Microstructure and mechanical behavior of stir-cast Zn–27Al based composites reinforced with rice husk ash, silicon carbide, and graphite. *Journal of King Saud University-Engineering Sciences, 29*, 172–177.

Alo, O. A., Umoru, L. E., Ajao, J. A, & Oluwasegun, K. M. (2012). Thermal, hardness and microstructural characterization of Al-Si-SiCp composites. *Journal of Minerals and Materials Characterization and Engineering, 11*, 159–168.

Arun Kumar, M. B., & Swamy, R. P. (2011). Evaluation of mechanical properties of Al6061, fly ash and E-glass fiber reinforced hybrid metal matrix composites. *ARPN Journal of Engineering and Applied Sciences, 6*, 40–44.

ASTM D638–14. (2014). *Standard test method for tensile properties of plastics*. ASTM International.

ASTM E10–18 Standard Test Method for Brinell Hardness of Metallic Materials, Annual Book or ASTM Standards, American Society for Testing and Materials (vol. 3.01).

Cubides-Herrera, C. S., Villalba-Rondón, D. A., & Rodríguez-Baracaldo, R. (2019). Charpy impact toughness and transition temperature in ferrite–perlite steel. *Scientia et Technica Year, 24*.

Devapiriam, R., Karthik, S., & Santhy, K. (2022). Evaluation of mechanical and thermal properties of fabricated Sansevieria and Kaans fiber/egg shell powder reinforced with polyester matrices. *Pigment & Resin Technology, 51*, 692–299.

Ejiofor, J. U., & Reddy, R. G. (1997). Developments in the processing and properties of particulate Al–Si composites. *Journal of Materials Science, 79*, 31–34.

Eswara Kumar, G., & Seshu Kumar, G. S. V. (2018). A study on the fabrication and testing of aluminum metal matrix composites using stir casting techniques. *IRJET, 5*(6), 2416–2420.

Gobalakrishnan, B., Rajaravi, C., Udhayakumar, G., & Lakshminarayanan, P. R. (2021). Effect of ceramic particulate addition on aluminium based ex-situ and in-situ formed metal matrix composites. *Metals and Materials International, 27*, 3695–3708.

Gobalakrishnan, B., Rajaravi, C., Udhayakumar, G., & Lakshminarayanan, P. R. (2023). A comparative study on ex-situ & in-situ formed metal matrix composites. *Archives of Metallurgy and Materials*, 171–185.

Gugulothu, B., Nagarajan, N., Pradeep, A., Saravanan, G., Vijayakumar, S., & Rao, J. (2022). Analysis of mechanical properties for Al-MMC fabricated through an optimized stir casting process. *Journal of Nanomaterials*, 1–7.

Madakson, P., Yawas, D., & Apasi, A. (2012). Characterization of coconut shell ash for potential utilization in metal matrix composites for automotive applications. *International Journal of Engineering Science, 4*.

Manohar, M. P. B., & Mukunda, P. G. (2014). Mechanical properties of aluminum metal matrix composites reinforced with SiC and coconut shell ash. *Journal of Materials Engineering and Performance, 23*, 2445–2450.

Pal, D., Vijayakumar, S., Rao, T. V. J., & Babu, R. S. R. (2022). An examination of the tensile strength, hardness and SEM analysis of Al 5456 alloy by addition of different percentage of SiC/fly ash. *Materials Today: Proceedings, 62*, 1995–1999.

Rajan, T. P. D, Pillai, R. M, Pai, B. C., Satya Narayana, K. G, & Rohatgi, P. K. (2007). Fabrication and characterisation of Al–7Si–0.35Mg/fly ash metal matrix composites processed by different stir casting routes. *Composites Science and Technology, 67*, 3369–3377.

Raju Siva Rao, G., & Siva, B. (2018). Experimental studies of mechanical properties and tribological behaviour of aluminium composite reinforced with coconut shell ash particulates. *International Journal of Materials Engineering Innovation, 9*.

Siva, D. P, & Krishna, A. R. (2011). Production and mechanical properties of A356.2/RHA composites. *International Journal of Advanced Science, 33*, 51–58.

ASTM E23–16b. (2016). *Standard test methods for notched bar impact testing of metallic materials*. ASTM International.

Surappa, M. K., & Ramesh, N. M. (2004). Processing and characterization of aluminium-coconut shell ash composites. *Composites Science and Technology, 64*, 2329–2336.

Venkataramana, S., Sudarshan, T., & Bhanu Prakash, K. (2017). Mechanical properties of aluminum hybrid composites reinforced with coconut shell ash and SiC. *Materials Science Forum, 879*, 80–87.

Vijayakumar, S., Satheesh Kumar, P. S., Sampath kumar, P., Manickam, S., Ramaiah, G. B., & Pydi, H. P. (2022). The effect of stir-squeeze casting process parameters on mechanical property and density of aluminum matrix composite. *Advances in Materials Science and Engineering*, 1–10.

Vijayakumar, S., Manickam, S., Seetharaman, S., Rao, T. V. J., Pounraj, D., & Pydi, H. P. (2022). Examination of friction stir-welded AA 6262/5456 joints through the optimization technique. *Advances in Materials Science and Engineering*, 1–11.

A Review on Rolling Process for Polymer Nanocomposites with Different Nanofillers to Enhance Properties

Aamir M. Shaikh and Pravin R. Kubade

Abstract

To improve mechanical properties of polymer nanocomposites, different techniques are used by researchers. One of the critical tasks in front of the industry is to strengthen the polymer nanocomposites. Weight reduction with better mechanical properties are required in many automotive applications. Polymer nanocomposites with dissimilar types of nanofillers or nanomaterials show a significant part in controlling mechanical properties. Ultimately the shape, dimensions, and orientation of the nanoparticles with proper dispersion in the polymer matrix are also considered in the different research studies by many researchers. The rolling process is followed by melt mixing, ultrasonication, injection molding, and extrusion of the polymer nanocomposites, where rolling parameters like compression ratio or reduction ratio also govern the associated properties. In the literature, it has been considering the manufacturing of polymer nanocomposites by rolling process with various nanofillers like graphene (G), Halloysite nanotube (HNT), carbon nanotube (CNT), multi-wall carbon nanotube (MWCNT), hexagonal boron nitride (h-BN), and effect on properties of different polymers like PP, epoxy by comparing other processing techniques.

Keywords

Nanocomposites • Nanofillers • Rolling process • Mechanical properties

A. M. Shaikh
Shivaji University, Kolhapur, Maharashtra, India

P. R. Kubade (✉)
Shri Guru Gobind Singhji Institute of Engineering and Technology, Nanded, Maharashtra 431606, India
e-mail: pravinrkubade@gmail.com

1 Introduction

The creation of nanocomposite substantially differentiates the claim of automotive, for nanoscale fillers may efficiently rise performance, counting thermal and mechanical properties (Svoboda et al., 2002). Polypropylene (PP) is a polymer in a rage of commodity which offers the better performance characteristics with best price among the thermoplastics. As the clay nanoparticles are dissolved or hydrophilic in nature and mismatched with polypropylene. For two or more un-mixable material, compatibilizer is needed to mix it properly (Sharma & Nayak, 2009). Dropping the car's mass is an active portion of decreasing vehicle products of combustion/emission smog and progressing automobile fuel effectiveness. A polymer Polypropylene has significant aids to weightiness drop, protection, then energy savings, subsequently it has progressively converted mostly usable of vehicle materials. Toward additional recover the characteristics of PP, many research scholars take spread around nanofillers/nanoparticles into plain PP so that nearly distinct parameters of nanomaterials, for example, small particle extent and its surface effect. Due to the availability and suitability of polymer polypropylene is finding many of applications in various fields. In automotive applications, the use of PP is increased (Sui et al., 2022). Nanomaterials are evolving by means of a novel class of materials and are on stage a significant part in technical applications, primarily as a result of desirable characteristics. The core motives these nanomaterials show irregular and advanced properties are major niche area, significant aspect ratio, low density, and behavior of faces (Verma et al., 2017).

Graphene—G is the elementary part of graphite-like resources by a sp2 mixed structure and 2-d sole atomic film, and it takes involved increased attention as a result of its admirable electro-mechanical, thermo-mechanical, and biological-related characteristics. The utilization of graphene in polymer improves properties when dispersed correctly in the polymer matrix. Graphene is the thinnest and tough

M. Sumesh et al. (eds.), *2nd International Conference on Smart Sustainable Materials and Technologies (ICSSMT 2023)*,
Advances in Science, Technology & Innovation, https://doi.org/10.1007/978-3-031-49826-8_6

nanomaterial (Zhang et al., 2015). Crucial 2D material is available as Hexagonal-boron nitride (h-BN) nanosheets. These are also recognized as white graphene (G). It consists of good physical and chemical characteristics like surface area, conductivity, and constancy. Enhanced mechanical properties were achieved for modified hBN nanocomposites rather than PP. The results are obtained for incorporation of less amount of modified nanofillers (Yola & Atar, 2018). The researcher Yu et al. found high thermal conductivity i.e., 9 W/meter ^{0}K by reason of the vertically associated arrangement for hBN + epoxy (EP) nanocomposites through 44 volumes percentage hBN loading. Researcher prepared the compounds, by cutting them into the pieces. For that authors use the vacuum filtration method to align hBN micro-platelets (Bayır et al., 2021).

The carbon nanotube (CNT); properties make it an ideal strengthening for polymers. The interface of the nanofiller with polymer matrix transfers the characteristics of wonder material CNTs to the polymer. SEM:—Scanning Electron Microscopy confirmed the result of uniform scattering of MWCNTs used in matrix, which tends to improve tensile strength with toughness, and storage modulus through 35%, 53%, also 35% respectively, on loading of less weight percentage (0.75 wt.%) of several walled carbon nanotubes say MWCNTs in polymer polyepoxides (Kumar et al., 2017). HNTs (Halloysite nanotubes) are modified successfully using PEI (polyethyleneimine). For nanocomposites of PP/ABS 80/20 (wt./wt.) are melt mixed and HNTs and modified HNTs at different concentrations were used. Morphography and topography of nanocomposites remained observed (Kubade & Tambe, 2016).

Researchers investigated different polymer matrix and their blend-based nanocomposites with different nanofillers like graphene, CNT, MWCNT, HNT, h-BN, etc., and processing procedures used as melt compounding/mixing, in situ bulk polymerization techniques, sonication, extrusion, cold rolling, hot rolling, etc. This review focuses on the processing of rolled polymer nanocomposites compared with other processes.

2 Processing Techniques

Researcher uses various techniques for processing polymer nanocomposites (like melt mixing, ultrasonication, extrusion, rolling, injection molding, etc.) which are displayed in Table 1. And methods for polymer nanocomposites characterization (like SEM:—Scanning Electron Microscopy; TEM:—Transmission electron microscopy; FTIR:—Fourier Transform Infrared; DSC:—Differential Scanning Calorimetry; XRD:—X-Ray diffraction etc.).

3 Morphological Study/Characterization of Polymer Nanocomposites

Researcher used Epoxy and hBN. Incorporation of hBN leads to improved mechanical properties. Altered h-BN nanosheet reinforced epoxy (polyepoxides) nanocomposites lead to improved flexural characteristics. Epoxy/hBN nanocomposite detected under TEM were h-BN (at lower i.e., 0.08 wt.%) nanosheet used in epoxy (polyepoxides) are revealed in Fig. 1 TEM image having distance between nanoparticles with its size shows in diagram Fig. 1a, while proper dispersion will be observed in Fig. 1b (Kulkarni et al., 2019a).

Saerom Kong et.al. used hBN and modified hBN (Alkyl-BN). Surface modification process of hBN leads to amended mechanical characteristics and thermal characteristics. TEM images of hBN also Alkyl (BN) (modified by surface treatment) give the morphology in Fig. 2. Both shows stacked structure with some layers. Agglomerates are observed in Fig. 2a while as proper dispersion due to modified hBN and color of modified hBN i. e. Alkyl-BN is faint than that of hBN as it has a lesser number of coatings/layers seen in Fig. 2b (Kong et al., 2019; Kubade et al., 2017).

Chan, K et.al. used nanoparticles hBN with PP. Figure 3a shows images (SEM) of PP + 5% h-BN and Fig. 3b shows PP + 10% h-BN nanocomposites. The amount of supporter cells observed in these samples appears to increase by growing h-BN percentage. Likewise, hybridization gives effective results (Chan et al., 2015; Patil et al., 2021).

Researcher found for cold rolling, does not have a big impact on the melting behavior of polymer nanocomposites, as in Fig. 4. It has been observed that polybutylene succinate (PBS) at unrolled condition, melting temperature (T_m) at x location is 112.4 °C and for PBS at rolled condition (compression ratio = 2.50) melting temperature T_m at x location is 112.9 °C. Similarly, it is found that PBS (unrolled condition) temperature T_m at y location is 102.2 °C and for PBS at rolled condition (compression ratio = 2.50) temperature T_m at y location is 102.4 °C (Phua, 2011).

4 Effect on Properties with Different Nanofillers

Due to structural resemblance with graphene, Boron nitride: BN is similarly recognized as white g: graphene. Hexagonal, cubic, and tubular structures are available for BN. In comparison with all structures, h-BN: hexagonal boron nitride is most commonly termed as lubricant as well as thermo-conductive nano-filler in polymer nanocomposites.

Table 1 Different processing techniques used by researchers for nanocomposites

Processing of polymer nanocomposites	Schematic arrangement	References
Melt mixing: Melt mixing by internal mixer technology is appealing because of its speed, availability, and ease of use, particularly in the common plastics business	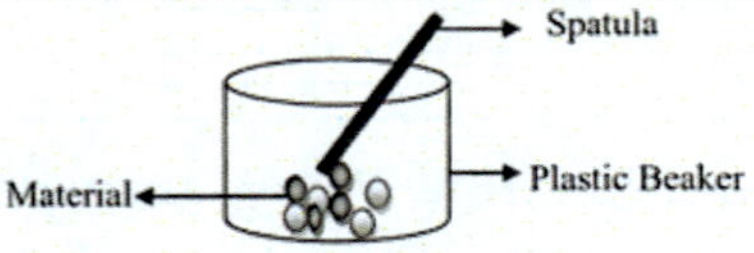	Breuer and Sundararaj (2004)
In situ polymerization method: In this method, nanometer-scale inorganic fillers or reinforcements are scattered in the monomer first. This is then polymerized using a method similar to bulk polymerization	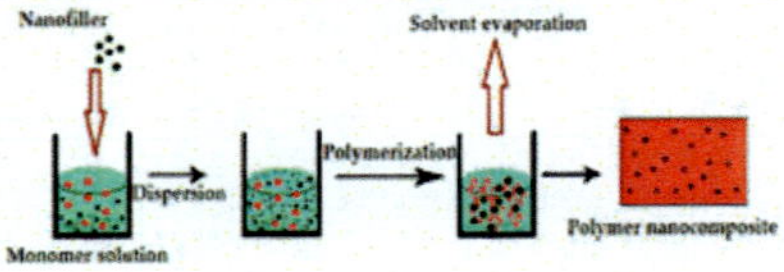	Thomas et al. (2007)
Ultrasonication: In ultrasonication-mechanical stirrer equipment set-up. The nanofiller-like carbon nanotube, cleaned in the beaker, was then dried overnight at 80 °C in the drying oven. Later powder was crushed by using an agate mortar. At this stage, it was assumed that MWCNT is separate from each other	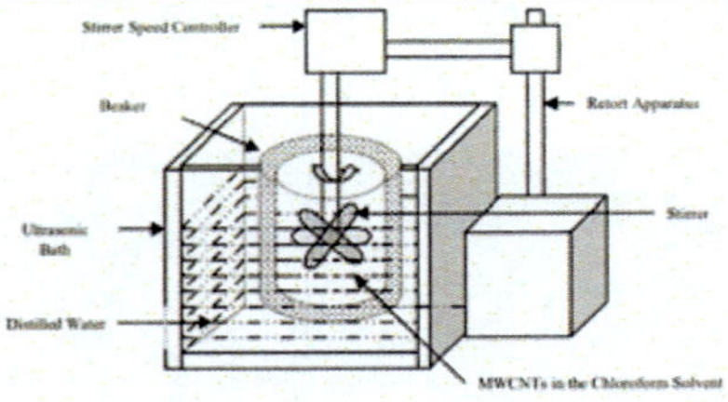	Salleh et al. (2008)
Extrusion: An extruder essential has a feeding system, a melting with plasticizing system, an impelling & pressing arrangement that produces a mixing result with a mechanism for shaping the melted substantial. Three primary regions can be distinguished created on the pressure applied along the drum or extrusion barrel	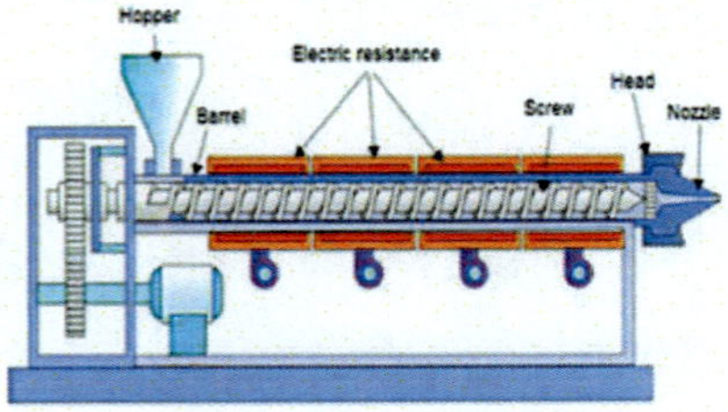	Avila-Orta et al. (2019)
Injection molding: In industry widely used technique is injection molding. With the help of injection molding arrangement, generally soft things also packages are formed. It is used for mass production The raw material (polymer + nanofillers) is fed into the hopper of the machine which is heated further and turned in a screw cylinder at the time of mixing	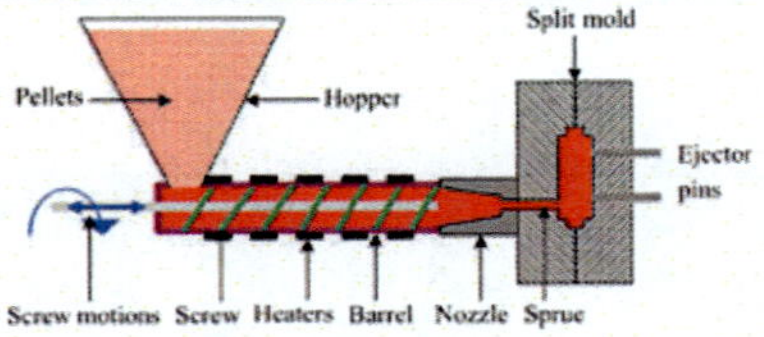	Kapłon et al. (2014)

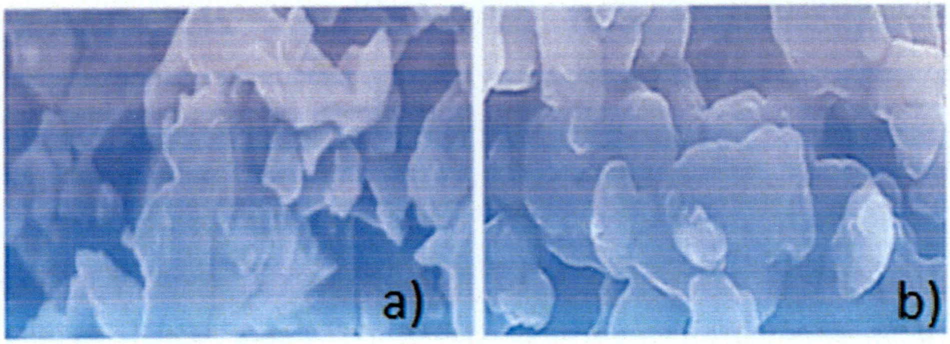

Fig. 1 Images at lower hBN (0.08 wt.%) (Cahill et al., 2020)

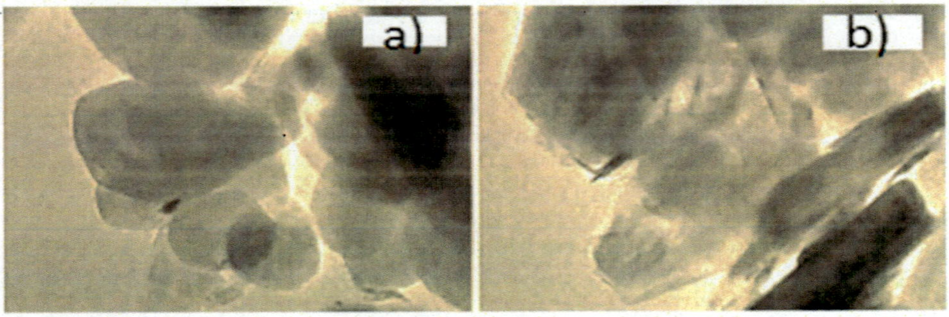

Fig. 2 Microscopic images; **a** hBN and **b** modified hBN [Alkyl(BN)] at lower wt.% (Lei et al., 2021)

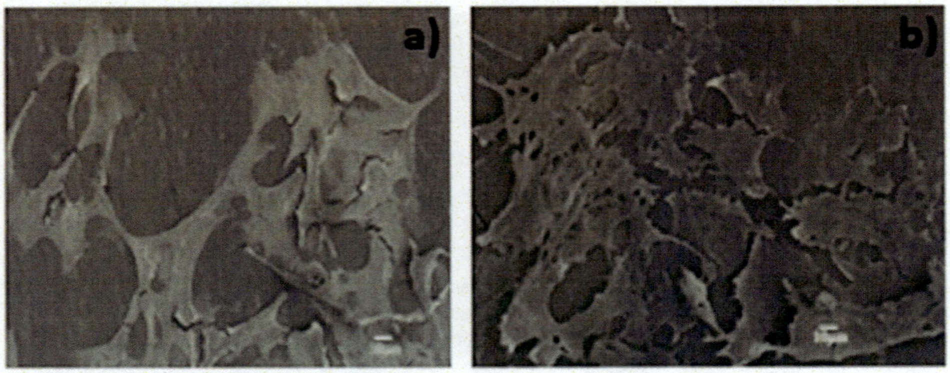

Fig. 3 Microscopic diagrams of: **a** PP + 5% h-BN and **b** PP + 10% h-BN (Chan et al., 2015)

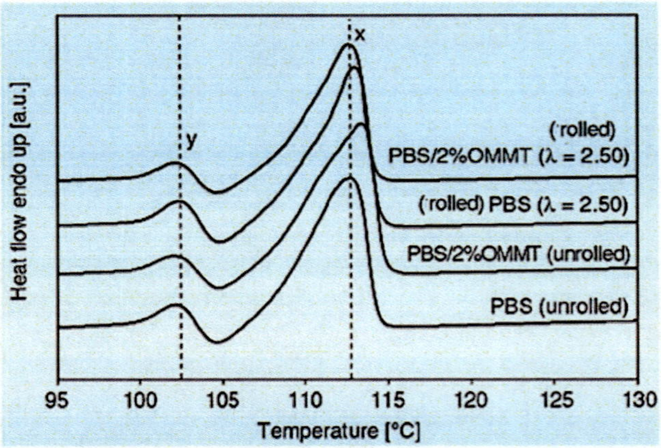

Fig. 4 DSC thermogram before and after rolling (Phua, 2011)

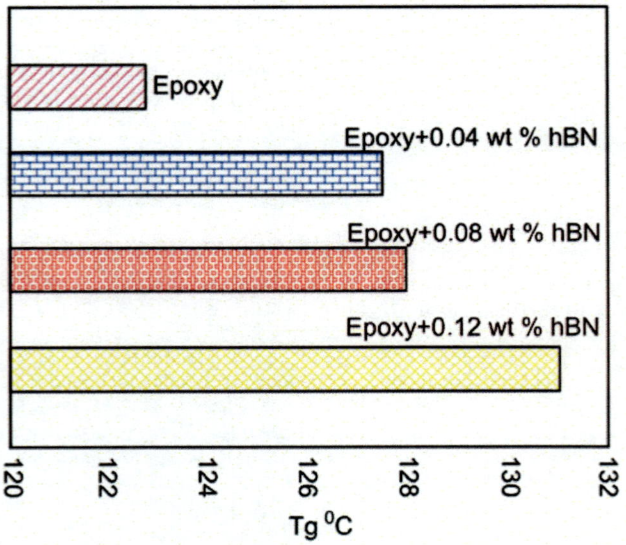

Fig. 5 Tg; Glass transition temp. of pristine epoxy and epoxy/hBN (Kulkarni et al., 2019b)

Figure 5 depicts the second order phase transition temperature of plain epoxy along with epoxy + h-BN. Initially, the (T_g) was 122.4 °C which varies to 131 °C when incorporated h-BN (Kulkarni et al., 2019b).

The result on flexural modulus and strength by the accumulation of nanofiller (h-BN) interested in the epoxy has been indicated in Fig. 6a, b. However, Flexural-strength observed was $107^{(\pm 3)}$ MPa which enhances after the incorporation of 0.04 wt.% h-BN up to $111^{(\pm 2)}$ MPa (Kulkarni et al., 2019b).

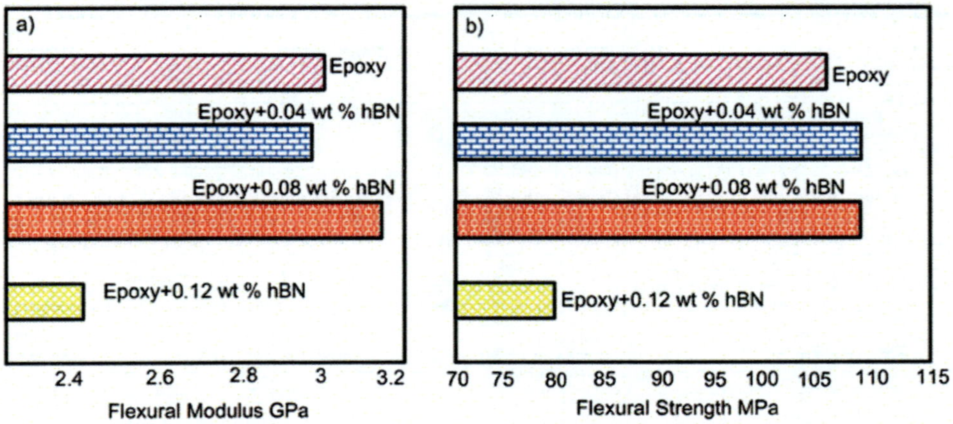

Fig. 6 Characteristics of nanocomposites (epoxy + hBN) (Kulkarni et al., 2019b)

Elastic modulus of PP grows when incorporated hexagonal boron nitride (h-BN) nanofiller, however tensile elongation decreases. Hybridization of hBN through nano-hydroxyapatite (nHA) additional boosts the elastic modulus of PP. It is beneficial for up to 20 wt.% hBN incorporation into PP composites, enhancing elastic modulus of component. Additionally, hBN-nHA nanofiller hybridizing further rises the strength of component, as predicted. Hybridization set of (PP) + 15% (h-BN) + 20% (nha) shows 2.38 GPa maximum modulus, which enhances 69 percentage instead of pristine PP (Chan et al., 2015). Improvement in thermal conductivity is obtained for CNT/hBN hybrid composition associated with single filled nanocomposites (TabkhPaz et al., 2016).

Researchers investigated different polymer matrix (PP, epoxy, PE etc.) blend-based nanocomposites through various compatibilizer (like PP-(g)-MAH, SEBS-g-(MA), (POE)-(g)-(MA) and nanofillers (like organoclay, graphene, CNT, MWCNT, HNT, h-BN, etc.) and these are processed with melt compounding/mixing, in situ bulk polymerization technique, sonication, extrusion, rolling which are summarized in Table 2.

5 Rolling Process

Cold rolling is one of the processes in which a polymer sheet is kept between two rollers or three rollers and then compressed. Rolling processes are followed by injection molding and extrusion. Due to compressive load act on polymer nanocomposites, dimension control will be maintained (Shaikh & Kubade, 2023). Researcher used initial thickness d_1 as 1 mm, after rolling of polymer nanocomposites final thickness d_2 would be 0.8, 0.6, 0.4. Accordingly, the low, medium, and high compression/reduction of component is found as illustrated in Fig. 7a–c. After rolling process the nanoparticle orientated properly and well defined so that enhancement in terms of mechanical qualities/properties was observed. For better result, component is passed 2 or 3 times. Properties are measured along transverse and rolling/machine directions (TD and MD). A compression ratio of 2.50, elongation at break and strength together improved at machine or rolling direction as compared to transvers direction (Shaikh & Kubade, 2023).

So as to maintain exact compression ratio, gap between the roller is adjusted. Table 3 shows compression or reduction ratio as 1.25, 1.67, 2.25. It has been revealed that the better result obtained at a high reduction ratio that is at 60% reduction referring to the polymer nanocomposite sheet (Phua, 2011).

6 Conclusion

There are several processes for the research of the polymer nanocomposite (PNC) materials, and separately has its advantages and weaknesses; hence, the appropriate methods should be applied to the target application, orientation, spreading performance, etc. Finally, it has been noted that the mechanical and moreover thermal properties are enhanced in rolling process as compared to other processing techniques like melt mixing, ultrasonication, injection molding and extrusion. Higher the reduction ratio or compression, better the properties. Melting point of the composites slightly increases after cold rolling for PBS. Number of passes, roller speed is affecting the mechanical properties. In case of cold rolling process for preparation of polymer nanocomposites, elongation at break and strength together enhanced at machine direction or rolling direction. Not only can the nanoparticles clarify the detected effects, but the impact of the contact between the matrix and particle is also important. There is scope to do study on polymer nanocomposites with suitable nanofiller for rolling process by performing different sets of reduction ratio.

Table 2 Current research on different materials and processes for nanocomposites

Sr. no.	Material	Compatibilizer	Nanofiller	Process used	Properties enhanced	Concluding remark	References
1	PMMA/BN	–	Graphene-like boron nitride (BN) 3 wt.%	BN layers	Hardness: 125%, elastic modulus: 130%	Incorporation of nanomaterial enhances mechanical/thermal properties	Kiran et al. (2011)
2	PP/h-BN	–	h-BN (35 wt.%)	Compression, FFF	Thermal property	Improved thermal conductivity with large size particle of hBN	Li et al. (2021)
3	Epoxy/mG/h-BN@PDA	–	mG/h-BN (5–25 wt.%)	Sonication	TS: 5.50 MPa	Combination of fillers, the orientation of these fillers enhanced thermo-mechanical, electrical characteristics. This mixture could be applied in thermal management area	Su et al. (2018)
4	PE/h-BN	–	h-BN 5 wt.%	Sonication, compression molding	TS: 27% YM: 64.1%	h-BN exhibits good reinforcing capability	Achaby et al. (2012)
5	PP/GNs	–	Graphene (0.2–3 wt.%; 0.08–1.2 vol%)	Melt mixing, twin screw co-rotating extruder	TS: 33.98–61.57 MPa (81%) influence on the rheological properties	Enhancement of mechanical properties significantly. Load transfer is efficient for PP + GN composites	Achaby et al. (2012)
6	PP/IFR/CNTs/RGO	–	IFR/CNTs/RGO (IFR 18 wt.%, CNTs and graphene 1 wt.%)	Altered Hummer's method With vacuum-dried	PHRR reduced by about 83%	PP's heat stability and char yields have been improved	Huang et al. (2014)
7	(PP)/G	–	Graphene (0.1–5.0 wt.%)	Melt-blending	YS: ~75% increase YM: ~74% increase	By adding graphene (g) sheets into PP, the stability of thermal oxidative is enhanced significantly. It has barrier effect on composites. Also, mechanical characteristics of PP are increased after the mixing of graphene	Song et al. (2011)
8	PP/fGO/PP-g-MA	PP-g-MA, MAPP, MAPP-g-fGO	fGO (0.1–1.0 wt.%)	Ultrasonication	EB of nanocomposites is decreased	Significant enhancement of thermal stability	Yuan et al. (2014)
9	PP/graphene/SiO_2 hybrid (GOS)	–	Graphene/SiO_2 hybrid (GOS) 0.25 wt.%	Melt-blending	TS: 27.95% Impact strength: 43.87%	Graphene + nano SiO2 gives better property of PP instead of pure PP	Yuan et al. (2014)
10	PP/MWCNT	PP-g-MA	MWCNT, ATH 1%	Twin screw extruder	Increase in thermal properties upto some extent	Well scattering observed at 1 wt.% ATH and 1 wt.% SCNT. Nanofiller + Polymer matrix bonding was enhanced	Sonawane et al. (2020)
11	Epoxy/graphene/basalt	–	Hybrid graphene/basalt fiber 0.3 wt.%	Ultra sonication, hand lay-up	Wear rate: −38%	Wear rate and Friction Coefficient reduces. At 0.3 wt.% (GNPs), better wear properties were found	Kazemi-Khasragh et al. (2019)
12	POM/h-BN	–	h-BN 3%	Mixer	Wear rate: −85%	Low friction and wear	Gao et al. (2017)
13	PP/GNPs	–	GNPs (1–5 wt.%)	Twin screw mixing and injection molding	Maximum tensile ~33 MPa FS: ~58.81 MPa	Improvement of mechanical strength is observed with GNP. Better spreading of nanofiller is observed at smaller thickness and the lower is the agglomeration	Sutar et al. (2021)

(continued)

Table 2 (continued)

Sr. no.	Material	Compatibilizer	Nanofiller	Process used	Properties enhanced	Concluding remark	References
14	PP/PP-g-MAH	PP-g-MAH	Nanoclay 5 wt.%	Screw extrusion	TM: 130% Good ductility, high stiffness	mAH gives flake-off structure and a functional enrichment of mechanical characteristics for PP + nanoclay composites as compared to pure PP	Modesti et al. (2005)
15	(PP) + xGnP + CNT	–	xGnP/CNT 3, 5, 10 wt.% 0.1, 1, 5 wt.%	Twin screw mini extruder (co-rotating arrangement)	Enhanced	An average size of approx. 30 nm at interphase. 5–12 GPa is obtained modulus	Mehdi et al. (2010)
16	PP/FGO	–	FGO 2 wt.%	Melt mixing	YM: 31% TS: 15%	Better exfoliation and dispersion compared with that of PP/GO or PP/RGO	Lee et al. (2020)
17	PP/CB	–	CB 1–10 wt.%	Melt compounding	Enhanced thermal stability	Enhanced flame retardancy characteristics nearly at scope because it acts as a barrier to the transfer of heat flow	Wen et al. (2012)
18	PP/CNF and PP/GRP	–	CNF GRP 45 wt.% 60 wt.%	Injection molding	Maximum flexural modulus: 15 GPa	Incorporating GRP and CNF with very high loading improves electrical and mechanical performance of PP nanocomposites compared with neat PP	Pathak et al. (2020)
19	PLA/BNNS	Modifier: chemical hydroxyl	Surface-treated BNNS 3 wt.%	Chemical and physical surface treatments	YM: 24.5%, toughness: 42%	Surface treated nanofiller were slightly better. Energy absorbed by PLA per unit volume is about 131%	Rastegar and Montazeri (2022)
20	EP/PEI-BNNS	–	PEI-BNNS	Ultrasonication		PEI grafted (EP + BNNS) results in a better anticorrosive nature	Wu et al. (2020)
21	PP/ABS	PP-g-MA SEBS-g-MA	PEI-modified HNTs 80/20 (wt./wt.)	Twin screw extruder with counter rotating	TM:28.8% TS: 26.6% IS:38.5%	Thermal stability also enhances for PEI-modified HNTs-filled polymer blend as compared to pure blends and HNTs-filled polymer blend	Kubade and Tambe (2016)
22	(PP) + (GNP)	(PP)-g-(MA), (POE)-g-(MA)	(GNP) 1–5 wt.%	Twin screw extruder by melt compounding	Increased TS: up to 4 MPa	(POE)-g-(MA) compatibilizers somewhat reduced the strength. No major effect on both melting temp. And crystallization. Plain PP enhances thermal stability slightly	Al-Saleh et al. (2019)
23	PP/GNPs	–	GNP (9–12 wt.%)	Melt extrusion	Thermal conductivity 23.8% higher	To achieve the improved thermal conductivity at little. The bigger diameter through moderate spreading will be observed as more conducive at GNPs (6 wt.%) content for same system	Xu et al. (2019)
24	PP/N6	PP-g-MAH	–	Twin screw extruder	TS: 25%	The incorporation of (PP)-(g)-(MAH) increases modulus. It also enhances the adhesion among the two different components	Shashidhara et al. (2019)
25	PP/graphene	–	Graphene (0.5, 1.5, 2.5, 5 wt.%)	Ultrasonication	Thermal, mechanical properties	Thermal stability of pure PP was lesser compared to PP/graphene PnCs	Bafana et al. (2017)
26	PP/EPR	PP-g-MAH	GO 30 wt.%	Melt mixing	Improved TS and YM		Guezzout et al. (2017)

(continued)

Table 2 (continued)

Sr. no.	Material	Compatibilizer	Nanofiller	Process used	Properties enhanced	Concluding remark	References
						Incorporation of (GO) and used compatibilizers progresses thermal stability	
27	PTFE/hBN/AlN	–	hBN/AlN	Ultrasonication	Thermal conductivity is 3.8 times that of neat PTFE	Hybrid fillers improve the thermal conductivity	Pan et al. (2018)
28	PLA	–	MWCNTs 1 wt.%	Cold rolling	TS: 89%	At reduction ratios of 30 and 50%, it is found that increased fracture strain. The values changes from 17.9–27.5 to 120%	Gu et al. (2018)

[TS: tensile strength, YM: Young's modulus; YS: Yield strength; IS: Impact Strength; EB: Elongation at break; TM: tensile modulus; PMMA: polymethylmethacrylate; FFF: Fused filament fabrication; mG: multi-layer grapheme; h-BN: hexagonal boron nitride; PDA: polydopamine; CER: Cycloaliphatic epoxy resin; PE: polyethylene; IFR: intumescent flame retardant; PHRR: heat release rate peak; PP-g-MA: maleic anhydride grafted polypropylene; MWCNT: Multiwall carbon nanotube; ATH: Aluminum Hydroxide; GNP: graphene nanoplatelets; POM: polyoxymethylene; xGnP: exfoliated graphite nanoplatelets; CNT: carbon nanotube; FGO: fluorinated graphene oxide; RGO:- reduced-graphene oxide; CB:- carbon black; CNF:- carbon nano-fiber; GRP: granulated graphitized graphene; PLA: polylactic acid; EP: epoxy; PEI: polyethyleneimine; ABS: acrylonitrile butadiene styrene; GNP: graphene nanoplatelets; POE-g-MA: ethylene-octene elastomer grafted maleic anhydride; EPR: ethylene-propylene rubber; PTFE: polytetrafluoroethylene; PLA: poly(lactic acid)]; (Kubade et al., 2017; Patil et al., 2021; Shaikh & Kubade, 2023)

Fig. 7 Rolling of polymer nanocomposites; **a** low **b** moderate **c** high reduction

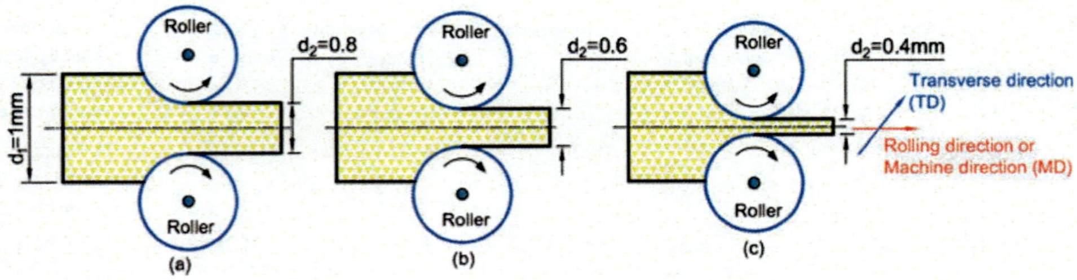

Table 3 Thickness reduction/compression ratio (Phua, 2011)

Initial thickness d_1 (mm)	1	1	1
Final thickness d_2 (mm)	0.8	0.6	0.4
Compression ratio (reduction ratio) = d_1/d_2	1.25	1.67	2.50
Thickness reduction (%) = $(d_1 - d_2)/100$	20	40	60

Acknowledgements Aamir M. Shaikh would like to thank Dr. Pravin R. Kubade for supervising research activities.

References

El Achaby, M., Arrakhiz, F.-E., Vaudreuil, S., el Kacem Qaiss, A., Bousmina, M., & Fassi-Fehri, O. (2012). Mechanical, thermal, and rheological properties of graphene-based polypropylene nanocomposites prepared by melt mixing. *Polymer Composites, 33*(5), 733–744.

Al-Saleh, M. A., Yussuf, A. A., Al-Enezi, S., Kazemi, R., Wahit, M. U., Al-Shammari, T., & Al-Banna, A. (2019). Polypropylene/graphene nanocomposites: Effects of GNP loading and compatibilizers on the mechanical and thermal properties. *Materials, 12*(23), 3924.

Avila-Orta, C., González Morone, P., Valdez, D., González-Sánchez, A., Martinez, G., Mata, J., & Cruz, V. (2019). Ultrasound-assisted melt extrusion of polymer nanocomposites. https://doi.org/10.5772/intechopen.80216

Bafana, A. P., Yan, X., Wei, X., Patel, M., Guo, Z., Wei, S., & Wujcik, E. K. (2017). Polypropylene nanocomposites reinforced with low weight percent graphene nanoplatelets. *Composites Part B: Engineering, 109*, 101–107. https://doi.org/10.1016/j.compositesb.2016.10.048

Bayır, S., Semerci, E., & Erdogan Bedri, T. (2021). Preparation of novel thermal conductive nanocomposites by covalent bonding between hexagonal boron nitride nanosheet and well-defined polymer matrix. *Composites Part a: Applied Science and Manufacturing, 146*, 106406.

Bian, J., Wang, Z. J., Lin, H. L., Zhou, X., Xiao, W. Q., & Zhao, X. W. (2017). Thermal and mechanical properties of polypropylene nanocomposites reinforced with nano-SiO_2 functionalized graphene oxide. *Composites Part a: Applied Science and Manufacturing, 97*, 120–127.

Breuer, O., & Sundararaj, U. (2004). Big returns from small fibers: A review of polymer/carbon nanotubes composites. *Polymer Composites, 25*(6).

Cahill, J. T., Du Frane, W. L., Sio, C. K., King, G. C. S., Soderlind, J. C., Lu, R., Kuntz, J. D., et al. (2020). Transformation of boron nitride from cubic to hexagonal under 1-atm helium. *Diamond and Related Materials*, 108078. https://doi.org/10.1016/j.diamond.2020.108078

Chan, K., Wong, H., Yeung, K., & Tjong, S. (2015). Polypropylene biocomposites with boron nitride and nanohydro xyapatite reinforcements. *Materials, 8*(3), 992–1008.

Gao, C., Guo, G., Zhang, G., Wang, Q., Wang, T., & Wang, H. (2017). Formation mechanisms and functionality of boundary films derived from water lubricated polyoxymethylene/hexagonal boron nitride nanocomposites. *Materials & Design, 115*, 276–286.

Gu, L., Qiu, J., Yao, Y., Sakai, E., & Yang, L. (2018). Functionalized MWCNTs modified flame retardant PLA nanocomposites and cold rolling process for improving mechanical properties. *Composites Science and Technology, 161*, 39–49. https://doi.org/10.1016/j.compscitech.2018.03.033

Guezzout, Z., Doufnoune, R., & Haddaoui, N. (2017). Effect of graphene oxide on the properties of compatibilized polypropylene/ethylene-propylene-rubber blend. *Journal of Polymer Research, 24*(8). https://doi.org/10.1007/s10965-017-1291-7

Huang, G., Wang, S., Song, P., Wu, C., Chen, S., & Wang, X. (2014). Combination effect of carbon nanotubes with graphene on intumescent flame-retardant polypropylene nanocomposites. *Composites Part a: Applied Science and Manufacturing, 59*, 18–25.

Kapłon, Ł., Kochanowski, A., Molenda, M., Moskal, P., Wieczorek, A., Bednarski, T., Bialas, P., Czerwiński, E., Korcyl, G., Kowal, J., Kowalski, P., Kozik, T., Krzemien, W., Niedźwiecki, S., Palka, M., Pawlik, M., Raczyński, L., Rudy, Z., Salabura, P., & Zon, N. (2014). Plastic scintillators for positron emission tomography obtained by the bulk polymerization method. *Bio-Algorithms and Med-Systems, 10*. https://doi.org/10.1515/bams-2013-0108

Kazemi-Khasragh, E., Bahari-Sambran, F., Siadati, S. M. H., Eslami-Farsani, R., & Chirani, S. A. (2019). The effects of surface-modified graphene nanoplatelets on the sliding wear properties of basalt fibers-reinforced epoxy composites. *Journal of Applied Polymer Science, 136*, 47986.

Kiran, M. S. R. N., Raidongia, K., Ramamurty, U., & Rao, C. N. R. (2011). Improved mechanical properties of polymer nanocomposites incorporating graphene-like BN: Dependence on the number of BN layers. *Scripta Materialia, 64*(6), 592–595. https://doi.org/10.1016/j.scriptamat.2010.12

Kong, S., Seo, H., Shin, H., Baik, J.-H., Oh, J., Kim, Y.-O., & Lee, J.-C. (2019). Improvement in mechanical and thermal properties of polypropylene nanocomposites using an extremely small amount of alkyl chain-grafted hexagonal boron nitride nanosheets. Polymer, 121714.

Kubade, P. R., Tambe, P., & Kulkarni, H. B. (2017). Morphological, thermal and mechanical properties of 90/10 (WT.%/WT.%) PP/ABS blends and their polymer nanocomposites. *Advanced Composites Letters, 26*(6). https://doi.org/10.1177/096369351702600602

Kubade, P., & Tambe, P. (2016). Influence of surface modification of halloysite nanotubes and its localization in PP phase on mechanical and thermal properties of PP/ABS blends. *Composite Interfaces, 24*(5), 469–487.

Kulkarni, H. B., Tambe, P. B., & Joshi, G. M. (2019). Influence of surfactant assisted exfoliation of hexagonal boron nitride nanosheets on mechanical, thermal and dielectric properties of epoxy nanocomposites. *Composite Interfaces*. https://doi.org/10.1080/09276440.2019.1663115

Kulkarni, H. B., Tambe, P. B., & Joshi, G. M. (2019). Influence of surfactant assisted exfoliation of hexagonal boron nitride nanosheets on mechanical, thermal and dielectric properties of epoxy nanocomposites. *Composite Interfaces*.

Kumar, A., Ghosh, P. K., Yadav, K. L., & Kumar, K. (2017). Thermo-mechanical and anticorrosive properties of MWCNT/epoxy nanocomposite fabricated by innovative dispersion technique. *Composites Part B*.

Lei, Y., Pakhira, S., Fujisawa, K., Liu, H., Guerrero-Bermea, C., Zhang, T., Dasgupta, A., Martinez, L. M., Singamaneni, S. R., Wang, K., Shallenberger, J., Elías, A. L., Cruz-Silva, R., Endo, M., Mendoza-Cortes, J. L., & Terrones, M. (2021). Low temperature activation of inert hexagonal boron nitride for metal deposition and single atom catalysis. *Materials Today, 51*, 108–116. ISSN 1369-7021, https://doi.org/10.1016/j.mattod.2021.09.017

Lee, M. G., Lee, S., Cho, J., Bae, S., & Jho, J. Y. (2020). Effect of the fluorination of graphene nanoflake on the dispersion and mechanical properties of polypropylene nanocomposites. *Nanomaterials, 10*(6), 1171.

Li, J., Leng, J., Jiang, Y., & Zhang, J. (2021). Experimental characterization of 3D printed PP/h-BN thermally conductive composites with highly oriented h-BN and the effects of filler size. *Composites Part a: Applied Science and Manufacturing, 150*, 106586. https://doi.org/10.1016/j.compositesa.2021.106586

Mehdi, K., Pucha, R. V., Bhuiyan, Md. A., & Kyriaki, K. (2010). Effect of interphase modulus and nanofiller agglomeration on the tensile modulus of graphite nanoplatelets and carbon nanotube reinforced polypropylene nanocomposites. *Carbon Lettres, 11*(4), 325–331.

Modesti, M., Lorenzetti, A., Bon, D., & Besco, S. (2005). Effect of processing conditions on morphology and mechanical properties of compatibilized polypropylene nanocomposites. *Polymer, 46*(23), 10237–10245.

Pan, C., Kou, K., Zhang, Y., Li, Z., & Wu, G. (2018). Enhanced through-plane thermal conductivity of PTFE composites with hybrid fillers of hexagonal boron nitride platelets and aluminum nitride particles. *Composites Part B: Engineering, 153*, 1–8. https://doi.org/10.1016/j.compositesb.2018.07.019

Pathak, A. K., Zhou, Y., Lecointre, L., & Yokozeki, T. (2020). Polypropylene nanocomposites with high-loading conductive carbon nano-reinforcements for multifunctional applications. *Applied Nanoscience*. https://doi.org/10.1007/s13204-020-01594-6

Patil, J., Patil, H., Sankpal, R., Rathod, D., Patil, K., Kubade, P. R., & Kulkarni, H. B. (2021). Studies on mechanical and thermal performance of carbon nanotubes/polypropylene nanocomposites. *Materials Today: Proceedings, 46*, Part 17, 7182–7186. ISSN 2214-7853, https://doi.org/10.1016/j.matpr.2020.11.452

Phua, Y. J. (2011). Mechanical properties and structure development in poly(butylene succinate)/organo-montmorillonite nanocomposites under uniaxial cold rolling. *eXPRESS Polymer Letters, 5*, 93–103.

Rastegar, S., & Montazeri, A. (2022). Atomistic insights into the toughening role of surface-treated boron nitride nanosheets in PLA-based nanocomposites. *European Polymer Journal, 168*, 111071. ISSN 0014-3057.

Salleh, A., Abd Razak, J., Ibrahim, N., Fakhru'l-Razi, A., & Abdul Rashid, S. (2008). The influences of melt-compounding parameters on the tensile properties of low filler loading of untreated-MWCNTs-polypropylene (PP) nanocomposites. *Journal of Engineering Science and Technology, 3*.

Shaikh, A. M., & Kubade, P. R. (2023). *Materials Today: Proceedings*. https://doi.org/10.1016/j.matpr.2023.07.346

Sharma, S. K., & Nayak, S. K. (2009). Surface modified clay/polypropylene (PP)nanocomposites: Effect on physico-mechanical, thermal and morphological properties. *Polymer Degradation and Stability, 94*(1), 132–138.

Shashidhara, G. M., Biswas, D., Shubhalaksmi Pai, B., Kadiyala, A. K., Wasim Feroze, G. S., & Ganesh, M. (2019). Effect of PP-g-MAH compatibilizer content in polypropylene/nylon-6 blends. *Polymer Bulletin, 63*(1), 147–157. https://doi.org/10.1007/s00289-009-0074-7

Sonawane, S., Thakur, P., & Paul, R. (2020). Study on thermal property enhancement of MWCNT based polypropylene (PP) nanocomposites. *Materials Today: Proceedings*. https://doi.org/10.1016/j.matpr.2019.12.018

Song, P., Cao, Z., Cai, Y., Zhao, L., Fang, Z., & Fu, S. (2011). Fabrication of exfoliated graphene-based polypropylene nanocomposites with enhanced mechanical and thermal properties. *Polymer, 52*(18), 4001–4010.

Su, Z., Wang, H., Ye, X., Tian, K., Huang, W., He, J., & Tian, X. (2018). Synergistic enhancement of anisotropic thermal transport flexible polymer composites filled with multi-layer graphene (mG) and mussel-inspiring modified hexagonal boron nitride (h-BN). *Composites Part a: Applied Science and Manufacturing, 111*, 12–22.

Sui, Y., et al. (2022). Rheological and mechanical properties of automobile polypropylene/silica (PP/SiO_2) nanocomposites. Journal of Physics: Conference Series (Vol. 2324, p. 012004).

Sutar, H., Mishra, B., Senapati, P., Murmu, R., & Sahu, D. (2021). Mechanical, thermal, and morphological properties of graphene nanoplatelet-reinforced polypropylene nanocomposites: Effects of nanofiller thickness. *Journal of Composites Science, 5*(1), 24. https://doi.org/10.3390/jcs5010024

Svoboda, P., Zeng, C., Wang, H., et al. (2002). Morphology and mechanical properties of polypropylene/organoclay nanocomposites. *Journal of Applied Polymer Science, 85*, 1562–1570.

TabkhPaz, M., Shajari, S., Mahmoodi, M., Park, D.-Y., Suresh, H., & Park, S. S. (2016). Thermal conductivity of carbon nanotube and hexagonal boron nitride polymer composites. *Composites Part B: Engineering, 100*, 19–30.

Thomas, S., Stephen, R., Bandyopadhyay, S., & Thomas, S. (2007). Polymer nanocomposites: Preparation, properties and applications. *Rubber Fibers Plastics International., 2*, 49–56.

Verma, A., Parashar, A., & Packirisamy, M. (2017). Atomistic modeling of graphene/hexagonal boron nitride polymer nanocomposites: A review. *Wiley Interdisciplinary Reviews: Computational Molecular Science, 8*(3), e1346.

Wen, X., Wang, Y., Gong, J., Liu, J., Tian, N., Wang, Y., & Tang, T. (2012). Thermal and flammability properties of polypropylene/carbon black nanocomposites. *Polymer Degradation and Stability, 97*(5), 793–801.

Wu, Y., He, Y., Zhou, T., Chen, C., Zhong, F., Xia, Y., & Zhang, C. (2020). Synergistic functionalization of h-BN by mechanical exfoliation and PEI chemical modification for enhancing the corrosion resistance of waterborne epoxy coating. *Progress in Organic Coatings, 142*, 105541.

Xu, Z., He, S., Zhang, J., Huang, S., Chen, A., Fu, X., & Zhang, P. (2019). Relationship between the structure and thermal properties of polypropylene/graphene nanoplatelets composites for different platelet-sizes. *Composites Science and Technology*, 107826. https://doi.org/10.1016/j.compscitech.2019.107826

Yola, M. L., & Atar, N. (2018). A Novel detection approach for serotonin by graphene quantu dots/two-dimensional (2D) hexagonal boron nitride nanosheets with molecularly ımprinted polymer. *Applied Surface Science*.

Yuan, B., Bao, C., Song, L., Hong, N., Liew, K. M., & Hu, Y. (2014). Preparation of functionalized graphene oxide/polypropylene nanocomposite with significantly improved thermal stability and studies on the crystallization behavior and mechanical properties. *Chemical Engineering Journal, 237*, 411–420.

Zhang, M., Li, Y., Su, Z., & Wei, G. (2015). *Polymer Chemistry*. https://doi.org/10.1039/C5PY00777A

Synthesis and Characterization of Co-precipitated Hematite-α-Fe_2O_3 Nanoparticles (AIO-NPs)

Dinesh Uthra, Krishna Patel, Jai Narayan Sahu, and M. Z. Khan

Abstract

In the present work, we used a co-precipitation-calcination approach to synthesized Hematite-Fe_2O3 Nanoparticles (AIO-NPs). In this technique we have taken 10 g of $FeCl_3 \cdot 6H_2O$, 150 ml of pure water was used to dissolve with stirring. After that, at room temperature, a 2 ml NH_4OH solution was added drop by drop (drop rate = 1 ml min^{-1}) to the stirring mixture. Throughout the synthesis, the pH remained at a constant one. After being continuously stirred for one hour at room temperature, the black dispersion was heated to evaporate for two hours at 80 °C to produce a brown powder. After being brought down to the room temperature, the product was then calcined for four hours at 500 °C. The as-prepared samples were subjected to calcination at various temperatures ranging from 400 °C to 800 °C. To study the structural properties in details, characterized by XRD. The lattice constant so obtained for α-Fe_2O_3 Nanoparticles (AIO-NPs) were a = b = 5.03421 Å and c = 13.746502 Å. The size of the particles was calculated from the XRD spectra with the help of Scherrer formula around 30 nm. Having increase in the annealing temperature phase transformation from magnetite (Fe_3O_4) to hematite (α-Fe_2O_3) is observed. The Raman study exhibited the lattice expansion and line broadening optical modes were observed. The modes observed at 217 and 490, 821 cm^{-1} are corresponding to A1g phonon modes corresponding to Fe–O symmetrical structure whereas the modes observed at wave numbers 278, 391 and 589 cm^{-1} are pointing the Eg phonon modes correspond Fe–O symmetrical bend. All these results are discussed in this paper. The morphology of hematite particle was not uniform and there was agglomeration.

Keywords

Hematite • Co-precipitation method • Iron oxide nanoparticles

1 Introduction

The unique structure for a nanomaterial ranges from 1 to 100 nm in dimension which is mainly in the form of powder, and is made up of nanoparticles. Due to the different chemical and physical properties, the size of the particles is less as compared to the macroscopic structure of the identical compound. Nanotechnology is nothing but the mixture of patterns and the characterization of small-sized substances. One can say that technology is a fine art of particles which is by manipulation. These produced nanostructured materials, having new properties with unique purposes. Due to all of these this technology is now growing in the field of research science and nanoscience (Adinaveen et al., 1721; Bahari, 2165; Bumbudsanpharoke & Ko, 2015; Bumbudsanpharoke et al., 2015; Carbone et al., 2016; Cha et al., 2019; Huang et al., 2018; Rezić et al., 2017; Tsagkaris et al., 2870; Whitney et al. 1993). Dimensional classification of nanomaterials is shown in Fig. 1.

As this technology produces nano sized materials that having a new material properties and for different purposes because of this it is called technology at nano scale. Most of the research and development in nanotechnology and nanoscience is moving toward this technology due to its unique size and features. Application of nanotechnology can be in areas like environmental protection, pharmaceutical industry, new material evolution, agriculture food processing and packaging.

D. Uthra (✉) · K. Patel
Department of Pure and Applied Physics, G.G.V. Central University, Bilaspur, Chhattisgarh 495009, India
e-mail: dkuthra45@gmail.com

J. N. Sahu · M. Z. Khan
Department of Physics, Dr. C.V. Raman University, Kota, Bilaspur, Chhattisgarh 495113, India

M. Sumesh et al. (eds.), *2nd International Conference on Smart Sustainable Materials and Technologies (ICSSMT 2023)*, Advances in Science, Technology & Innovation, https://doi.org/10.1007/978-3-031-49826-8_7

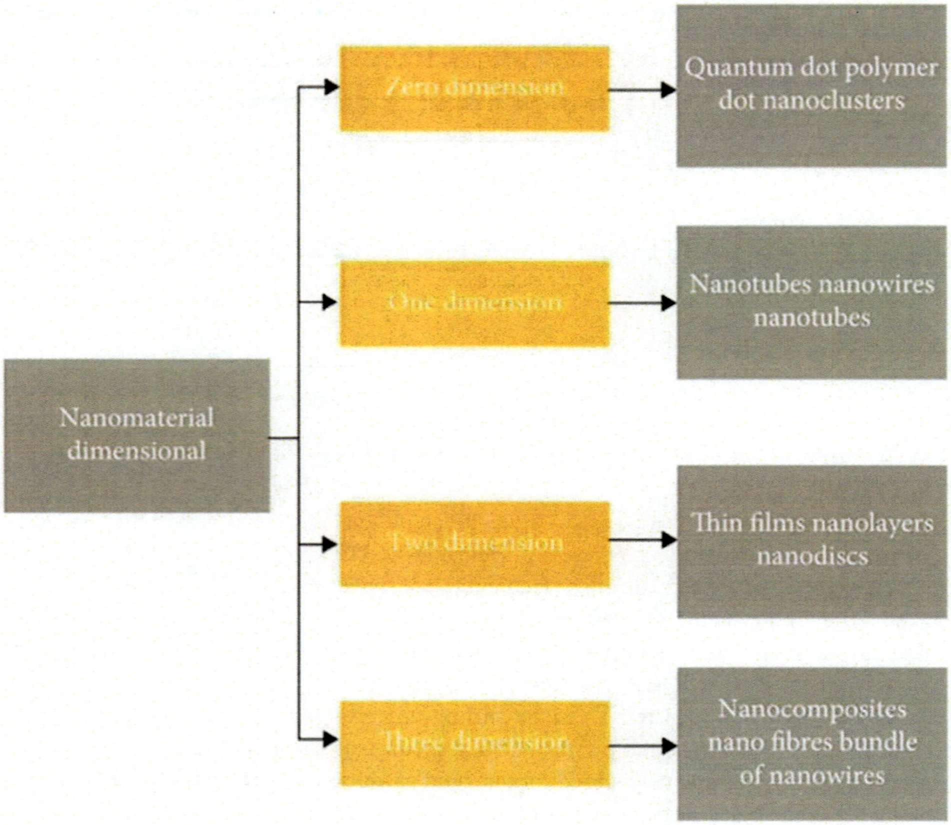

Fig. 1 Dimensional classification of nanomaterials

Nanomaterial is well known as it might be pragmatic like in food production, nanoemulsion pesticides, in the agriculture sectors, for large surface area nanoceramic devices as well as the maintenance sector; for food packing as well as many incorporated food products like titanium oxide and silver oxide which can be used as colour, flavours, food agent and test. Most developed and advanced technology is the nanotechnology in recent times. For food industry more than 400 companies have adopted the nanotechnology for better products. A report predicted in 2013 that by the end of 2018 the global market supply of nano-enabled packages of food will be around 6.5 billion and that also be predicted or expected to reach approximately 20 billion by the end of 2025 (Cao et al., 2005; Guodong et al., 2021; Kusior et al., 2019; Li et al., 2019a, 2019b; Lian et al., 2009; Qiu et al., 2019; Rahman et al. 2022; Sarangi et al., 2009; Zhang et al., 2019). Because of proper food packaging, the food can provide healthy and nutrition food products. Many nanometers are used for food packaging, such Silver nanoparticles are one of the most productive because they possess properties like antibacterial and antiviral, as compared to different nanometrics like zinc and titanium. Nanomaterials possess deeper properties of chemical structures and characteristics because this nanometer has different applications in code packaging as compared to micro and MacroParticles. Silver Nanoparticles starch and nano zinc oxide have a larger surface area.

Hematite α-Fe_2O_3 Nanoparticles (AIO-NPs) is the most stable iron oxide and the semiconductor with most friendly environmental with an energy band gap of Eg = 2.1 eV. Iron oxide nanoparticle preparation such as hematite Fe_2O_3 has attracted significant attention in past recent years because of its properties and application in numerous fields of technology. For sensible photocatalytic and electrochemical properties also in applications like organic pollutants degradation, as an anode, in various methods that include hydrothermal, solvothermal and co-precipitation, direct calcination of the ferric salt in air atmosphere, liquid phase-based ultrasonication (Baratto et al. 1998), chemical deposition and sol–gel, were used to prepare this inexpensive and environmentally friendly semiconductor device with a narrow band gap (2.0–2.2 eV). Altering the concentration, time, and temperature of the raw materials as well as the thermal routes can also alter the dimensions and morphology of the α-Fe_2O_3 nanoparticles. Kusior et al. (Kusior et al., 2019) utilized an ion-mediated hydrothermal method and reported distinct α-Fe_2O_3 shapes. Li et al. (2019b) used the hydrothermal method, various morphologies of α-Fe_2O_3 were prepared by varying the time and temperature. Rahman et al. (2020) made uniform nanodisks of hematite catalysts through a straightforward hydrothermal process. Tadic et al. (2019) developed rhombohedron and plate-like haematite (α-Fe_2O_3) as potential biomedical applications for MRI.

Researchers have used a variety of methods, including combustion, hydrothermal, precursor, solvent evaporation, and others, to produce homogeneous iron oxide nanoparticles. However, these methods typically necessitate the use of specialized equipment, high temperatures, and laborious impurity removal—all of which take a lot of time and cost a lot of money. Using inorganic salts as precursors, the co-precipitation method has been utilized to produce nano or sub-micro powders in a variety of metal oxides. The synthesis of managed-dimension magnetite nanoparticles has long been of interest to scientists and engineers. However, the synthesis method greatly affects the uniform physical and chemical properties of magnetite nanoparticles, making it difficult to devise a straightforward and efficient method for producing magnetite particles with a restricted dimension distribution and high dispersion. α-Fe_2O_3 can be found in two well-known crystalline forms: Hematite, the phase, has a rhombohedral structure, while maghemite, the -phase, has a cubic shape. During calcination, around 400 °C, the transition of α-Fe_2O_3(AIO) occurs. Using the iron chloride $FeCl_3{\cdot}6H_2O$ precursor, ferric oxide was synthesized via co-precipitation in this study. Using XRD and RAMAN, the surface morphological properties as well as the structural properties are discussed (Avasthi et al., 2020; Baratto et al., 1998; Chen et al., 2020; De Faria et al., 1997; Jesus et al., 2018; Lagodi & Waal, 2007; Rahman et al., 2020; Tadic et al., 2019; Zhang et al., 2022).

2 Experimental Details

The procedure for producing iron oxide nanoparticles was a straightforward co-precipitation synthesis.

First at room temperature, 10 g of $FeCl_3{\cdot}6H_2O$ were dissolved in 150 ml of pure water with stirring. After that, at room temperature, a 2 ml NH_4OH solution was added drop by drop (drop rate = 1 ml min^{-1}) to the stirring mixture. Throughout the synthesis, the pH remained at a constant one. After being continuously stirred for one hour at room temperature, the black dispersion was heated to evaporate for two hours at 80° to produce a brown powder. After being brought down to room temperature, the product was then calcined for four hours at 500 °C as shown in Fig. 2.

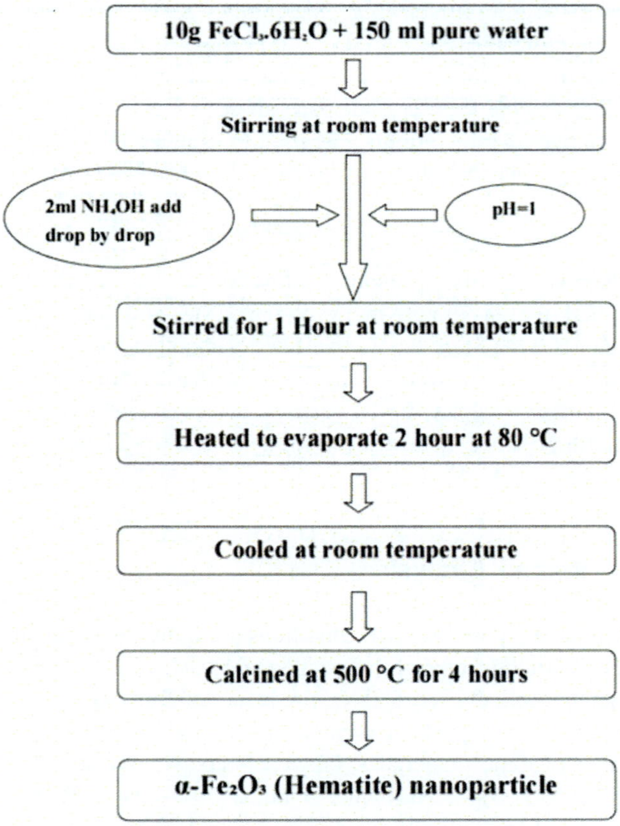

Fig. 2 Flow chart of α-Fe_2O_3 nanoparticles

3 XRD Analysis

All of the samples were analysed without being washed or purified. The as-synthesized and annealed α-Fe_2O_3 nanoparticles (AIO-NPs) size, structure, and surface morphological properties were specified. The crystalline phase was identified and the crystalline size was estimated using an X-Ray diffractometer (XRD). With type XPert Pro MPD, Cu-Kα λ = 1.54 Å., the XRD pattern was recorded in the range of 4–85°.

The crystalline phases and their sizes were determined using X-ray diffraction (XRD) at 40 kV. The sample's XRD pattern prior to annealing is depicted as the heat-treated XRD pattern of the sample. Between 300 and 600 °C, $\gamma \rightarrow \alpha$-$Fe_2O_3$ phase transformation occurred during calcination. When the temperature of the calcination rose above 400 °C, there was a sudden increase in the amount of a phase. The powder that was calcined at temperatures greater than 500 °C only contained the α-Fe_2O_3(AIO) phase. A rhombohedral structure of α-Fe_2O_3 is identified using standard data and belongs to the R-3c crystal space group.

Debye-Sherrer formula was used to estimate the ordered α-Fe_2O_3 nanoparticles (AIO-NPs) mean sizes, as shown

$$D = 0.89\lambda/(b\cos\theta)$$

where θ = the Bragg angle, b = line broadening in Radians at half the maximum intensity (FWHM), and the shape factor is 0.89. From this Debye-Sherrer equation the average size of annealed α-Fe_2O_3 nanoparticles was approximately 30 nm. For alpha Fe_2O_3 nanoparticles (AIO-NPs), the resulting lattice constants were a = b = 5.03421 A^0 and c = 13.746502 A^0.

4 Raman Analysis

Homogeneity and surface conditions are quite fast to quantify crystalline quality by Raman analysis of a semiconductor. Actually crystalline products exhibit way sharp peaks. Raman peaks are rather broad for amorphous or polycrystalline materials. Previous reports (Gouadec & Colomban, 2007; Nasibulin et al., 2009) showed Raman spectra of hematite are in between 200 and 800 cm^{-1}. In the present sample annealed at 500 °C, six optical modes were observed.

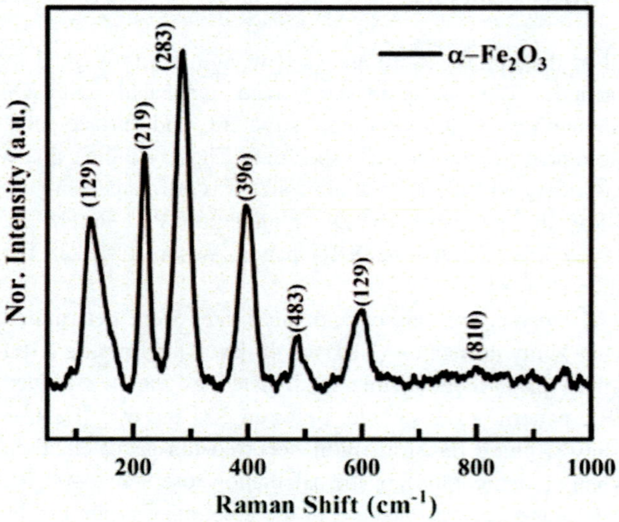

Fig. 3 Raman spectra of α-Fe_2O_3 Nanoparticles (AIO-NPs)

The modes observed at 217 and 490, 821 cm^{-1} correspond to Eg phonon modes corresponding to Fe–O symmetrical structure whereas the modes observed at wave numbers 278, 391 and 589 cm^{-1} are pointing the A1g phonon modes corresponding to Fe–O symmetrical bend. All the observed phonon modes are in good agreement with the group theory (Gouadec & Colomban, 2007) and other reports (Nasibulin et al., 2009). No other Raman peaks corresponding to other phases of iron oxide were observed in the present Raman spectra in Fig. 3. It confirmed the formation of α-Fe_2O_3(AIO) phase of iron oxide. These results also support the conclusions obtained from the XRD analysis mentioning the phase transformation from Fe_3O_4 to α-Fe_2O_3(AIO). The Eg mode observed at 217 and 490, 821 cm^{-1} is corresponding to Fe–O symmetrical structure whereas modes are related to Fe–O symmetrical bend. Thus Raman shift and line broadening effect has observed which effect is attributed to a size-dependent effect on nanoparticles and which is in good agreement as compared to previous report (Jesus et al., 2018; Rahman et al., 2020). Figure 3 shows the Raman spectra of α-Fe_2O_3 Nanoparticles (AIO-NPs).

5 SEM Analysis

A scanning electron microscope (SEM) was used to examine the Hematite α-Fe_2O_3 Nanoparticles (AIO-NPs) surface morphology. The morphology of Hematite α-Fe_2O_3(AIO) calcined for four hours in Fig. 4 revealed that the sample's particle shape is irregular and produces agglomerates.

6 Conclusion

The rhombohedral structure of Hermites-α-Fe_2O_3(AIO) was obtained through the Coprecipitation-calcination method, belongs to the R-3c crystal space group. Hermites −Fe_2O_3(AIO) are used to make for heavy media separation and also to produce pigments and put in homes for positive energy also to make jewellery. Besides of all these it has got healing powers. Future prospects of Hermites−Fe_2O_3(AIO) NPs in the MRI, biomolecules detection, hyperthermia, photothermal therapy, chemotherapy, antimicrobial activity and also their role for diagnosis and nanomedicines as a multifunctional agent. The hematite crystallite with the largest size was found after four hours (30 nm). When hematite is synthesized, the parameters directly correlate with the

Fig. 4 Morphology of α-Fe_2O_3 nanoparticles

Fig. 4 (continued)

four-hour redness that results from the highest chroma. There is agglomeration and irregularity in the morphology of the hematite α-Fe_2O_3 Nanoparticles.

References

Adinaveen, T., Judith Vijaya, J., & John Kennedy, L. (2014). Studies on the structural, morphological, optical, and magnetic properties of αFe_2O_3 nanostructures by a simple one-step low temperature reflux condensing method. *Journal of Superconductivity and Novel Magnetism*, *27*, 1721–1727. https://doi.org/10.1007/s10948-014-2497-0

Avasthi, A., Caro, C., Pozo-Torres, E., et al. (2020) Magnetic Nanoparticles as MRI Contrast Agents. *Topics in Current Chemistry (Z)*, *378*, 40. https://doi.org/10.1007/s41061-020-00302-w

Bahari, A. (2017). Characteristics of Fe_3O_4, αFe_2O_3 and γ-Fe_2O_3 nanoparticles as suitable candidates in the field of nanomedicine. *Journal of Superconductivity and Novel Magnetism*, *30*(1), 2165–2174. https://doi.org/10.1007/s10948-017-4014-8

Baratto, C., Lottici, P. P., Bersani, D., & Anionioli G. (1998). Sol-Gel preparation of α-Fe_2O_3 thin films: Structural characterization by XAFS and Raman. *Journal of Sol-Gel Science and Technology*, *13*, 667–671. https://doi.org/10.1023/a:1008694519106

Bumbudsanpharoke, N., & Ko, S. (2015) Nano-food packaging: An overview of market, migration research, and safety regulations. *Journal of Food Science*, *80*(5), R910–R923. https://doi.org/10.3390/nano8100830

Bumbudsanpharoke, N., Choi, J., & Ko, S. (2015). Applications of nanomaterials in food packaging. *Journal of Nanoscience and Nanotechnology*, *15*(9), 6357–6372. https://doi.org/10.1166/jnn.2015.10847

Cao, M., Liu, T., Gao, S., Sun, G., Wu, X., Hu, C., et al. (2005) Singlecrystal dendritic micro-pines of magnetic α-Fe_2O_3: Largescale synthesis, formation mechanism, and properties. *Angewandte Chemie International Edition*, *44*, 41974201. https://doi.org/10.1002/anie.200500448

Carbone, M., Donia, D. T., Sabbatella, G., & Antiochia, R. (2016). Silver nanoparticles in polymeric matrices for fresh food packaging. *Journal of King Saud University-Science*, *28*(4), 273–279. https://doi.org/10.1016/j.jksus.2016.05.004

Cha, H. G., Kim, C. W., Kim, Y. H., Jung, M. H., Ji, E. S., Das, B. K., et al. (2019). Preparation and characterization of α-Fe_2O_3 nanorodthin film by metal–organic chemical vapor deposition. *Thin Solid Films*, *517*, 1853. https://doi.org/10.1016/j.tsf.2008.10.023

Chen, T., Jiang, W., Sun, X., Ning, W., Liu, Y., Xu, G., et al. (2020). Size-controlled synthesis of hematite α-Fe_2O_3 nanodisks closed with (0001) basal facets and {11-20} side facets and their catalytic performance for CO2 hydrogenation. ChemistrySelect, *5*, 430. https://doi.org/10.1002/slct.201904490

De Faria D. L. A., Silva, S. V., de Oliveria, M. T. J. (1997). Raman micro-spectroscopy of some iron oxides and oxyhydroxides. *Raman Spectroscopy*, *28*, 873–878. https://doi.org/10.1002/(SICI)1097-4555(199711)28:113.0.CO;2-B

Gouadec, G., & Colomban, P. (2007). Raman spectroscopy of nanomaterials: How spectra relate to disorder, particle size and mechanical properties. *Progress in Crystal Growth and Characterization of Materials*, *53*, 1–56. https://doi.org/10.1016/j.pcrysgrow.2007.01.001

Huang, Y., Mei, L., Chen, X., & Wang, Q. (2018). Recent developments in food packaging nanomaterials. *Nanomaterials*, *8*(10), 830. https://doi.org/10.3390/nano810080

Jesus, J. R., Lima, R. J. S., Moura, K. O., Duque, J. G. S., & Meneses, C. T. (2018). Anisotropic growth of α-Fe_2O_3 nanostructures. *Ceramics International*, *44*, 3585–3589. https://doi.org/10.1016/j.ceramint.2017.11.068

Kusior, A., Michalec, K., Jelen, P., & Radecka, M. (2019). Shaped α-Fe_2O_3 nanoparticlesa—synthesis and enhanced photocatalytic degradation towards RhB. *Applied Surface Science*, *476*, 342–352. https://doi.org/10.1016/j.apsusc.2018.12.113

Lagodi, M. A., & de Waal, D. (2007). The preparation of magnetite, goethite, hematite and maghemite of pigment quality from mill scale iron waste. *Dyes and Pigments*, *74*(1), 161–168. https://doi.org/10.1016/j.dyepig.2006.01.038

Li, Z., Mao, Y., Tian, Q., Zhang, W., & Yang, L.: Extremely facile preparation of high-performance Fe_2O_3 anode for lithiumion batteries. *Journal of Alloys and Compounds*, *784*, 125–133 (2019). https://doi.org/10.1016/j.jallcom.2018.12.392

Li, D., Xu, R., Jia, Y., Ning, P., & Li, K. (2019). Controlled synthesis of αFe2O3 hollows from β-FeOOH rods. *Chemical Physics Letters*, *731*, 1336623. https://doi.org/10.1016/j.cplett.2019.136623

Lian, J., Duan, X., Ma, J., Peng, P., Kim T., & Zheng W. (2009). Hematite (α-Fe_2O_3) with various morphologies: Ionic liquid-assisted synthesis, formation mechanism, and properties. *ACS Nano*, *3*, 3749–3761. https://doi.org/10.1021/nn900941e

Nasibulin, A. G., Rackauskas, S., Jiang, H., Tian, Y., et al. (2009). Simple and rapid synthesis of α-Fe_2O_3 nanowires under ambient conditions. *Nano Research*, *2*, 373–379. https://doi.org/10.1007/s12274-009-9036-5

Qiu, M., Wang, R., & Qi, X. (2019). Hollow polyhedral α-Fe_2O_3 prepared by self-assembly and its photocatalytic activities in degradation of RhB. *Journal of the Taiwan Institute of Chemical Engineers*, *102*, 394–402. https://doi.org/10.1016/j.jtice.2019.05.024

Rahman, G., Najaf, Z., Ali Shah, A. H., & Mian, S. A. (2020). Investigation of the structural, optical, and photoelectrochemical properties of α-Fe_2O_3 nanorods synthesized via a facile chemical bath deposition. *Optik, 200*, 163454. https://doi.org/10.1016/j.ijleo.2019.163454

Rahman, A., Kang, S., McGinnis, S., & Vikesland, P. J. (2022). Life cycle impact assessment of iron oxide (Fe_3O_4/γ-Fe_2O_3) nanoparticle synthesis routes.ACS Sustainable Chem. Eng. 10, 10, 3155–3165, https://doi.org/10.1021/acssuschemeng.1c05763

Rezić, I., Haramina, T., & Rezić, T. (2017) Metal nanoparticles and carbon nanotubes—perfect antimicrobial nano-fillers in polymer-based food packaging materials. In *Food packaging* (pp. 497–532). Academic Press. https://doi.org/10.1016/B978-0-12-804302-8.00015-7

Sarangi, P. P., Naik, B., & Ghosh, N. N. (2009). Low temperature synthesis of single- phase α-Fe_2O_3 nano-powders by using simple but novel chemical methods. *Powder Technology*, *192*(3), 245–249. https://doi.org/10.1016/j.powtec.2009.01.002

Tadic, M., Panjan, M., Tadic, B. V., Lazovic, J., & Damnjanovic, V. (2019). Magnetic properties of hematite (α-Fe_2O_3) nanoparticles synthesized by sol-gel synthesis method: The influence of particle size and particle size distribution. *Journal of Electrical Engineering*, *70*, 71–76. https://doi.org/10.2478/jee-2019-0044

Tsagkaris, A. S., Tzegkas, S. G., & Danezis, G. P. (2018). Nanomaterials in food packaging: State of the art and analysis. *Journal of Food Science and Technology*, *55*(8), 2862–2870. https://doi.org/10.1007/s13197-018-3266-z

Whitney, T. M., Searson, P. C., Jiang, J. S., & Chien, C. L. (1993). Fabrication and magnetic properties of arrays of metallic nanowires. *Science*, *261*(1), 316–319. https://doi.org/10.1126/science.261.5126.1316

Zhang, H. J., Meng, F. N., Liu, L. Z., & Chen, Y. J. (2019). Convenient route for synthesis of alpha-Fe_2O_3 and sensors for H_2S gas. *Journal of Alloys and Compounds, 774*, 1181–1188. https://doi.org/10.1016/j.jallcom.2018.09.384

Zhang, Z., Zhou, Y., Zhao, S., Ding, L., Chen, B., Chen, Y. (2022). Nanomedicine-enabled/augmented cell pyroptosis for efficient tumor nanotherapy. *Advanced Science*, *9*(35), 2203583. https://doi.org/10.1002/advs.202203583

Zhao, G., Wang, E., & Tong, R. (2021). From reactive oxygen species to reactive brominating species: Fenton chemistry for oxidative bromination. *ACS Sustainable Chemistry & Engineering*, *9*(17), 6118–6125. https://doi.org/10.1021/acssuschemeng.1c01709

Influence of the Bio-lubricant Along with Nano Compounds as Additives on the Properties of the Lubricant

S. Nithya Poornima and V. Shantha

Abstract

Lubricant oils play a vital role in all areas with the purpose of reducing wear, smooth functioning of the system, reducing cutting temperature, absorbing heat, and overall improving machining efficiency. General impact of the commercial lubricant that has been used in automobiles has adverse effect on the environment. With the advancement in nanomaterials, the nanomaterials have proved to perform under extreme temperature and pressure and have shown some significance in the lubrication areas as well. With the growing demands of lubricant oil and the pollution caused by them, there has been a need for utilizing a lubricant oil that is available in abundance and at the same time trying to utilize the naturally available resource. In order to enhance its characteristic performance a small addition of nano compound is added to improve its performance and efficiency. As it is already proved that non-edible vegetable oils can effectively be used as a lubricant in machines for its smooth functioning. The present work focuses on collaborating the nano compound with the non-edible vegetable oil and studying its physical properties. The obtained results were compared with the SAE 20W40 which is a petroleum-based lubricant oil.

Keywords

Nanomaterials • Neem oil • Pongamia oil • Blending

S. N. Poornima (✉)
Department of Mechanical Engineering, BMS Institute of Technology and Management, Bengaluru, 560064, India
e-mail: nithyapmech@bmsit.in

V. Shantha
Department of Mechanical Engineering, Sir M. Visvesvaraya Institute of Technology, Bengaluru, 560064, India
e-mail: shantha_mech@sirmvit.edu

1 Introduction

Recent advancements in nanomaterials have proved to be promising for all sectors of people. There has been constant research on nanomaterials for a quite long period of time and numerous studies have proved that nanomaterial can enhance the lubrication properties of oils. The efficiency and stability of any machine depends on its degree of resistance to wear and to protect it from wear under all circumstances. Today's trends are to have ecofriendly lubricant that does not cause any environmental adverse effects. Every responsible citizen's duty is to focus on saving the environment and avoiding pollution to the maximum. Commercially available lubricants are proven to be highly toxic and have contributed to the pollution crisis. The disposal of these commercially available lubricants is also found to be tedious process and these deposits are harmful debris to the environment (Chandrakar & Suhane 2014; Khan et al., 2009; Kolawole & Odusote, 2013; Satheesh Kumar et al. 2015; Upadhyay 2015).

Wear and friction are the two most inherent properties of the machines. Friction and wear have to be closely monitored to overcome damages in machines. Hence at the same time it has become necessary to familiarize the concept of using nanomaterials as lubricant or in the process of lubrication thereby acknowledging the machines to show consideration to explore the range of nanomaterials application in the area of lubrication. The use of nanomaterials in the various blends has resulted in indicating different sound qualities as possible in studies of wear and friction. The main intension of experimenting with all these nanomaterials was to ensure environment-friendly and use of natural available resources (EmelKuram et al., 2012; Fairuz et al., 2015; Jyothi et al., 2017; Kuram et al., 2010; Vaibhav Koushik et al., 2012).

With this being the primary focus the naturally available vegetable nonedible lubricant oil with a small percentage of nano compound is considered for the study. Our country is

M. Sumesh et al. (eds.), *2nd International Conference on Smart Sustainable Materials and Technologies (ICSSMT 2023)*,
Advances in Science, Technology & Innovation, https://doi.org/10.1007/978-3-031-49826-8_8

rich with almost 300 types of non-edible oil seed. These have not been exploited to maximum potential and Neem tree and Pongamia tree are abundantly found in all regions. Their physical properties are determined and compared with the commonly used lubricant oil. The use of nanoparticles/nanomaterials as lubricant additives are known as nanolubricants. Typically, their diameter particle size is between 1 and 100 nm. In laboratory tests, the use of nanolubricants in base or coatings promotes a significant reduction of friction and wear, which exhibits interesting tribological properties This research aims in adding additives to the lubricant base oil whose primary aim is to reduce friction and wear. In previous work, base oil such as Neem oil and Pongamia oil was used as a lubricant oil (Gui et al., 2008; Jyothi et al., 2017) and has already shown excellent tribological behavior. The idea of introducing zinc borate (Zhao et al., 2014) and graphene (Deepika 2020; Yang et al. 2019) as additives to the lubricant added to the enhancement of its wear properties. The nanographene application as an additive to the base oil has already shown their excellent tribological properties (Akbulut 2011; Baskara & Srirama, 2014; Bofardi, 1987; Koppula et al., 2016; Seah et al., 1997; Xiandong et al. 1997; Zhmud & Pasalskiy, 2013).

2 Experimentation and Methodology

2.1 Need for the Lubricant

The frictional heat that is generated by any machine is generally absorbed by the lubricant and there is a need for the lubricant. The lubricant is very necessary as it absorbs heat and tries to fill the airline gap between two surfaces thereby reducing friction and having a relative motion with each other.

Wear is bound to happen due to the continuous friction force acting along the surfaces that are relative to each other. Hence lubrication is a continuous process which will help in decreasing the heat around the surfaces that could be generated during machining process and reducing wear and increasing the life of the component. All the lubrication has several functions, some of them which plays a major role in reducing friction between any moving components so as to reduce wear and tear of the components. It enhances the life of the component life and machine life. It also can act as a coolant and remove any depositions along the surface of the machine during the process of machining. Lubricant's important property is to reduce the noise that is like to be produced during friction. Therefore, for every machine lubricant plays a vital role in the life and service of the component. The lubricant that has to be used has to be carefully examined and its utilization factor should be observed or studied so that it has a very good result in reducing wear. It has to reduce the rate of abrasion during the process.

2.2 Methods of Lubrication

An effective method of lubrication has been chosen for the process. There are several methods of lubrication such as manual lubrication, gravity feed lubrication, forced lubrication and splash method of Lubrication. Manual lubrication is a process in which the oil flows freely, in this category oil can that is used is poured at different points for lubrication. Gravity Feed Lubrication is a process in which the components are lubricated by using the natural property of gravity for lubrication. Forced lubrication is a process in which the oil is made to flow along the surfaces or at the parts at high pressure by using an oil pump and splash method of lubrication is a process that works by splashing the lubricant oil along the surfaces of the engine parts that is immersed in the oil tank. This method has the advantage as the oil after splashing gets deposited back in the oil well so there is no loss of the oil and it can be reused.

2.3 Types of Lubricants

Lubricants are generally classified into solid lubricants and liquid lubricants. Solid lubricants are those materials which are in the form of powders, and they are basically applied on the surface for lubrication. Liquid lubricants are those that are in the liquid form and are generally obtained by refining of petroleum. The various different types of liquid lubricants are mineral oils, animal oils and synthetic oils. The synthetic oils are in wide variations such as SAE 10, SAE 20, SAE 30 and SAE 40. These are widely used in industry applications and in automotive engines. Semi-liquid lubricants are obtained by mixing petroleum products with soap mixture. These materials are generally where adherence is at most required. It is also suitable for components where a large clearance is required.

2.4 Properties of Good Lubricant

A good lubricant can be justified if it satisfies the following properties. The properties are volatility, adhesiveness, oiliness, surface tension, flashpoint and fire point. Any lubricant, if it has been introduced or a variation in the lubricant has been made, has to satisfy the conditions of the following properties for it to be widely used. Volatility is the property when the lubricant can vaporize when it is subjected to heat, hence it should always be low. Adherence is an important property so as to settle on the surface and to maintain a certain gap between two rubbing mediums.

Oiliness is the capacity of the lubricant to maintain a lining along the rubbing surfaces, generally it should be high. Flashpoint is the important property of a lubricant which basically has the ability to give vapors. Fire point is the ability to catch fire when introduced to flame. Generally, a good lubricant has to have a high flash point and high fire point. There are several experiments that can be used to study flashpoint and fire point such as open cup and closed cup methods. In open cup Cleveland's Open Cup Apparatus is used and whereas in closed cup Pensky-Marten's Closed Cup Apparatus is used. Saybolt viscometer is used to study fire point and flash point of the lubricant.

Surface tension is the property of avoiding the formation of bubbles along the surface of the lubricant. Other properties are cloud point and pour point. Cloud point is the property at which the liquid tends to solidify, that is it changes its state from liquid to solid and pour point is at the lowest temperature at which it can stop flowing. Generally, a good lubricant has to have a high-power point and a lower pour point.

2.5 Methodology

The blending of the oils and analysis of these were carried out at R and D lab chemistry and experimentation was carried out at energy conversion lab at B.M.S Institute of Technology and Management. Nanoparticles and base oil lubricants can be mixed thoroughly in the first procedure which is magnetic stirrer for about 30 min and the second procedure is ultra sonification for about 25 min, so that a uniform mixture can be obtained. The flow of activities that are involved in the experimentation is clearly explained with the help of flow chart as shown in Fig. 1. Next comes the process of choosing the nano compound. This compound was chosen with the intension to reduce wear and maintain smooth finish on the surface of the material. Then the process of blending both of them together so that no residue will be deposited, and a uniform mixture could be obtained. Then different physical properties of the lubricant were tested and analyzed.

2.5.1 Identifying the Right Alternative for Lubricants

The alternative lubricant that was considered for my study is bio-lubricant. Bio-lubricant is something that is easily available and is not processed with chemicals. The bio-lubricants considered for the study are Pongamia oil and Neem oil. These lubricants are naturally available and have less impact on the environment. The oil is extracted from non-edible seeds. Neem oil is extracted in three different methods such as pressing, steam distillation and solvent distillation. The quality of the oil obtained varies appropriately according to the methods chosen. Pongamia oil is Latin name is Pongammia Pinnata throughout the globe it is referred to as Pongammia Pinnata Perry. The oil is obtained from the Pongammia seed. Pongamia oil is produced on a large scale. These bio-lubricants possess a higher superior anti-corrosive property, high flash, and fire point as well as feeble aquatic toxicity (Kuram et al., 2010).

2.5.2 Recognizing the Right Nano Lubricant that Can Blend with Bio Lubricant Thereby Enhancing Its Properties

The nano compound chosen for the work is Zinc Oxide and Graphene. These nano compounds are chosen in small percentage and added to the base oil and hence are termed as nano additives. Zinc Deposits have proved to have sacrificial protection for any base. Several studies have shown that addition of zinc deposition has enhanced Corrosion Resistance. They also prevent the formation of white rusty products. They also have an anti-oxidation property (Gui et al., 2008). The molecular weight of zinc oxide is 81.406 g/mol and density of zinc oxide is 5.6 g/cm^3. Graphene oxide is an intermediate product that contains hydrophilic and reactive oxygen. It has shown excellent dispersion stability in water and in most solvents. These materials have been shown to have highly corrosion-resistant composite coatings (Zhao et al., 2014). Graphene is one of the strongest materials and at the same time it is also very lightweight (0.77 mgm^{-2}). These compounds are added in very low quantity as there is a foundation of a continuous protective layer (Zhmud & Pasalskiy, 2013). A stable suspension of nano sized particles in oils is added as repulsive forces are larger over the attractive forces (Zhmud & Pasalskiy, 2013). Hence these nano additives were chosen to enhance the performance of lubricants.

2.5.3 To Obtain Different Samples by Blending Them in Different Proportions

The lubricant Neem oil and Pongamia oil with nano additives such as Graphene and zinc oxide were considered for my study. Here three samples are considered for my study. The first samples were prepared by blending half litre of Neem oil with nano compounds such as 10 gms of nano Zinc oxide (GRZO) that was synthesized by using germinated Rajma seeds (GRZO) as a catalyst and 10 gms of Graphene. The second samples were prepared by blending half litre of Pongamia oil with nano compounds such as 10 gms of nano Zinc oxide (GRZO) that was synthesized by using germinated Rajma seeds as a catalyst and 10 gms of Graphene and the third samples were prepared by blending half liter. The third sample was prepared using Pongamia oil with nano

Fig. 1 Shows the schematic view of our experimental procedures

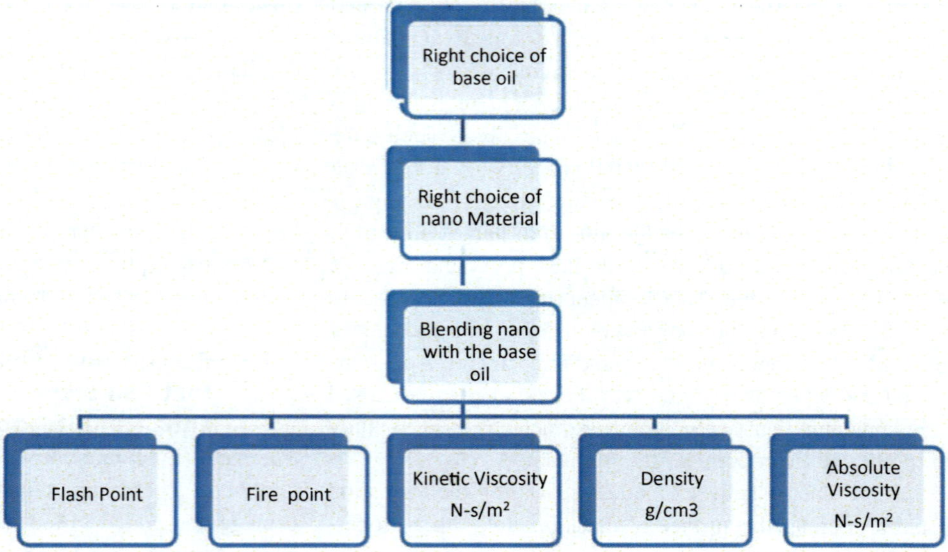

compounds such as 10 gms of nano Zinc oxide (RZO) that was synthesized by using Rajma seeds without germination as a catalyst and 10 gms of Graphene. These samples were blended together by using an ultrasonicator. The ultrasonicator basically agitates the compounds by using sound energy so that a uniform mixture can be obtained. Each sample was blended for about 20 min.

2.5.4 Studying Various Physical Properties of This Nano Lubricant

The physical properties of this Bio-nano lubricant were examined. The Flash and Fire points of this bio-nano lubricant were determined by using the Able Pensky martin experiment. 70 ml of these various samples were taken for the conduction of the experiment. The flash point and fire point of various samples are obtained and discussed.

The physical properties such as Density, Absolute viscosity and Kinematic viscosity were examined using Saybolt viscometer. The Saybolt viscometer was chosen as its thickness is high. At two different temperatures the samples were examined at 90 and 80 °C. Averaged values have been taken for discussion at these two temperatures.

3 Results and Discussion

The important property of any lubricant is viscosity, viscosity of any fluid is the resistance offered by the fluid against the rate of flow. It is determined by the researchers that if the viscosity of the lubricant is high then it will not carry away the heat and thereby contributing to wear.

Results obtained from the experiment show that the bio-Nano lubricants Pongamia oil, GRZO, Graphene and Pongamia oil, RZO and Graphene the kinetic viscosity is low thereby enhancing the property of the lubricant. The Viscosity of the samples was determined at 90 and 80 °C. The Kinetic Viscosity and Dynamic Viscosity values are tabulated in Table 1 and are plotted on the graph indicated in Figs. 1 and 2 respectively. Similarly, when absolute viscosity is considered for the sample Pongamia oil, GRZO, Graphene and Pongamia oil, RZO, Graphene shows a higher value; thereby allowing it to consume less and contributing less wear. However, the samples have been shown to possess good physical properties when compared with the conventional oil used.

An ideal Fire point should be high and Flash point should be low for a good lubricant. However, Bio-nano lubricant must of flash point could be considered because if it is low, there would be a possibility of fire in storage. Flash point and fire of the samples are tabulated in Table 1 and graph for the same is plotted and represented in Fig. 3. The results below show that the fire and flash point of sample 1: Neem oil, GRZO and Graphene, sample 2: Pongamia oil, GRZO and Graphene and sample 3: Pongamia oil, RZO and Graphene is high when compared with SAE20W40 [34].

Density is also determined for the samples and compared with the ideal lubricant. It is known that a good lubricant oil density will range between 700 and 950 kg per cubic meter (kg/m^3). From Table 1 for the samples Neem oil, GRZO, Graphene, Pongamia oil, GRZO, Graphene and Pongamia oil, RZO, Graphene shows that density is quite close with when compared with the ideal lubricant (Jyothi et al., 2017) [25, 32]. The values plotted are represented in Fig. 4.

Figure 5 shows the graph of various blends with their densities. Although the natural lubricant has high density, when the values are again compared with blends with nano additives Neem oil, GRZO and Graphene have been found to have high density.

Table 1 Tested Lubricants and their properties

Sl. no.	Type of cutting fluid	Flash point	Fire point	Density g/cm^3	Kinetic viscosity N-s/m^2	Absolute viscosity N-s/m^2
1	Neem oil+GRZO +Graphene	250	265	0.88	10.99	$9.67*10^{-3}$
2	Pongamia oil+GRZO +Graphene	270	320	0.83	4.77	$4.77*10^{-3}$
3	Pongamia oil+RZO +Graphene	250	300	0.85	6.14	$6.14*10^{-3}$
4	Neem Oil	248	285	0.92	24.14	$3.45*10^{-3}$
5	Pongamia oil	220	245	0.90	27.12	$2.66*10^{-3}$
6	SAE20W40	210	215	0.88	21.72	$1.97*10^{-3}$

Fig. 2 Bio-nano lubricants

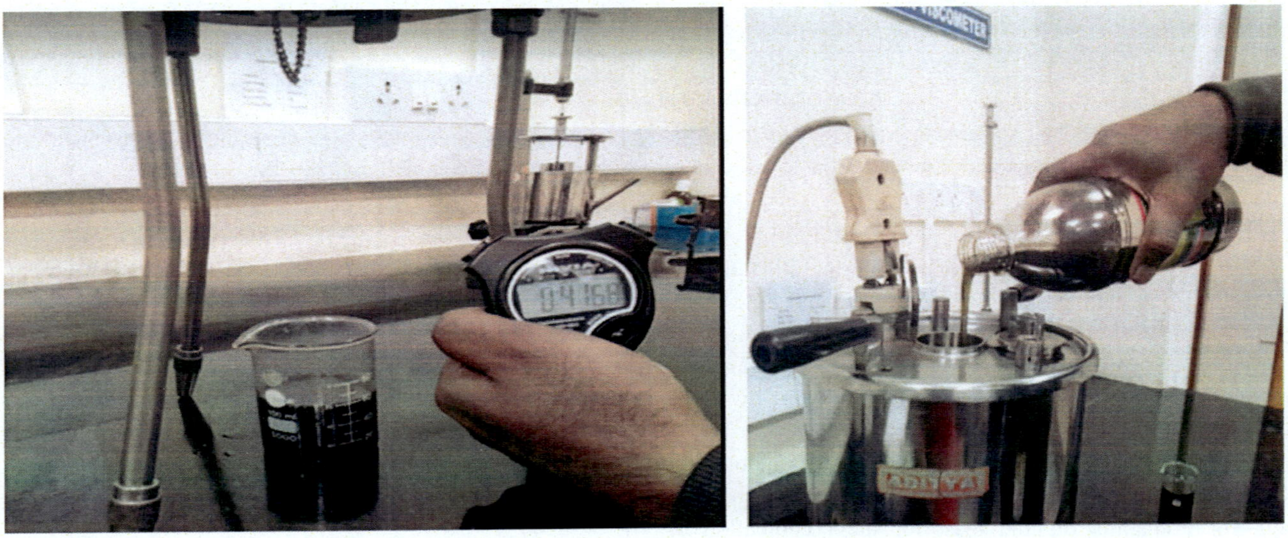

Fig. 3 Physical properties determined using Saybolt viscometer

Fig. 4 Physical properties determined using Able Pensky martin experiment

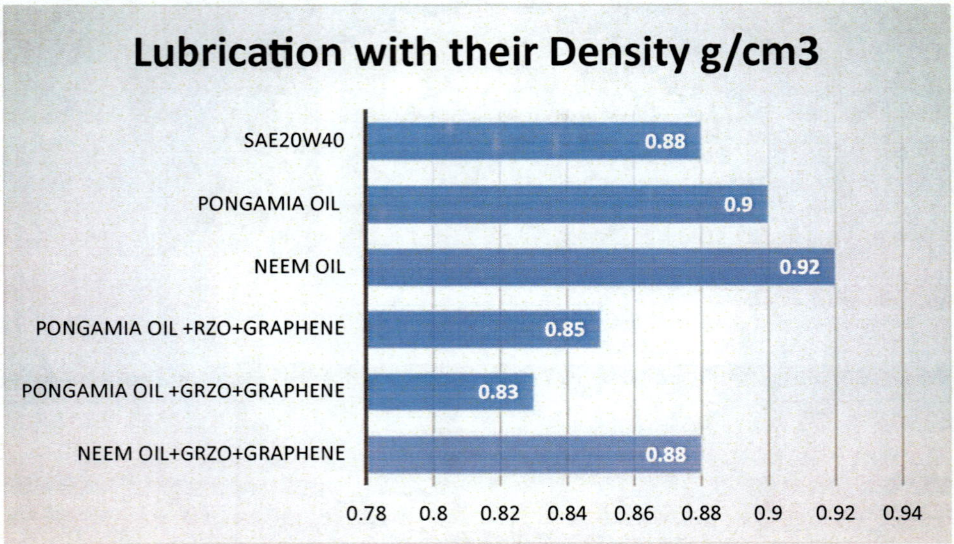

Fig. 5 Lubrication with their density

From Fig. 6 the result recommends that the utilization of Neem oil along with the additives Graphene and Zinc oxide as its Absolute viscosity is high.

Figure 7 of various lubricants along with their Flash and fire point are plotted. All samples have shown low flash point and high fire point.

Figure 8 shows the values of kinetic viscosity along with all the lubricants. It is observed that the kinetic viscosity is high for Pongamia oil and least for the combination of Pongamia oil, GRZO and Graphene.

4 Conclusion

Bio-nano lubricants have been initiated for enhancing physical properties of the lubricant but still there is a long way to go. Bio-nano lubricant could significantly improve the friction-reducing and antiwear properties. The Neem and Pongamia oil that was chosen could be advantageous as they are widely available in India and act as cost-efficient. Furthermore, the addition of nano compounds has improved the

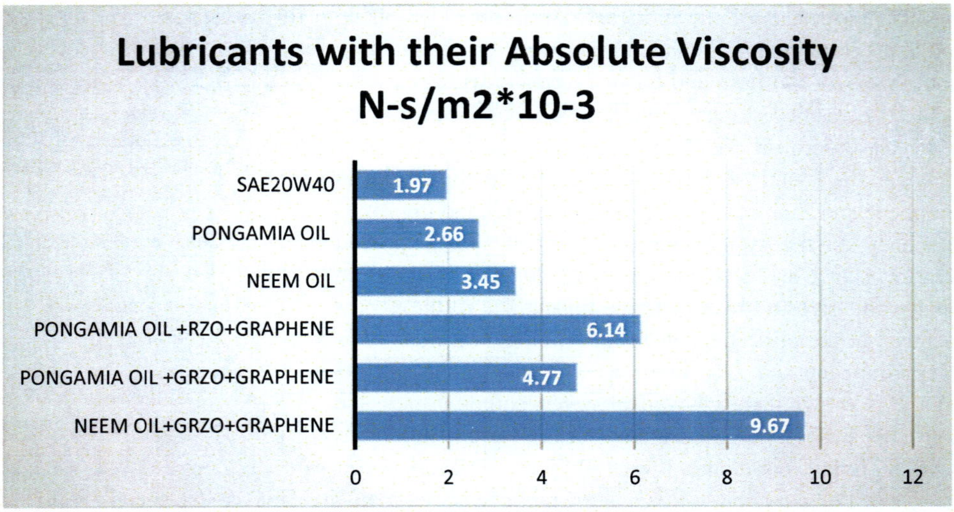

Fig. 6 Lubrication with absolute viscosity

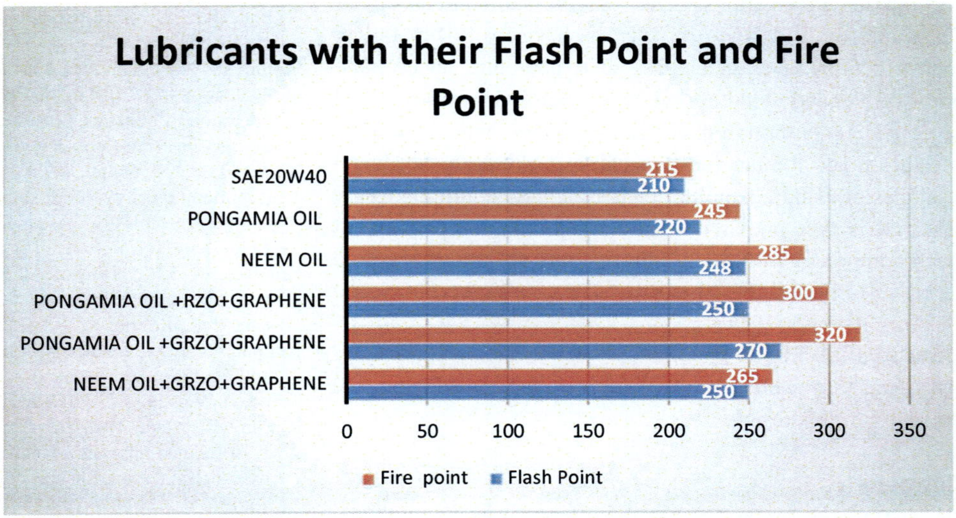

Fig. 7 Lubricants along with their flash and fire point

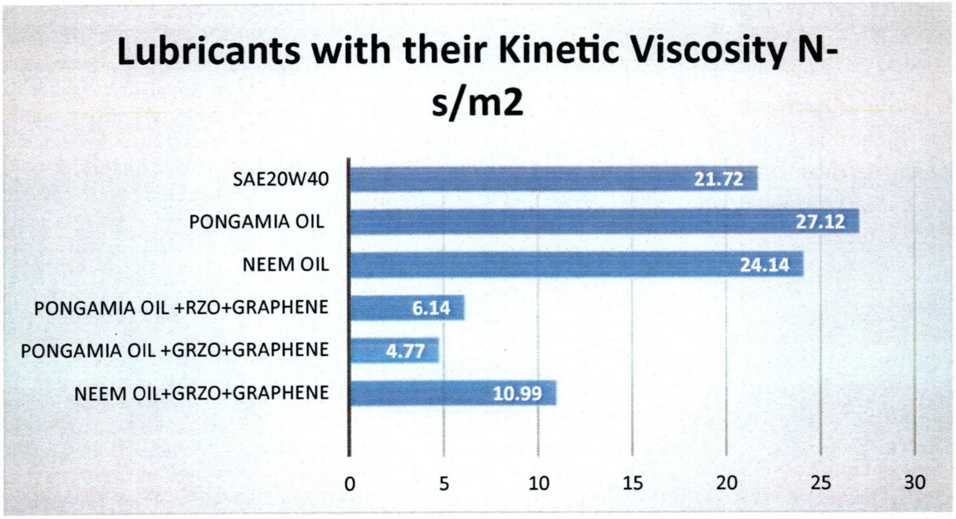

Fig. 8 Lubrication with kinetic viscosity

physical properties of the lubricant. Addition of graphene and Zinc oxide as Nano material can work on maintaining smoothness of the base and has reduced the friction coefficient. The results show that the densities of Neem oil are high. The Neem oil could be chosen as a natural lubricant however the blend of Neem oil along with the additives Graphene and Zinc oxide has greater potential and improved efficiency on the system because graphene has improved wear resistance and acts as a self-lubricator; and while using Zinc oxide gives a greater surface finish. The result recommends that the utilization of Neem oil along with the additives Graphene and Zinc oxide as its dynamic viscosity is high. It is known that for any ideal system should possess the high. The flash point and fire point of the various lubricants have high fire point and low flash point. Although this sample of Pongamia oil, GRZO, and Graphene shows a high fire point and a low flash point, it could be chosen for the appropriate applications. The natural lubricant such as Pongamia oil and Neem oil has high viscosity but fail to maintain oiliness with increase in temperature hence nano additive blended with base oil Neem and Pongamia oil could be considered although their viscosity is low which could be a right choice for the most machines. Although the disadvantages of using a nanolubricant are many, confining to my research is that preparation of homogenous suspension of nano compound with the bio lubricant is not always assured to be uniform and the cost of lubricants is higher.

Data Availability

All data that were used to support the conclusion are included in this article.

Citations

All citations in the text are in the reference list.

Conflicts of Interest

The authors declare that there are no conflicts of interest.

Acknowledgements

The author is grateful in general for the support of the Principal, Head of the Department, and the department of Mechanical Engineering for providing a conducive Environment.

References

Akbulut, M. (2011). Nanoparticle-based lubrication systems. *Journal of Powder Metallurgy & Mining*, *1*, 77843–3122.

Baskara, S., & Srirama, G. (2014). Tribological behavior of journal bearing material under different lubricants. *Tribology in Industry*, *36*, 127–133.

Bofardi, B. P. (1987). Control of environmental variables in water reticulating systems. In *Metals Handbook*, Vol. 13 of Corrosion (9th ed., p. 487). ASM International, Materials Park, Ohio, USA.

Chandrakar, J. K., & Suhane, A. (2014, May). The prospects of vegetable based oil as metal working fluid in manufacturing application-a review. *International Journal of Engineering Research and Technology*, *3*(5). ISSN 2278-0181.

Deepika. (2020). Nanotechnology implications for high performance lubricants. *SN Applied Sciences* (springer.com).

Fairuz, M. A., Nurul Adlina, M. J., Azmi, A. I., Hafieza, M. R. M., & Leong, K. W. (2015). Investigation of chip formation and tool wear in drilling process using various types of vegetable-oil based lubricants. *Applied Mechanics and Materials*, *799–800*, 247–250.

Gui, M. M., Lee, K. T., & Bhatia, S. (2008). Feasibility of edible oil vs. nonedible oil vs. waste edible oil as biodiesel feedstock. *Energy, 33* (11), 1646–1653.

Jyothi, P. N., Susmitha, M., & Sharan, P. (2017). Performance evaluation of NEEM oil and Pongamia oil as cutting fluid in drilling operation of mild steel. In *IOP conference series: Materials science and engineering* (Vol. 191, No. 1, p. 012026).

Khan, M. M. A., Mithu, M. A. H., & Dhar, N. R. (2009). Effects of minimum quantity lubrication on turning AISI 9310 alloy steel using vegetable oil-based cutting fluid. *Journal of Materials Processing Technology*, *209*, 5573–5583

Kolawole, S. K., & Odusote, J. K. (2013). Performance evaluation of vegetable oil-based cutting fluids in mild steel machining. *Chemistry and Material Research*, *3*(9).

Koppula, S. B., & Sudheer, N. V. V. S. (2016). A review on effect of adding additives and nano additives on thermal properties of gear box lubricants, *11*, 3509–3526.

Kuram, E., Cetin, M. H., Ozcelik, B., & Demirbas, E. (2012). Performance analysis of developed vegetable-based cutting fluids by d-optimal experimental design in turning process. *International Journal of Computer Integrated Manufacturing*, *25*(12).

Kuram, E., Ozcelik, B., Demirbas, E., & Şık, E. (2010). Effects of the cutting fluid types and cutting parameters on surface roughness and thrust force. In *Proceedings of the World Congress on Engineering 2010 Vol II WCE 2010, June 30–July 2, 2010*, London, U.K.

Satheesh Kumar, B., Padmanabhan, G., & Vamsi Krishna, P. (2015). Experimental investigations of vegetable oil based cutting fluids with extreme pressure additive in machining of AISI 1040 steel. *Manufacturing Science and Technology*, *3*(1), 1–9.

Seah, K. H. W., Sharma, F. S. C., & Girish, B. M. (1997). Corrosion characteristics of ZA-27-graphite particulate composites. *Corrosion Science*, *39*(1), 1–7.

Upadhyay, N. (2015). Environmentally friendly machining: Vegetable based cutting fluid. *SJPSET*, *7*(2). ISSN: 2229-7111.

Vaibhav Koushik, A. V., Narendra Shetty, S., & Ram Prasad, C. (2012). *International Journal on Theoretical and Applied Research in Mechanical Engineering (IJTARME), 1*(1). ISSN (Print): 2319-3182.

Xiandong, S., Chengping, L., Zhuoxuan, L., & Liuzhang, O. (1997). The fabrication and properties of particle reinforced cast metal matrix composites. *Journal of Materials Processing Technology, 63* (1–3), 426–431.

Yang, B., Zhang, P., Wang, G., Wang, A., Chen, X., Wei, S., & Xie, J. (2019). Effect of graphene oxide concentration in electrolyte on corrosion behavior of electrodeposited Zn–electrochemical reduction graphene composite coatings.

Zhao, C., Chen, Y. K., Jiao, Y., Loya, A., & Ren, G. G. (2014). The preparation and tribological properties of surface modified zinc borate ultrafine powder as a lubricant additive in liquid paraffin. *Elsevier Ltd./Tribology International, 70*, 155–164.

Zhmud, B., & Pasalskiy, B. (2013). Nanomaterials in lubricants: An industrial perspective on current research. *Lubricants, 1*, 95–101.

Experimental and Heat Transfer Analysis Using Nanofluid in Cylindrical Heat Pipe Heat Exchanger

P. Ramkumar, C. M. Vivek, P. Latha, and S. P. Manikandan

Abstract

In waste heat management systems for the utilization of heat, many new techniques are used currently, by taking into account an effective one is heat pipe heat exchanger. Three heat pipes are placed inside the heat exchanger for analyzing the heat transfer performance by using acetone and copper oxide with DI water mixture nanofluid. The influence of nanofluid for various heat input temperature conditions is studied for horizontal and vertical orientation of heat pipe. The study reveals at horizontal orientation the maximum effectiveness is obtained as 66.7% for 100 LPH and Reynolds number ranges as 1173 for similar conditions. The influence of nanofluid as heat transport fluid shows maximum heat transfer capacity for given conditions and this shows good performance in the system.

Keywords

Heat pipe • Heat exchanger • Nanofluid • Effectiveness • Reynolds Number

1 Introduction

The requirement of efficient heat transfer equipment is increasing gradually for the industries, but the equipment needs to be cost-effective and less maintenance. This requirement leads to the introduction of heat pipe technique for several waste heat utilization applications in energy sectors. Thus industries get focused into significant techniques by utilizing the heat pipe in heat exchanger, thus leading to the development of heat pipe heat exchanger, this was economically effective in design & operation, compactness, and reliability. The new design in the functional system at the surface encounters performance development in the heat pipe (Tang, 2023). The study focuses on axially rotating heat pipe with angle of rotation the performance of the system gets improved (Wang et al. 2023). The numerical study using ANFIS for single heat pipe analysis has been performed and results are compared (Ramkumar et al., 2022). The heat pipe analysis is done for different orientations to investigate the performance of the system (Ramkumar et al., 2021a, 2021b). The pulsating heat pipe was studied using sodium stearate of 0.31, 0.44, and 0.57 as fill ratio. The observation reveals that at 0.57, there was 0.002 wt.% maximum evaporating temperature (Bao et al., 2023). The copper-pulsating heat pipe using ethylene glycol shows less thermal resistance at 55% fill ratio and at 600 W possesses high efficiency for less thermal resistance of 0.136 °C/W (Rudresha et al., 2023). The Copper—cylindrical heat pipe with copper oxide nanofluid represents good performance for given conditions (Venkatachalapathy et al., 2015). The flat heat pipe with capillary wick of metal fiber uses sodium fluid. The performance increased for 15° orientation for 600 W has 498.702 W/m^2 K (Xu et al. 2023). The solar water heater is investigated with flat and closed loop pulsating heat pipe. The flow rate of 0.6 l/min, the results show 31.4, 22.5, and 18.5 °C temperatures for in and out conditions (Siritan et al., 2022). The loop heat pipe with ammonia with sintered nickel powders as biporous wick. The heat

P. Ramkumar (✉)
Department of Mechanical Engineering, Kalasalingam Academy of Research and Education, Virudhunagar, Tamil Nadu, India
e-mail: rkmailmech@gmail.com

C. M. Vivek · S. P. Manikandan
Department of Mechanical Engineering, Periyar Maniammai Institute of Science and Technology, Thanjavur, Tamil Nadu, India

P. Latha
Department of Civil Engineering, Periyar Maniammai Institute of Science and Technology, Thanjavur, Tamil Nadu, India

M. Sumesh et al. (eds.), *2nd International Conference on Smart Sustainable Materials and Technologies (ICSSMT 2023)*,
Advances in Science, Technology & Innovation, https://doi.org/10.1007/978-3-031-49826-8_9

input load of 280 W, there was 0.103 °C/W of least thermal resistance (Zhang et al., 2023). Insulated Gate Bipolar Transistor induced heat pipe demonstrates 450 $m^3\ h^{-1}$ air flow rate has good performance (Jiazheng et al., 2019). The miniature loop copper wire heat pipe with 100 PPI and 37% permeability predicts the thermal resistance of 0.109–0.396 °C/W and 0.132–0.644 °C/W with 3.749 and 3.744% error, respectively has high efficiency (Kamran et al., 2021). The heat pipe designed for energy recovery applications shows maximum results for all given conditions and parameters (Chaudhry et al., 2012; Choling & Zou, 2015). The methanol—heat pipe was studied with 50, 65, and 75% of fill ratios with 5 − 15 W heat input. The optimized conditions show that 65% (Sonawane et al., 2023). The cylindrical micro-grooved heat pipe is studied for various angles, but the best results are obtained at horizontal angle and grooves make high-performance improvement (Zeghari et al., 2023). The composites are made with nano particles and results are analyzed (Deepak et al., 2011; Sadanand et al., 2016). The given literatures mainly analyzed on operational parameters, and limitations with single geometrical heat pipe, working parameters and fluids analyzed. This analysis is evaluated with multiple heat pipe induced in heat exchangers with different parameters and influence of nanofluid is also studied and observations are reported.

2 Experimental Fabrication and Procedure

2.1 Fabrication of Setup

The heat pipe is fabricated using three copper pipes with diameters of 19 and 17 mm and the heat exchanger shell is fabricated using galvanized iron and similarly heat pipe is placed inside the shell of the evaporator zone and it's given in Fig. 1. The fabricated setup possess the heat pipe heat exchanger geometry as 1000 mm long, by including evaporator zone as 700 mm length and diameter as 102 mm for circulating hot Copper oxide with De-Ionized water mixture nanofluid (CuO/DI water) and condenser zone of 300 mm with 35 mm diameter for circulating cold water. The heat pipe is filled with acetone for 50% fill ratio (Ramkumar et al., 2021a). The mess for heat pipe is used as stainless steel with 50 holes per inch. The fluid reservoir capacity of 5 l and 3 LPM rotameters for controlling the mass flow rates, and a water heater of 2 kW with ±10 W capacity is placed in hot fluid tank. In the evaporator region the heated nanofluid is circulated using a 1 HP monoblock pump. Each temperature node is measured using K-type thermocouples and the readings are recorded for the analysis.

2.2 Nanofluid Preparation Using CuO Nano Particles

In the preparation of Copper oxide with De-Ionized water mixture (CuO/DI water) nanofluid (Venkatachalapathy et al., 2015), the CuO nanoparticles of 2.4422 g are required for 100 ml base fluid of DI water for 0.4% concentration given in Fig. 2 (Ramkumar et al., 2022). This fluid is prepared using magnetic stirring and ultra-sonic vibration for 5 h for uniform mixing of particles. The prepared fluid is left over for 4 h to check the partical settlement, but no partical settlement was observed at bottom of flask. This shows the prepared nanofluid is isentropic and Newtonian in nature thus properties are given in Table 1 and distribution of CuO is warranted using Scanning Electron Microscope (SEM) represented at Fig. 3.

2.3 EDAX APEX Studies

The CuO powder presence is evaluated with ZESIS EDAX APEX and it's given in Fig. 4 (Ramkumar et al., 2022). There is a presence of 72.9% of Copper and 27.1% of Oxygen element, this peaks clearly confirm the pure CuO and no impurities in the material.

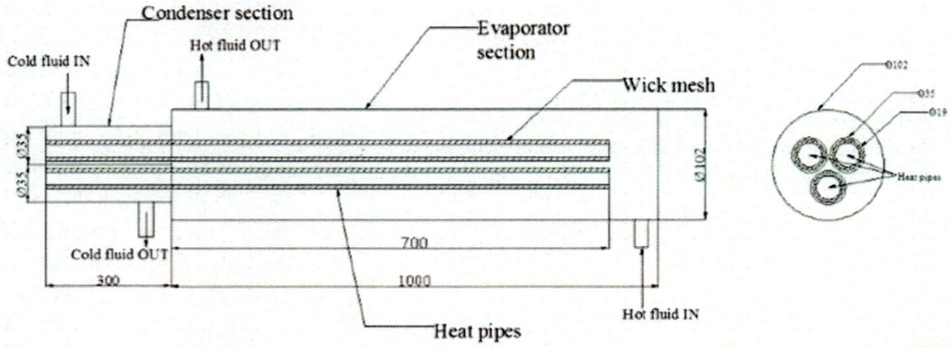

Fig. 1 Schematic diagram of heat pipe heat exchanger

Fig. 2 Photographic view of CuO nano particles and CuO/DI water nanofluid (Ramkumar et al., 2022)

Table 1 Thermo physical properties of CuO nano particles (Ramkumar et al., 2022)

Properties	Values
Particle size (nm)	10–50
Thermal conductivity (W/mK)	18
Purity (%)	99.0
Density (kg/m^3)	6510
Specific heat (J/kg K)	540

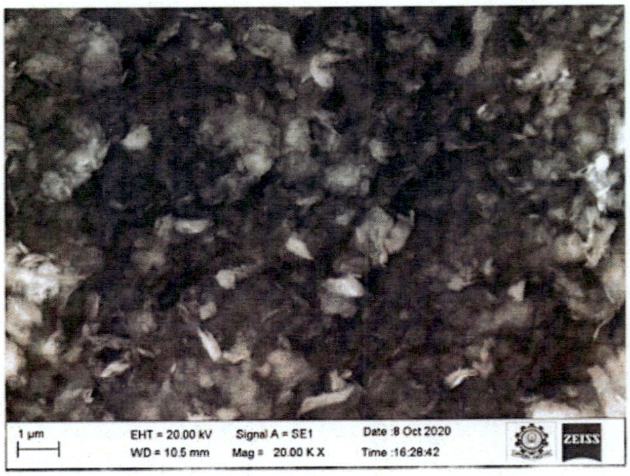

Fig. 3 Distribution of CuO nano particles (SEM) (Ramkumar et al., 2022)

2.4 XRD Analysis

The CuO nano particle powder is analyzed using Rigaku X-ray diffractometer (XRD) by Cu-kα radiation which ranges as 20–80° and is represented in Fig. 5. The intensity variation is represented as count per second (cps). The image shows two major peaks in between the range of 35–40°. The major peak for the particle is compared with the CuO synthesized. Hence it is undoubtedly the CuO of the nano scale. The Sherrer formula is used for calculating the grain size, the observed average size is 15.3 nm, and it's given in Eq. (1) (Ramkumar et al., 2022).

$$\begin{aligned} \beta &= \text{Full width at half maximum height} \\ \text{d} &= \text{Grain size} \\ \theta &= \text{Angle at which peak occurs} \\ \lambda &= \text{Wavelength of Cu K}\alpha\,\text{radiation} = 1.54\,\text{A} \\ \text{d} &= [0.9\lambda/(\beta\cos\theta)] \\ \text{d} &= 15.3\,\text{nm.} \end{aligned} \quad (1)$$

2.5 Procedure of Experimentation

In the fabricated heat pipe heat exchanger the study analysis is carried out for horizontal (0°) and vertical (90°) orientations and are shown in Fig. 6. The study is initially carried out for 0° angle with hot nanofluid mass flow rate (m_{hi}) of 40–120 LPH with 20% increment and cold water flow rates (m_{ci}) as 20–60 LPH as 10% increment, this hot & cold fluid circulations are indicated as red and blue this is shown in Fig. 7. The hot nanofluid temperature ranges (T_{hi}) as 50–

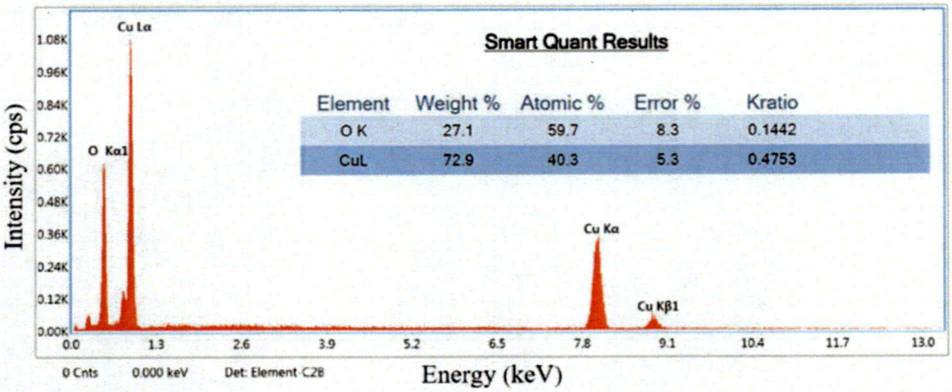

Fig. 4 EDAX APEX of CuO nano particles

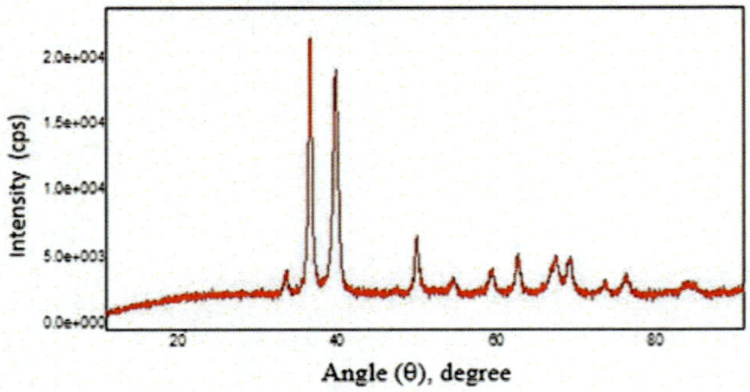

Fig. 5 CuO nano particles presence in XRD (Ramkumar et al., 2022)

70 °C with 10% increment and cold water T_{ci} as 32 °C. The acetone (Washburn, 1930) absorbs the heat from nanofluid and converts to vapour and phase transfer occurs.

The same above conditions are repeated for 90° angle and observations are analyzed and recorded. The temperatures of all nodes are measured and recorded using K-type thermocouples at surface & wall of evaporator and condenser (T_{wc}, T_{we} and T_{se}, T_{sc}) and similarly losses due to environmental interaction are prevented using Styrofoam materials (20 mm thickness).

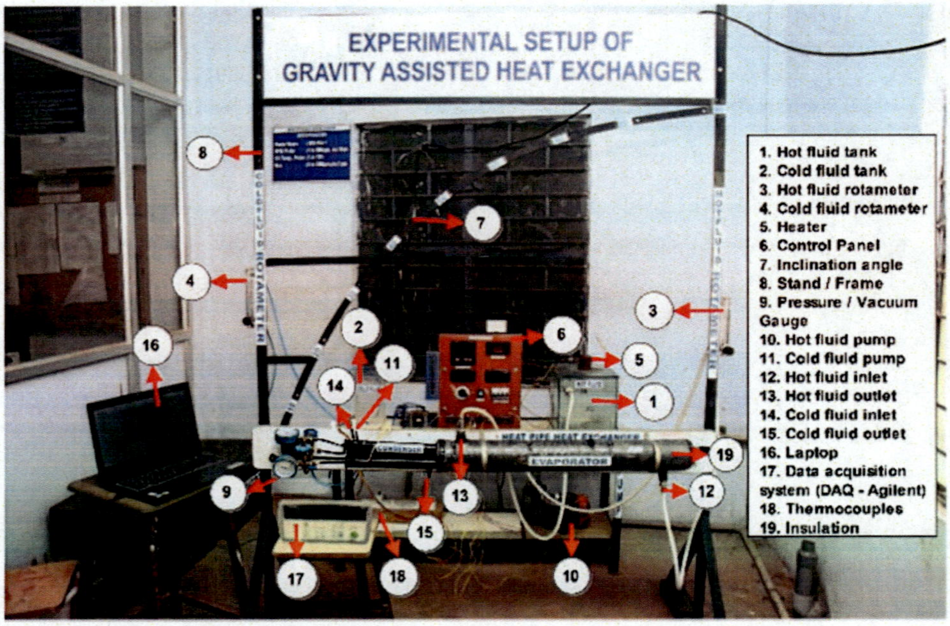

Fig. 6 Experimental setup photograph

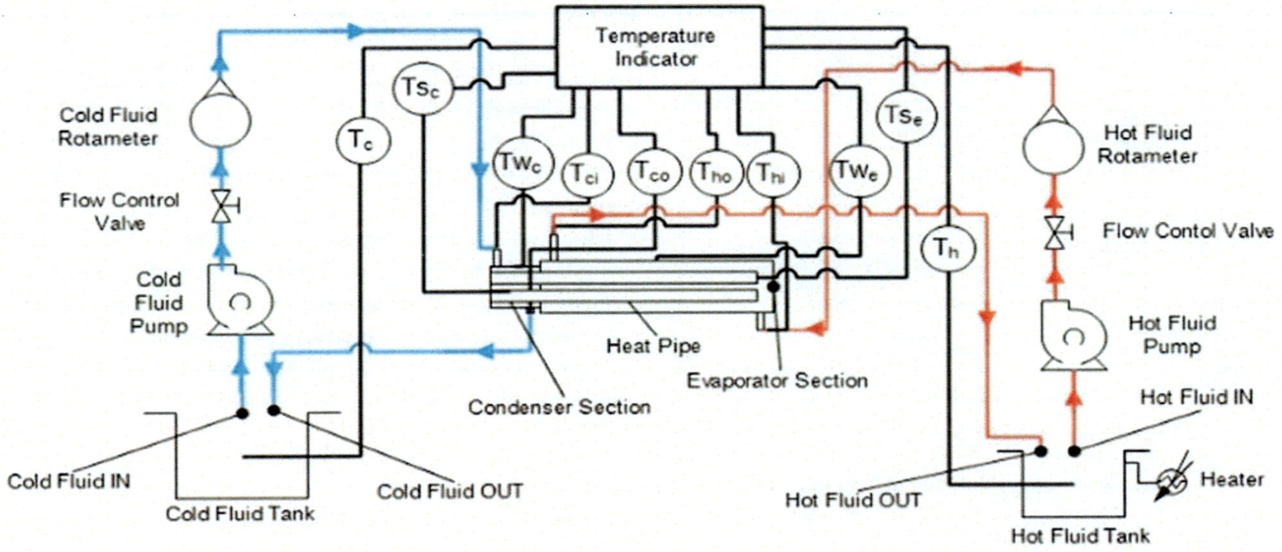

Fig. 7 The line diagram of heat pipe heat exchanger

2.6 Uncertainty Study

The Kline and McClintock's method (Holman, 2007) is used for investigating the uncertainty of the experimentation. The effectiveness possess 1.07%, heat transfer coefficient shows 1.79%, heat transfer rate denotes 1.56%, Reynolds number states 1.49% and thermal resistance is 0.76% of uncertainty. The instrumental uncertainties are calibrated to measure the sensitivities of fluid pumps, flow meters, thermocouples, heaters, temperature indicators. The periodical calibrations are done frequently for reducing the instrumentation errors.

3 Result and Discussion

3.1 Effectiveness Analysis

The investigation is carried out with above parameters and observations are noted and discussed. The maximum performance obtained for 100 LPH, 60 °C and 0° angle for CuO/DI water nanofluid, the Fig. 8 shows increase in values for 0° than 90° as 66.7 and 62.8%, there is an increment of 6.21% is observed for 0°. The nanofluid holds the obtained heat and transfers to the working fluid at evaporator and influence of wick in the heat pipes produces the capillary effect and pushes the working fluid movement by phase change along the pipe. This produces the uniform heat along the pipe and performance is recorded. When pivoting the heat pipe from 0° to 90° the downtrend in performance is achieved due to the sudden gravitational effect creating a thermal resistance in fluid flowing due to the frictional effect. This creates the lag in performance for 90°.

The rising trends of effectiveness observed from 50 to 60 °C beyond that decrease in heat absorbing for acetone, leads to downfall of results at 65 and 70 °C (Ramkumar et al., 2021b). During 120 LPH results for 0 and 90° represent 60.6–58.1%. The downtrend in effectiveness for 120 than 100 LPH is due to acetone possessing less heat absorbing and delivery capacity due to the dry-out condition.

3.2 Heat Transfer Coefficient Analysis

In the similar conditions for Fig. 9, infers the influence of nanofluid shows maximum heat transfer coefficient of

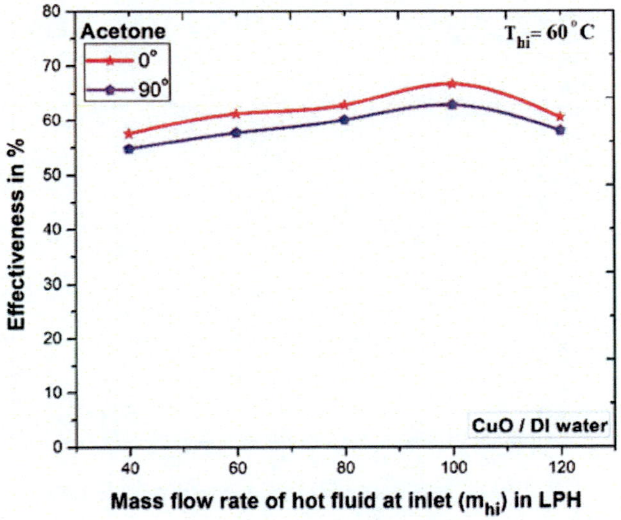

Fig. 8 Variations of effectiveness (0°–90°, Acetone)

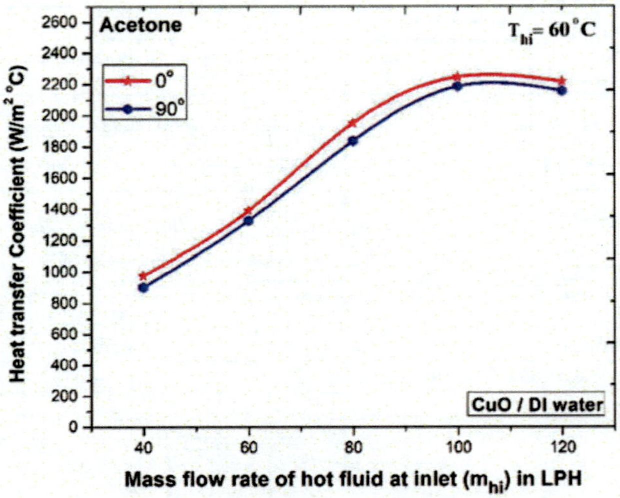

Fig. 9 Variations of heat transfer coefficient (0°–90°, Acetone)

2245.61 W/m^2 °C at 60 °C, 100 LPH and 0°. Meanwhile for 90° it is 2185.90 W/m^2 °C. This reveals that there is 2.73% of increment in heat transfer coefficient for 0° than 90°.

While analyzing for 120 LPH, the observation shows downtrend for 100 LPH as 2218.12 and 2156.85 W/m^2 °C for ψ as 0°–90°. This happens because of the less heat absorption capability of acetone from nanofluid. It results in negligible quantity of phase transfer which arises at evaporator and condenser regions.

3.3 Heat Transfer Rate Analysis

The heat transfer rate study is carried out for horizontal and vertical orientations and results are noted in Fig. 10, and the observation for 0° and 90° are 1639.62 and 1429.47 W. There is 12.81% increment in the value than 90° and this possesses the capillary action. Moreover, the pressure difference of working fluid induces maximum heat transfer rate. In 120 LPH, the obtained results are 1451.42 and 1348.36 W for 0° and 90° and it provides comparatively decreased results than 100 LPH.

The liquid acetone possesses the maximum heat energy absorption from hot CuO/DI water nanofluid and phase change occurs quickly and vapour acetone expels heat to the cold water and this process occurs simultaneously inside the heat pipe. This quick phase change leads to the maximum energy transfer at optimum conditions of 100 LPH, 0° and 60 °C.

3.4 Reynolds Number Study

The Reynold's number (Re) was studied using acetone and nanofluid for 0 and 90° angles and the observed values are 1173.26 and 1146.60, there is an increment of 2.27% given in Fig. 11. The (Re) increases from 40 to 100 LPH and similarly at 120 LPH as 1051.36–1031.65 for 0° and 90°, this decrease happens because of maximum fluid flow along the channel leads to reduction in performance. This discloses at 100 LPH, for 0° angle and 60 °C, the Re is higher than in other conditions. This increasing nature of Reynold's number leads to higher performance.

The movement of heat pipe at vertical position leads to the downfall in performance this happens due to sudden influence of gravitational effect, the density of the condensate formed is high than the vapour at evaporator. The influence of gravity acts along the heat pipe, this pushes the condensate against vapour acetone. This leads to the

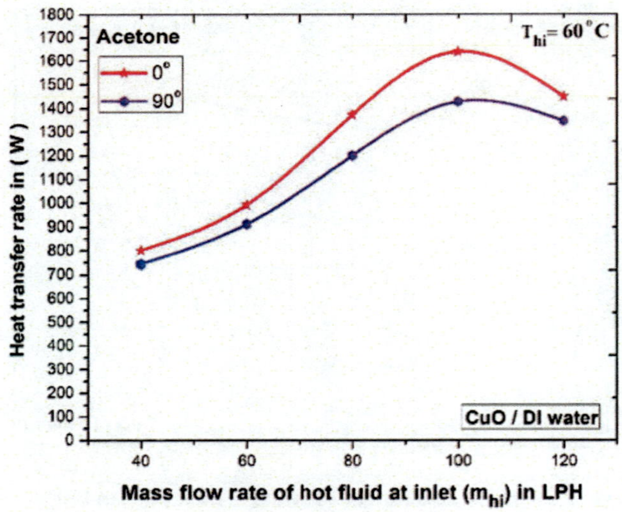

Fig. 10 Variations of heat transfer rate (0°–90°, Acetone)

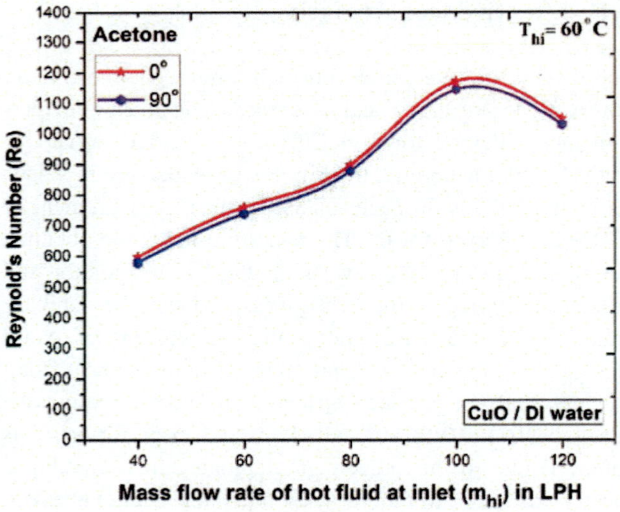

Fig. 11 Variations of Reynolds number (0°–90°, Acetone)

frictional effect along the pipe thus affecting the uniform heat transfer. In the evaporator zone, the condensate will be in immobile state, thus heat requirement is high for the vaporization of working fluid, but there is no heat transfer in the system. Thus performance depletion happens at vertical angle of heat pipe.

3.5 Thermal Resistance Study

The effect of thermal resistance with acetone and CuO/DI water nanofluid is depicted in Fig. 12. The optimum parameters identified are 0°, 60 °C and 100 LPH, with acetone and CuO/DI water nanofluid. The evaluation for 0 and 90° are 0.0005 and 0.0017 °C/W this is stated as downfall in trend of 70.58% than 90°. This downfall makes enhancement in efficiency of the system. The similar conditions at 120 LPH for 0 and 90° provide 0.0033 and 0.0038 °C/W. While comparing the results increasing trends are noted for 120 LPH rather than 100 LPH, thus performance is affected for heat pipe at 120 LPH for all conditions.

Inside the heat pipe the wick surface is accumulated with liquid formation and leads to the circumferential heat drop and induces the friction effect this creates the thermal resistance in fluid flow inside the heat pipe. To overcome this at evaporator zone there is a need for supplying the uniform heat by the hot fluid and mass flow rate leads to the removal of liquid layer and performance enhancement is achieved. Even though after the 100 LPH and 60 °C condition the acetone attains the saturation point and thus heat transfer performance gets decreased.

The analysis clearly states the influence of CuO/DI water nanofluid and acetone makes better heat transfer performance for all the conditions. The effect of inclination angle makes the influence in the results thus 0° position of heat pipe makes higher performance in the system. Thus 60 °C input temperature shows high heat transfer capability and 100 LPH flow rate also transfers heat in the system.

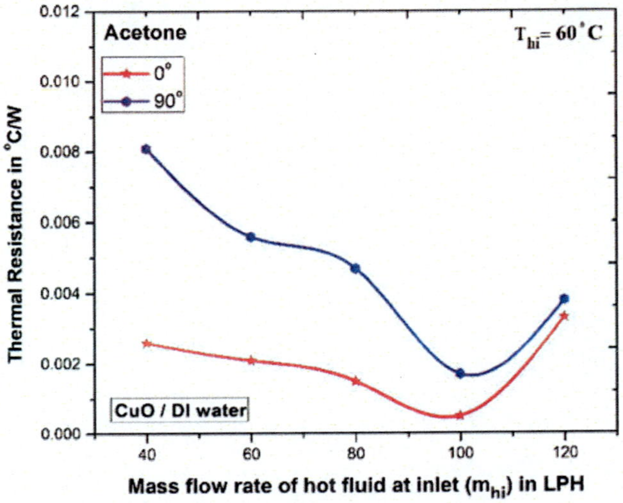

Fig. 12 Variations of Thermal resistance (0°–90°, Acetone)

4 Conclusion

The industrial requirement of effective waste heat recovery system, this proposed model is compatible for minimum heat input applications. The investigation convicts the acetone as working fluid for heat pipes and CuO/DI water nanofluid for hot fluid circulation along the heat exchanger for different input conditions. The evaluation warrants that for given parameters the maximum efficiency is obtained at 0°, 100 LPH and 60 °C. The results reveal that for effectiveness as 66.7%, heat transfer coefficient represents 2245.61 W/m^2 °C, heat transfer rate shows 1639.62 W, Reynold's number possesses 1173.26 and thermal resistance of 0.0005 °C/W. Influence of nanofluid and acetone-charged heat pipe heat exchangers are best suitable for this range of temperature operation. The acetone has boiling point of 56 °C, hence this can be used for minimum temperature range and for higher temperature conditions suitable working fluid can be choosed and tested for direct application systems.

References

Bao, K., Wang, X., Zhang, P. E., Han, X., & Tan, J. (2023). Evaporating temperature uniformity of the pulsating heat pipe with surfactant solutions at different concentrations. *Journal of Thermal Science*, *32*, 183–191. https://doi.org/10.1007/s11630-022-1756-9

Chaudhry, H. N., Hughes, B. R., & Ghani, S. A. (2012). A review of heat pipe systems for heat recovery and renewable energy applications. *Renewable and Sustainable Energy Reviews, 16*(4), 2249–2259.

Choling, H., & Zou, L. (2015). Study on the heat transfer characteristics of a moderate temperature heat pipe heat exchanger. *International Journal of Heat and Mass Transfer, 91*, 302–310.

Deepak, V., Umamaheshwaran, P. S., Guhan, K., Nanthini, R. A., Krithiga, B., Jaithoon, N. M. H., & Gurunathan, S. (2011). Synthesis of gold and silver nanoparticles using purified URAK. *Colloids and Surfaces b: Biointerfaces, 86*(2), 353–358.

Holman, J. P. (2007). *Experimental methods for engineers* (7th ed.) McGraw-Hil, New York.

Kamran, M. S., Naz, K., Umer, J., Sajjad, M., Saleem, M. W., Ibrahim, M., & Zeinelabdeen, M. (2021). Experimental evaluation of performance characteristics of a horizontal copper mesh wick-based miniature loop heat pipe. *Arabian Journal for Science and Engineering*, *46*, 2121–2132. https://doi.org/10.1007/s13369-020-05027-y

Lu, J., Shen, L., Huang, Q., Sun, D., Li, B., & Tan, Y. (2019). Investigation of a rectangular heat pipe radiator with parallel heat flow structure for cooling high-power IGBT modules. *International Journal of Thermal Sciences*, *135*, 83–93.

Ramkumar, P., Sivasubramanian, M., Rajesh Kanna, P., & Raveendiran, P. (2021a). An experimental inquisition of waste heat recovery in electronic component system using concentric tube heat pipe heat exchanger with different working fluids under gravity assistance. *Microprocessors and Microsystems, 83*, 104033.

Ramkumar, P., Sivasubramanian, M., Rajesh Kanna, P., & Raveendiran, P. (2021b). Heat transfer behaviour on influence of an adiabatic section on concentric tube shell assisted heat pipe heat exchanger. *International Journal of Ambient Energy, 42*(06), 672–681.

Ramkumar, P., Nair, A., Sivasubramanian, M., Buddhi, D., & Prakash, C. (2022). Effectiveness prediction of CuO nanofluid heat pipe system using fuzzy neuro approach. *International Journal on Interactive Design and Manufacturing (IJIDeM)*. https://doi.org/10.1007/s12008-022-01066-x

Rudresha, S., Babu, E. R., & Thejaraju, R. (2023). Experimental investigation and influence of filling ratio on heat transfer performance of a pulsating heat pipe. *Thermal Science and Engineering Progress, 38*, 101649. https://doi.org/10.1016/j.tsep.2023.101649

Sadanand, V., Rajini, N., Satyanarayana, B., & Varada Rajulu, A. (2016). Preparation and properties of cellulose/silver nanoparticle composites with in situ-generated silver nanoparticles using *Ocimum sanctum* leaf extract. *International Journal of Polymer Analysis and Characterization, 21*(5), 408–416

Siritan, M., Vafai, K., Kammuang-Lue, N., Terdtoon, P., & Sakulchangsatjatai, P. (2022). An innovative design for a solar water heating system utilizing a flat-shaped heat pipe. *Journal of Solar Energy Engineering*, *145*(5), 051002. https://doi.org/10.1115/1.4056624

Sonawane, C. R., Tolia, K., Pandey, A., Kulkarni, A., Punchal, H., Sadasivuni, K. K., Kumar, A., & Khalid, M. (2023). Experimental and numerical analysis of heat transfer and fluid flow characteristics inside pulsating heat pipe. *Chemical Engineering Communications*, *210*(4), 549–565. https://doi.org/10.1080/00986445.2021.1974413

Tang, H., Xie, Y., Xia, L., Tang, Y., & Sun, Y. (2023). Review on the fabrication of surface functional structures for enhancing heat transfer of heat pipes. *Applied Thermal Engineering*, *226*, 120337. https://doi.org/10.1016/j.applthermaleng.2023.120337

Venkatachalapathy, S., Kumaresan, G., & Suresh, S. (2015). Performance analysis of cylindrical heat pipe using nanofluids—an experimental study. *International Journal of Multiphase Flow, 72*, 188–197.

Wang, H., Tang, Y., Liu, M., Zhu, S., Zheng, K., & Du, X. (2023). Experimental study on heat transfer performance of axially rotating heat pipe in steady state. *International Journal of Thermal Sciences*, *184*, 107975. https://doi.org/10.1016/j.ijthermalsci.2022.107975

Washburn, E. W. (1930). *International critical tables of numerical data, physics, chemistry and technology*. National Research Council, The National Academies Press, Washington DC. https://doi.org/10.17226/20230

Xu, H., Yang, Y., Gan, K., Zhang, H., Gao, Y., Li, R., Jiang, Y., Qian, J., Wei, Z., Zheng, Y., & Ding, Q. (2023). Heat transfer performance of novel high temperature flat heat pipe (HTFHP) with heating power and inclination angles. *Applied Thermal Engineering*, *220*, 119679. https://doi.org/10.1016/j.applthermaleng.2022.119679

Zeghari, K., Louahlia, H., Tiffonnet, A.-L., & Schaetzel, P. (2023). Micro-grooved circular miniature heat pipe for thermal management: experimental and analytical investigations. *Thermal Science and Engineering Progress*, *40*, 101714. https://doi.org/10.1016/j.tsep.2023.101714

Zhang, Z., Zhao, R., Zhao, S., Zou, H., Liu, Z., Luo, X., & Liu, W. (2023). Performance characteristics of a two-phase pump-assisted loop heat pipe with dual-evaporators in parallel. *Thermal Science and Engineering Progress*, *38*, 101657. https://doi.org/10.1016/j.tsep.2023.101657

Two Au Atoms-Doped Silicene Nanoribbons in Unit Cell with an Electrical Field: A DFT Study

Hoang Van Ngoc

Abstract

The present investigation is centered on the exploration of the impact of Au doping on silicene nanoribbons (SNRs), which are systems with one dimension characterized by edges of hydrogenated silicene. Specifically, the research scrutinizes three distinct structures, namely meta, ortho, and para, each containing two Au atoms per unit cell, under the influence of an electrical field of 0.1 V/m. The doping process entails the substitution of two Si atoms with two Au atoms within each unit cell. By employing Density Functional Theory (DFT) methodology, the investigation computes the formation energy, electrical band structure, and states density for the doped structures. Notably, the optimized doping structures exhibit stability in the presence of an electrical field. The findings indicate that the meta configuration displays the minimum energy of formation, hence signifying its superior stability and optimal characteristics. Furthermore, the utilization of Au doping in conjunction with an external electric field triggers modifications in the energy band, thus signaling possible applications in the future.

Keywords

Silicene nanoribbons • Au atom • Doping configuration • DFT study

1 Introduction

The remarkable advancements in science and technology have given rise to devices that are smaller, lighter, more efficient, and energy-saving. The development of such technology heavily relies on the contributions of materials science, as new materials are constantly synthesized to cater to practical needs. Low-dimensional materials offer several advantages in the production of high-tech devices by making them more compact and efficient, and by harnessing the quantum theory and nanoscale confinement effects that are distinct from those observed in bulk materials. The confined electron in low-dimensional materials exhibits unique properties found only at the nanoscale, owing to energy disruption caused by confinement in all directions.

Silicon (Si) belongs to Group IVA in the periodic table, positioned just below carbon (C). It shares the same number of valence electrons (which determines electronic properties) as carbon, enabling it to adopt various forms and sizes. This suggests that when silicon atoms are arranged on a suitable surface layer, they can spontaneously organize into a honeycomb lattice, forming silicene. Clues about silicene's structure emerged in 2010 (Aufray et al., 2010). Through scanning tunneling microscopy, scientists examined self-assembled silicene nanoribbons and films on silver crystals Ag (110) and Ag (111) with atomic precision. Density functional theory (DFT) and calculations demonstrated that silicon atoms preferentially form honeycomb structures on silver with a slight buckling, similar to the configuration found in graphene.

Low-dimensional forms of silicon, such as 1D nanotubes and 0D fullerenes, were discovered early on, while the basic 2D structure of silicene was only predicted in 1994 (Takeda & Shiraishi, 1994). Silicene is known as a 2D material similar to graphene. In experimental terms, silicene can only be synthesized through a bottom-up epitaxial growth method since silicon does not naturally exist in layered structures like graphite (Cherukara et al., 2017). Silicene exhibits numerous novel physical properties with great potential for various applications, such as room-temperature field-effect transistors (FETs) based on silicene (Tao et al., 2015), 2D sensors (Hu et al., 2014), and energy storage devices (Galashev & Ivanichkina, 2020). Moreover, several other

H. Van Ngoc (✉)
Center for Forecasting Study, Institute of Southeast Vietnamese Studies, Thu Dau Mot University, Thu Dau Mot, Binh Duong 820000, Vietnam
e-mail: ngochv@tdmu.edu.vn

M. Sumesh et al. (eds.), *2nd International Conference on Smart Sustainable Materials and Technologies (ICSSMT 2023)*,
Advances in Science, Technology & Innovation, https://doi.org/10.1007/978-3-031-49826-8_10

special physical properties of silicene have been discovered, such as the wide bandgap induced by spin–orbit coupling at Dirac points (Karpas et al., 2019) and large quantum spin Hall effect (Liu et al., 2011).

However, the potential applications of silicene in nanoelectronic devices are limited due to its relatively small bandgap. Scientists have been continuously researching methods to widen the bandgap for silicene, making this topic an attractive and highly sought-after research area in recent times. Various methods have been employed to create energy bandgaps for silicene, such as atomic doping (Hernández Cocoletzi & Castellanos Águila, 2018), surface functionalization (Huang et al., 2013), different stacking structures (Padilha & Pontes, 2015), mechanical deformation (Qin et al., 2012), and the application of external fields (Wu et al., 2016).

From a geometric perspective, the simplest and most effective method to open the energy bandgap for silicene is to confine it to a finite size in 2D. The quantum confinement of 2D silicene results in 1D silicene nanoribbons (SiNRs) (Padova et al., 2012) with two typical edge structures, namely armchair SiNRs (ASiNRs) and zigzag SiNRs (ZSiNRs). However, ASiNRs and ZSiNRs possess different magnetic configurations due to the different spin orientations of the Si atoms at the edges. ASiNRs are non-magnetic semiconductors, whereas ZSiNRs are ferromagnetic semiconductors (Dong et al., 2015). The opened bandgap of 1D SiNRs offers a significant advantage in overcoming the main obstacles in 2D silicene for electronic device applications. Additionally, the unique magnetic configuration of ZSiNRs 1D could make them highly promising materials for spin-based electronic devices.

In addition to top-down methods from 2D silicene, bottom-up epitaxial growth is also an effective method for synthesizing 1D SiNRs with atomic-level precision, where SiNRs are grown on metal substrates or thin insulating films (Broek et al., 2016; Quertite, 2020). 1D SiNRs with diverse physical properties have been widely used in various applications, including gas sensors (Aghaei et al., 2016), electronic devices (Kharadi et al., 2020), and spin electronics (Zhou et al., 2017).

However, to be suitable for many practical applications, the essential physical properties of 1D SiNRs need to be diversified. Various methods have been employed to enrich the essential physical properties of 1D SiNRs, including edge functionalization (Mehdi Aghaei et al., 2016), atomic substitution or adsorption (Tran et al., 2020), creating network defects (Walia & Randhawa, 2018), forming heterostructures (Rosales et al., 2008), mechanical deformation (Mahmoudi et al., 2016), and applying external fields (Lu et al., 2020). Among these methods, the chemical doping and atomic impurity are effective ways to strongly modify the essential properties of 1D SiNRs. These methods can induce a strong diversity in different atomic concentrations and distributions. To date, many different types of atoms have been used to dope 1D SiNRs. The adsorption of Cu atoms on ZSiNRs 1D has been studied using first-principles calculations. The results showed that the charge transfer from Si to Cu atoms leads to ferromagnetic states (Ghasemi et al., 2019). Additionally, the adsorption of Ti atoms on ZSiNRs 1D has also been investigated, where Ti atoms optimally adsorb at the vacancy sites near the center of the 1D band (Xu et al., 2015). The doping of P in SiNRs 1D has also been studied, with P atoms optimally substituted at the edge positions of SiNRs 1D (Zhang et al., 2014). The doping of N or B in SiNRs has also been investigated, with N or B atoms showing optimal substitutions at the edge positions of 1D.

Modifying the elements on the edges generates SNRs with varying bandgaps. In this study, the author examines SNRs with hydrogen-modified edges placed in an electrical field of 0.1 V/m, generating a system with distinct electronic properties. The coexistence of Au and the electrical field in this system shows potential for future applications.

2 Theoretical Foundations

DFT is a powerful framework within quantum mechanics that is used to describe the properties of electron systems in atoms, molecules, solids, and other systems. In contrast to traditional methods that involve the complex wave function depending on 3N spatial coordinates for an N-electron system, DFT represents the system's properties solely through the electron density function, which depends only on three spatial coordinates. This approach simplifies the description and computation of electronic properties, making it computationally more efficient and widely applicable to various systems. Therefore, DFT has a significant advantage (and is currently the most widely used) in calculating physical properties for specific systems based on very fundamental equations of quantum physics.

$$\widehat{H}\Psi = E\Psi \tag{1}$$

In quantum theory, H represents the Hamiltonian operator that describes the nature of the system, and it is responsible for determining the energy of the system when applied to its wave function. The energy of the system is denoted by E, and ψ represents the corresponding wave function for that system. The Hamiltonian operator includes both the energy operator and the interaction operator among particles in the system and is considered a crucial part in solving the Schrödinger equations to find the states and energies of the quantum system.

Due to the extremely large number of electrons (on the order of Avogadro's number) and their interactions through the Coulomb potential, it is impossible to exactly solve Eq. (1). Instead, one must find approximate solutions to the equation using various approximations.

2.1 Electron Density

The electron density, denoted as $n(r)$, is a fundamental quantity in DFT and is defined as the number of electrons per unit volume at a specific position in space, represented by the spatial coordinate vector $r = (x, y, z)$. In quantum mechanics, electrons do not possess definite positions; instead, they are described by probability distributions. Thus, the electron density must be interpreted as a probability density function, representing the likelihood of finding an electron in a small volume element around the point r in space. The total electron density over all space is given by integrating the probability density function over the entire three-dimensional space. This approach provides a statistical description of electron distribution and is essential for the accurate representation of electronic properties in the DFT framework.

$$N = \int n(\vec{r})d\vec{r} \tag{8}$$

where

$$n(\vec{r}) = \mathrm{N}\int \ldots \int |\Psi(\overrightarrow{x_1}, \overrightarrow{x_2}, ..\overrightarrow{x_N}|^2 \, \mathrm{d}\sigma_1, d\overrightarrow{x_2}...d\overrightarrow{x_N} \tag{9}$$

However, this is the only term in Eq. (8) that can be explicitly written in terms of the electron density function. For the kinetic energy term K_e, the presence of the wave function's derivative makes it impossible to be explicitly written. In the electron–electron interaction energy term V_{ee}, the fractional position of electrons in the denominator complicates combining the terms together.

With these limitations, there cannot be a general expression for the energy functional of the density. Since both K_e and V_{ee} involve electrons, it is more convenient to combine them into a single term and write it as follows:

$$\mathrm{F}[\Psi_e] = K_e + V_{ee} \tag{11}$$

Then, the total energy of the system is:

$$E_e = F[\Psi_e] + V_{ee}(v, n) \tag{12}$$

The objective of the upcoming section is to express the electronic energy in the system as a function of the electron density. In other words, the aim is to transform $F[\psi_e]$ from a functional dependent on the wave function to a functional dependent on the electron density $F[n]$.

2.2 Thomas–Fermi Approximation

In the Thomas–Fermi approximation, the kinetic energy of the electron system is estimated by using an explicit functional of the electron density, which bears resemblance to the expression for the kinetic energy of a non-interacting electron system. Furthermore, the interaction energy between electrons is approximated exclusively by the static Coulomb interaction energy. As a consequence, the total energy functional can be explicitly expressed as follows:

$$E_{TF}[n(\vec{r})] = \frac{3}{10}3\pi^{\frac{2}{3}} \int n\left(\vec{r}^{\frac{5}{3}}\right)d\vec{r} + \frac{1}{2}\iint \frac{n(\vec{r}_1)n(\vec{r}_2)}{|\vec{r}_1 + \vec{r}_2|}d\vec{r}_1 d\vec{r}_2 + \int V_{ext}(\vec{r})n(\vec{r})d\vec{r} \tag{13}$$

The ground-state density can be taken from the variational condition of the energy functional, for example, using the Lagrange multiplier method. The results of this approximate method, when applied to electron systems in atoms and molecules, are quite modest. Although the overall shape of the electron density may be qualitatively reasonable, it is quantitatively inaccurate. As a result, unphysical consequences appear, such as the inability to describe the electronic orbital structure in atoms, the lack of chemical bonding in molecules, etc. This can be understood because the kinetic energy term approximation above is rather crude and only works well for systems with nearly constant electron density. Moreover, the electron–electron interaction part, which involves quantum effects and contributes to the total energy of the ground state as exchange and correlation energies, is neglected in the Thomas–Fermi approximation. These deficiencies are mostly addressed in the Kohn–Sham equations, which form the foundation for the success of Density Functional Theory (DFT) in accurately describing electronic properties of various systems.

2.3 Hohenberg–Kohn Theorems

The fundamental principle of DFT is to describe a many-particle system through its electron density rather than using a many-particle wave function. Employing the electron density as the sole variable offers a significant advantage since, regardless of the system's particle count, the electron density relies solely on the three spatial coordinates (x, y, and z), while the many-particle wave function depends on 3N spatial coordinates.

Though DFT was utilized in its early stages, it wasn't until 1964 that the idea of characterizing the fundamental properties of an electron system through its electron density was firmly established by the first Hohenberg–Kohn theorem. This theorem asserts: "For any interacting particle system in an external field represented by $V(r)$, the external potential is uniquely determined (up to an additive constant) by the ground-state electron density $n(r)$". Hence, it is impossible to have two different external potentials (differing by a constant) that result in the same ground-state electron density. This theorem has significant implications: the system's Hamiltonian and wave function are entirely determined by the electron density $n(r)$. In essence, knowing the ground-state electron density fully determines all properties of the system.

2.4 Kohn–Sham Equation

The Kohn–Sham equations can be effectively solved through a self-consistent approach. One approach involves constructing an effective potential using the equations and subsequently solving the equations to obtain the Kohn–Sham orbitals. Using these updated orbitals, the new electron density is then calculated using the following expression:

$$n(\vec{r}) = \sum_i |\Psi_i|^2 \quad (14)$$

This process is repeated until convergence is achieved, meaning the difference in the electron density (or effective potential) between two consecutive steps becomes smaller than a certain threshold (the convergence condition). In principle, the Kohn–Sham equation is exact, but in practice, approximate methods are used for the exchange–correlation energy component.

2.5 Pseudopotentials

The Kohn–Sham equations are a set of partial differential equations, and solving these numerical equations is typically done using either grid-based methods (finite difference) or finite element methods (expanding wave functions in basis sets). For crystalline solids, electrons move in periodic potentials, so Kohn–Sham wave functions follow the Bloch theorem and are classified according to Bloch states. The plane-wave basis set is convenient for expanding these wave functions. However, in reality, a major drawback of the plane-wave basis set is that a large number of plane waves are required to represent wave functions close to the nuclei. Pseudopotential methods have been developed to overcome this drawback and have been proven to be efficient in practical calculations.

The physical idea of the pseudopotential method is that most properties in the low-energy region of atomic, molecular, or solid-state systems are determined by valence electrons. The inner-core electrons in the atomic nucleus are tightly bound to the nuclei and contribute little. Thus, in the pseudopotential approximation, an electron can be considered "frozen' in the atomic core and the valence electrons move in an effective field created by the nuclei and the frozen core electrons. Mathematically, this can be achieved by replacing the strong, oscillatory Coulomb potential of the nucleus with a smooth and finite pseudo-potential.

3 Structure and Formation Energy

Figures 1, 2, 3, and 4 illustrate the pristine, ortho, meta, and para structures, respectively. Upon optimizing the configuration of the unit cell, its structure is found to be stable with a buckling of 0.538 Å (as shown in Fig. 1). In each doped configuration, two silicon atoms are replaced with two gold atoms, resulting in three distinct configurations that have all been found to be stable and remain unbroken even in the presence of an electrical field after optimization.

Table 1 displays the energy required to form the doped structures, which was calculated using DFT theory and VASP software after optimizing the system. The formation energy serves as an indicator of the stability of each doped configuration. Based on the table data, it can be observed that the meta configuration has a lower formation energy than the other two configurations, making it the most optimal option.

4 The Electrical Characteristics of the Structures with Doping

The original structure is a semiconductor with a narrow bandgap (Fig. 5), a band gap of about 0.33 eV, which is perfectly consistent with hydrogen-modified edges.

Multiple peaks can be observed in the bands, with the conduction band having the largest peak, as shown in Figs. 1, 2, 3, and 4. Figure 5 depicts the involvement of Si(s) and Si(p) states to the band formation, revealing that Si(p) is the primary contributor to states around the Fermi level. The band energy and density of states (DOS) of the doped structures are presented in Figs. 6, 7, and 8. All the configurations (ortho, meta, and para) are semiconductors with respective bandgaps of 0.29, 0.38, and 0.2 eV. The bandgaps of the doped structures exhibit significant variation, with direct and indirect bandgaps observed in Figs. 6, 7, and 8. Compared to the undoped structure, only the meta configuration displays a slightly wider bandgap, while the other two configurations have contracted bandgaps. Each configuration has unique characteristics that enable diverse applications in electronic

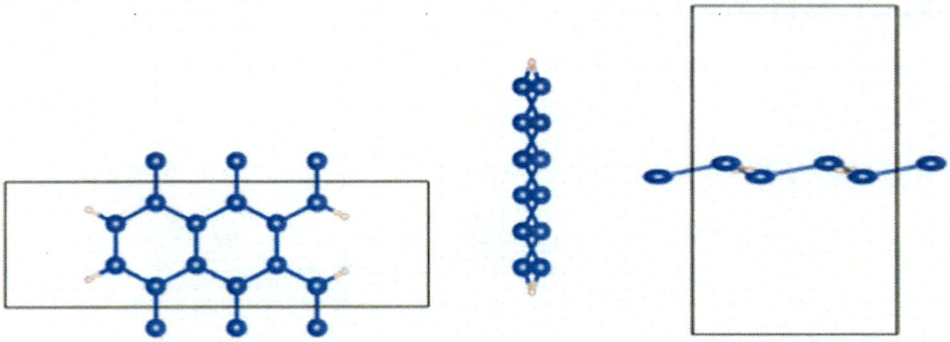

Fig. 1 The primordial system

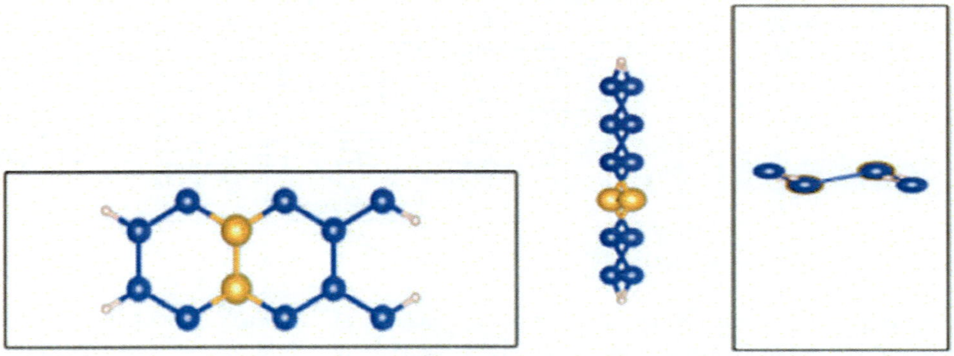

Fig. 2 The ortho doping structure

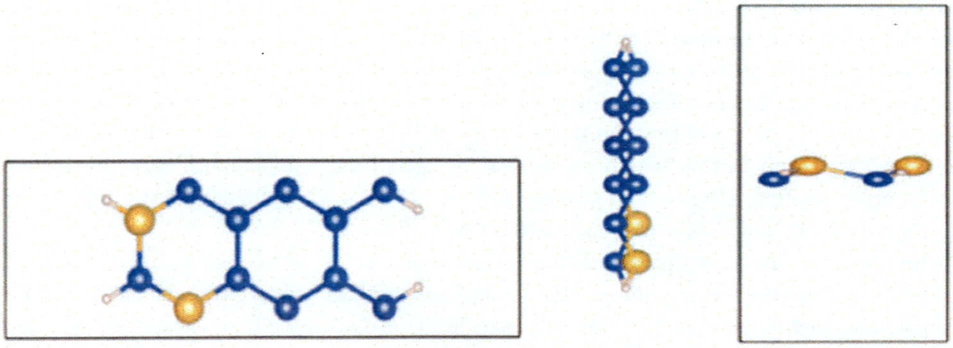

Fig. 3 The meta doping structure

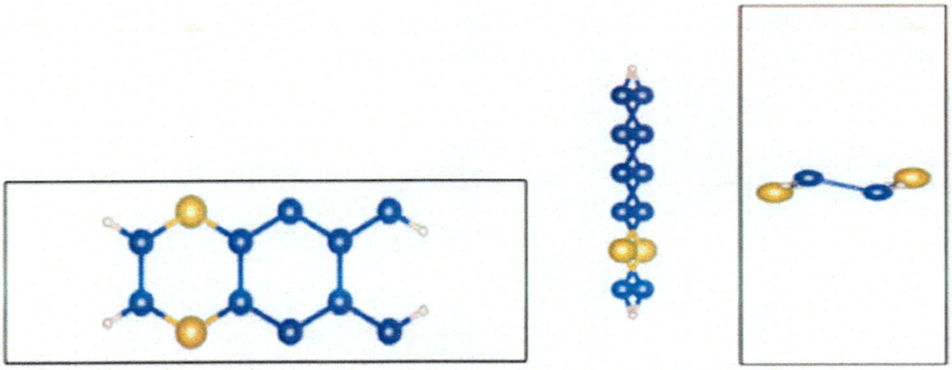

Fig. 4 The para doping structure

Table 1 The formation energy

Structure	E_t (eV)	E_p (eV)	E_{Si} (eV)	E_{Au} (eV)	E_f (eV)
Ortho	−69.501	−69.41	−0.135	−0.047	−0.267
Meta	−69.570	−69.41	−0.135	−0.047	−0.336
Para	−69.499	−69.41	−0.135	−0.047	−0.265

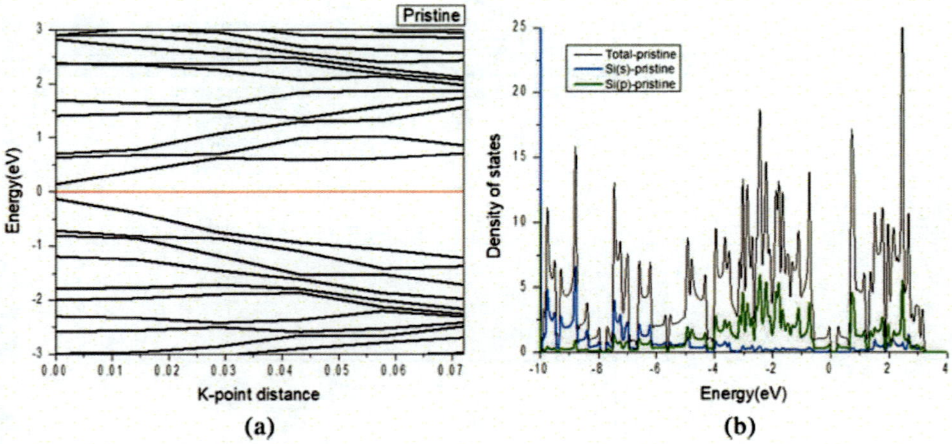

Fig. 5 **a** Band structure; **b** DOS (**b**) of the primordial system

technology. The meta configuration, with a bandgap of 0.38 eV, is suitable for use in field-effect transistors operating at room temperature. Furthermore, the doped configurations have fewer state peaks than the pristine configuration, with the highest peak of the doped configurations appearing in the valence band. The states in the doped configurations extend more towards the bands compared to the undoped structure. The state density plot confirms that states are more dispersed around the Fermi level, which is consistent since the post-doping configurations are all semiconductors.

Figures 9, 10, and 11 illustrate the s and p states of Si and Au in the ortho, meta, and para structures. The involvement of Si's s and p states is significant due to its high particle density and four outermost electrons, with two in the p-shell and two in the s-shell. Conversely, the s and p states of Au are minor because Au has a small density of particles, and only one outermost electron. The dominant involvement of Au is in the Au(d) state, as Au has up to 10 electrons in the d-shell.

Each Au atom bonds with three Si atoms next to it, but as gold has only one electron in the outermost shell, there is only one sigma bond with one Si atom, resulting in less stability for the Au atoms than in the doped configurations. The band structure plot reveals that the doped configurations are stable and do not break under the influence of an electrical field. However, if the electrical field is increased, the gold-doped structures will break, and the band gap of the doped configurations cannot be expanded further.

In the energy region from −6 eV to 0, especially from −2.5 eV to 0, hybridization between the s orbitals of Au and P of Si in the meta configuration is shown (Fig. 12). The red dots represent the s orbitals of Au, the green dots represent the p orbitals of Si. We see that there is overlap between the orbitals in these energy regions, these hybridizations create sigma bonds between Au and Si, while the remaining electrons of Si form π bonds, which keep gold stable in doped configurations.

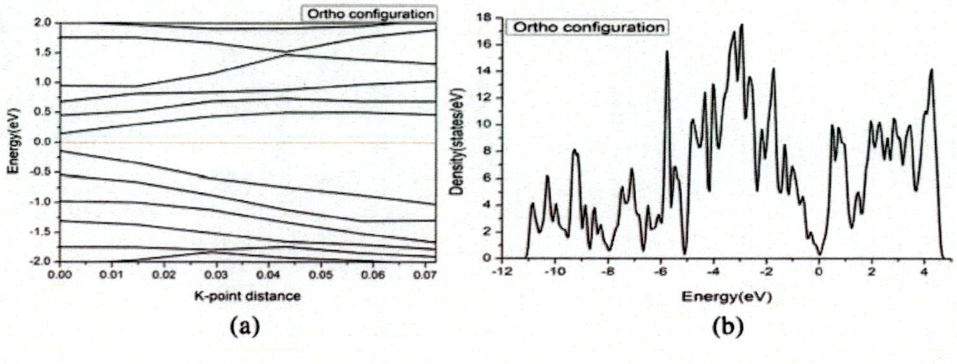

Fig. 6 **a** Band structure; **b** DOS of the ortho doping structure

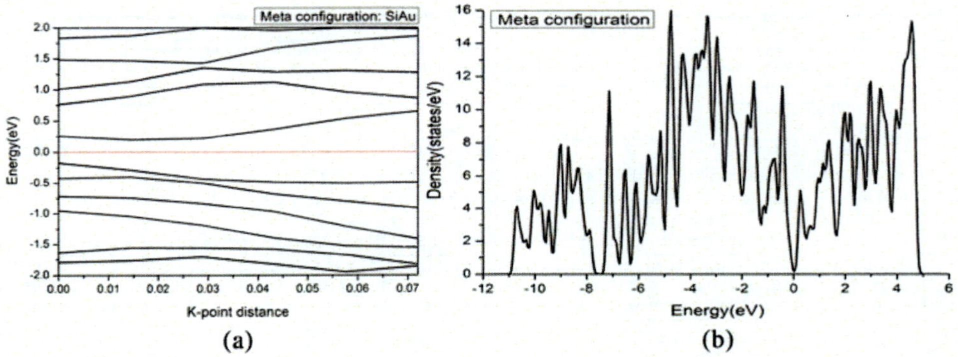

Fig. 7 **a** Band structure (**a**), DOS (**b**) of the meta doping structure

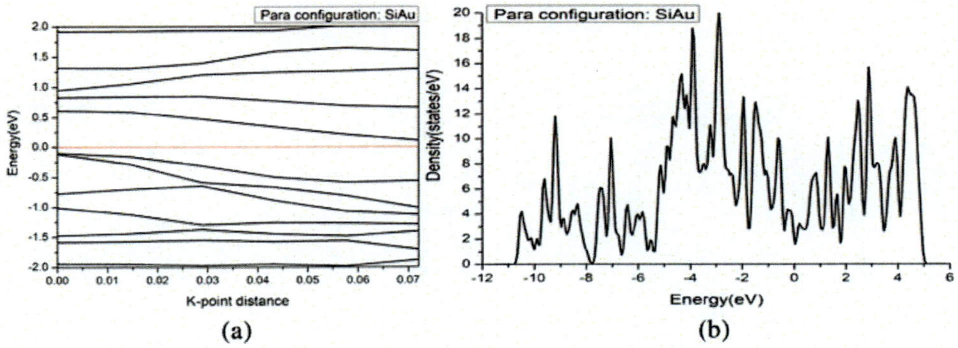

Fig. 8 **a** Band structure; **b** DOS of the para doping structure

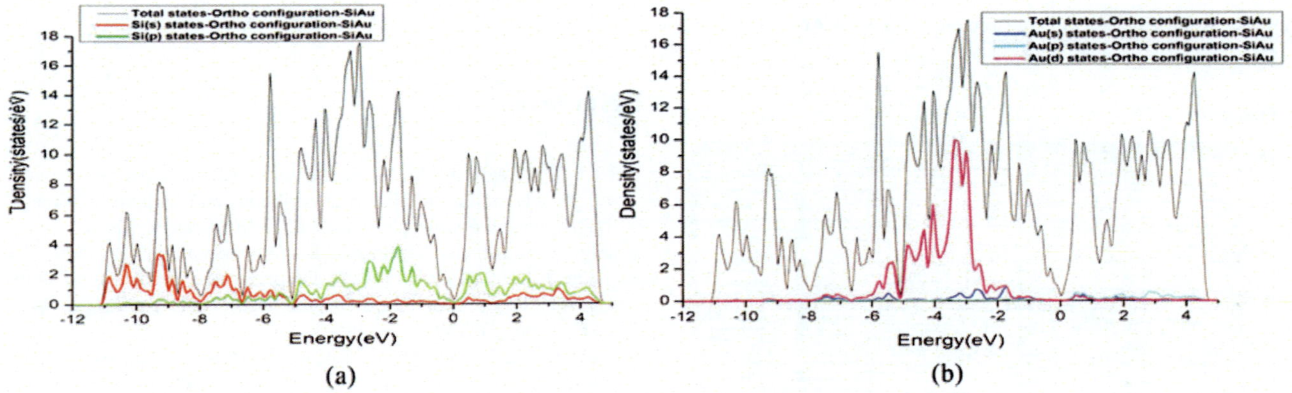

Fig. 9 **a** Partial states in ortho doping structure of: **a** Si; **b** Au

5 Conclusions

In conclusion, this study successfully introduced gold atoms into SNRs and investigated their effects on the system's properties. By replacing two Si atoms with two Au atoms in each unit cell, the resulting doped structures were found to be stable and able to withstand an electrical field of 0.1 V/m. The band structures of the doped systems were found to differ significantly from the original structure due to the presence of Au atoms and electrical field. The study also analyzed the s, p, and d states of the elements, providing insights into the electrical characteristics of the structures with doping. The findings of this study lay the foundation for future research in the field of SNRs and their potential applications in electronic technology.

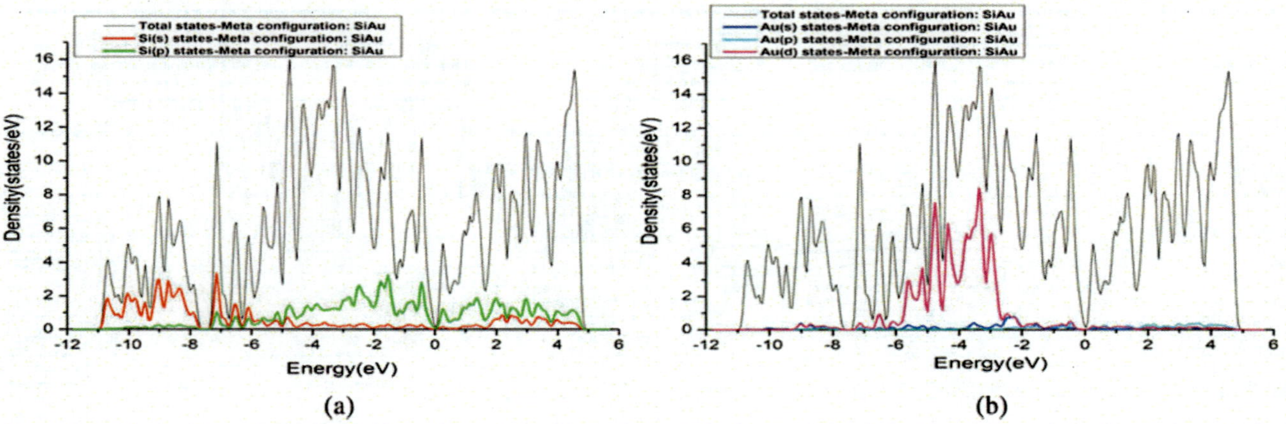

Fig. 10 a Partial states in meta structure of: **a** Si; **b** Au

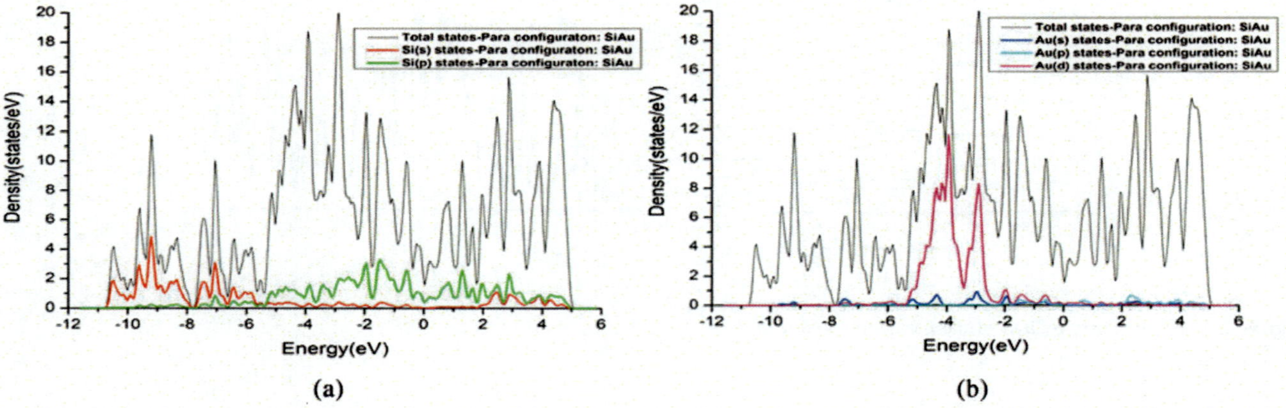

Fig. 11 a Partial states in para structure of: **a** Si; **b** Au

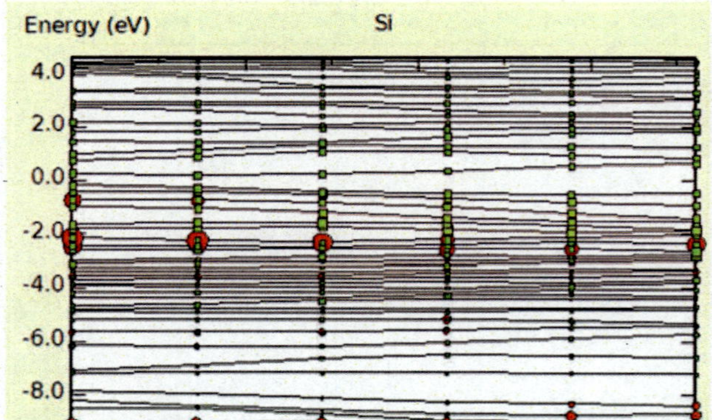

Fig. 12 Au(s)-Si(p) hybridization in meta configuration

Acknowledgements This research used resources of the high-performance computer cluster (HPCC) at Thu Dau Mot University (TDMU), Binh Duong Province, Vietnam.

References

Aghaei, S. M., Monshi, M. M., & Calizo, I. (2016). A theoretical study of gas adsorption on silicene nanoribbons and its application in a highly sensitive molecule sensor. *RSC Advances, 6*, 94417.

Aufray, B., Kara, A., Vizzini, S. B., Oughaddou, H., LéAndri, C., Ealet, B., & Le Lay, G. (2010). Graphene-like silicon nanoribbons on Ag(110): A possible formation of silicene. *Applied Physics Letters, 96*, 223109.

Van den Broek, B., Houssa, M., Lu, A., Pourtois, G., Afanas'ev, V., & Stesmans, A. (2016). Silicene nanoribbons on transition metal dichalcogenide substrates: Effects on electronic structure and ballistic transport. *Nano Research, 9*, 3394.

Cherukara, M. J., Narayanan, B., Chan, H., & Sankaranarayanan, S. K. R. S. (2017). Silicene growth through island migration and coalescence. *Nanoscale, 9*, 10186.

Dong, H., Fang, D., Gong, B., Zhang, Y., Zhang, E., & Zhang, S. (2015). Electronic and magnetic properties of zigzag silicene nanoribbons with Stone-Wales defects. *Journal of Applied Physics, 117*, 64307.

Galashev, A. Y., & Ivanichkina, K. A. (2020). Silicene anodes for lithium-ion batteries on metal substrates. *Journal of the Electrochemical Society, 167*, 50510.

Ghasemi, N., Ahmadkhan Kordbacheh, A., & Berahman, M.: Electronic, magnetic and transport properties of zigzag silicene nanoribbon adsorbed with Cu atom: A first-principles calculation. *Journal of Magnetism and Magnetic Materials, 473*, 306 (2019).

Hernández Cocoletzi, H., & Castellanos Águila, J. E. (2018). DFT studies on the Al, B, and P doping of silicene. *Superlattices and Microstructures, 114*, 242–250.

Hu, W., Xia, N., Wu, X., Li, Z., & Yang, J. (2014). Silicene as a highly sensitive molecule sensor for NH_3, NO, and NO_2. *Physical Chemistry Chemical Physics: PCCP, 16*, 6957.

Huang, B., Xiang, H. J., & Wei, S.-H. (2013). Chemical functionalization of silicene: Spontaneous structural transition and exotic electronic properties. *Physical Review Letters, 111*, 145502.

Karpas, M., Faria Junior, P. E., Gmitra, M., & Fabian, J. (2019). Spin-orbit coupling in elemental two-dimensional materials. *Physical Review B, 100*, 125422.

Kharadi, M. A., Malik, G. F. A., Khanday, F. A., Shah, K. A., Mittal, S., & Kaushik, B. K. (2020). Review-silicene: From material to device applications. *ECS Journal of Solid State Science and Technology, 9*, 115031.

Liu, C.-C., Feng, W., & Yao, Y. (2011). Quantum spin Hall effect in silicene and two-dimensional germanium. *Physical Review Letters, 107*, 76802.

Lu, W.-T., Sun, Q.-F., Tian, H.-Y., Zhou, B.-H., & Liu, H.-M. (2020). Band bending and zero-conductance resonances controlled by edge electric fields in zigzag silicene nanoribbons. *Physical Review B, 102*, 125426.

Mahmoudi, M., Fathipour, M., & Ahangari, Z. (2016). Improved double-gate armchair silicene nanoribbon field-effect-transistor at large transport band gap. *Chinese Physics B, 25*, 8.

Mehdi Aghaei, S., Torres, I., & Calizo, I. (2016). Structural stability of functionalized silicene nanoribbons with normal, reconstructed, and hybrid edges. *Journal of Nanomaterials*, 5959162.

De Padova, P., Perfetti, P., Olivieri, B., Quaresima, C., Ottaviani, C., & Le Lay, G. (2012). 1D graphene-like silicon systems: Silicene nano-ribbons. *Journal of Physics: Condensed Matter, 24*, 223001.

Padilha, J. E., & Pontes, R. B. (2015). Free-standing bilayer silicene: The effect of stacking order on the structural, electronic, and transport properties. *Journal of Physical Chemistry C, 119*, 3818.

Qin, R., Wang, C.-H., Zhu, W., & Zhang, Y. (2012). First-principles calculations of mechanical and electronic properties of silicene under strain. *AIP Advances, 2*, 22159.

Quertite, K., et al. (2020). Silicene nanoribbons on an insulating thin film. *Advanced Functional Materials*, 2007013.

Rosales, L., Orellana, P., Barticevic, Z., & Pacheco, M. (2008). Transport properties of graphene nanoribbon heterostructures. *Microelectronics Journal, 39*, 537–540.

Takeda, K., & Shiraishi, K. (1994). Theoretical possibility of stage corrugation in Si and Ge analogs of graphite. *Physical Review B, 50*, 14916.

Tao, L., et al. (2015). Silicene field-effect transistors operating at room temperature. *Nature Nanotechnology, 10*, 227.

Tran, N. T. T., Gumbs, G., Nguyen, D. K., & Lin, M.-F. (2020). Fundamental properties of metal-adsorbed silicene: A DFT study. *ACS Omega, 5*, 13760.

Walia, G. K., & Randhawa, D. K. K. (2018). First-principles investigation on defect-induced silicene nanoribbons-A superior media for sensing NH_3, NO_2 and NO gas molecules. *Surface Science, 670*, 33–43.

Wu, J.-Y., Chen, S.-C., Gumbs, G., & Lin, M.-F. (2016). Feature-rich electronic excitations of silicene in external fields. *Physical Review B, 94*, 205427.

Xu, L., Wang, X.-F., Zhou, L., & Yang, Z.-Y. (2015). Adsorption of Ti atoms on zigzag silicene nanoribbons: Influence on electric, magnetic, and thermoelectric properties. *Journal of Physics D. Applied Physics, 48*, 215306.

Zhang, J.-M., Song, W.-T., Xu, K.-W., & Ji, V. (2014). The study of the P doped silicene nanoribbons with first-principles. *Computational Materials Science, 95*, 429.

Zhou, J., Bournel, A., Wang, Y., Lin, X., Zhang, Y., & Zhao, W. (2017). Silicene spintronics: Fe(111)/silicene system for efficient spin injection. *Applied Physics Letters, 111*, 182408.

Enhancing Mechanical Characteristics and Cost-Efficiency of Composite Materials Through Hybridization and Nanoparticle Incorporation

G. Boopathy, G. Balaji, R. Bhoominathan, A. Bovas Herbert Bejaxhin, and N. Ramanan

Abstract

Lightweight materials are increasingly being used in engineering applications today. Composites are replacing traditional metallic materials in a variety of sectors, including aerospace, defence, and aircraft production, where there is a demand for structural materials with high strength-to-weight and stiffness-to-weight ratios. Natural fibre composites are also increasingly used in place of synthetic fibre composites in a variety of technical fields due to their affordability and environmental friendliness. In this particular study, Kevlar (Aramid fibres) is combined with other materials to improve the mechanical characteristics and impact resistance of composites. The only material that costs more than Kevlar is carbon fibres. The goal is to maximize mechanical qualities while utilizing the fewest amounts of pricey Kevlar fibres possible. The hand layup technique was used to create the hybrid composites, which included both natural and Aramid fibres. The performance of various mechanical properties was then assessed. In addition, a morphological examination was done to look at the interior structure of the composite materials that were examined. The results show that the hybrid composite, with or without the addition of nanoparticles, demonstrates great strength with little reliance on Kevlar fibres.

Keywords

Hybridization • Nanoparticle • Composites • Aramid fibre

1 Introduction

Extensive research has been conducted on the mechanical behaviour of unidirectional Kevlar/epoxy composites under tensile strength and various failure modes, with a particular emphasis on minimizing deviations (Asumani et al., 2012). By utilizing modified high-speed drill bits operating at operational speeds, the drilling capabilities of Kevlar composites were experimentally investigated. The study revealed that the application of liquid nitrogen at the drilling site significantly improved tool life, hole quality, and surface finish (Azmir & Ahsan, 2008). Experimental assessments have been carried out to examine the effects of treatment on Kevlar 29 fibre. The study focused on investigating the influence of temperature on the mechanical characteristics. The results indicated a decrease in mechanical parameters such as tensile strength and tensile strain as the treatment temperature increased. Furthermore, it was observed that vacuum treatment had no impact on Young's modulus and tensile strength (Azmir & Ahsan, 2009). The impact of using Kevlar fibre as reinforcement with bismaleimide, along with the effects of chemical treatments, was thoroughly examined. Two categories of bismaleimides were subjected to thermal analysis using techniques such as thermogravimetry, differential scanning calorimetry, and thermo mechanical analysis. The results demonstrated that interfacial strength was significantly improved through cholosulphuric acid

G. Boopathy (✉)
Aeronautical Engineering, Vel Tech Rangarajan Dr. Sagunthala R&D Institute of Science and Technology, Chennai, India
e-mail: boopathyg81@gmail.com

G. Balaji
Aeronautical Engineering, Hindustan Institute of Technology and Science, Chennai, India

R. Bhoominathan
Aeronautical Engineering, Bharath Institute of Higher Education and Research, Chennai, India

A. B. H. Bejaxhin
Department of Mechanical Engineering, Saveetha School of Engineering, SIMATS, Chennai, India

N. Ramanan
Mechanical Engineering, Sri Jayaram Institute of Engineering and Technology, Chennai, India

M. Sumesh et al. (eds.), *2nd International Conference on Smart Sustainable Materials and Technologies (ICSSMT 2023)*, Advances in Science, Technology & Innovation, https://doi.org/10.1007/978-3-031-49826-8_11

treatment. Scanning electron microscopy was employed to investigate the fracture surface (Anuar & Zuraida, 2011).

Researchers investigating the filtration characteristics of ground kenaf core. A comparison was made between the constant pressure pre-coat filtration properties of kenaf and diatomaceous earth (DE). Three challenge solutions were specifically analysed in detail. The results demonstrated that both kenaf and DE were capable of efficiently removing silica particles from the solution without any degradation in flux during the filtration process. As a result, the study concluded that kenaf and DE displayed similar filtration characteristics (Akil et al., 2011). The characteristics of marble kerf were experimentally investigated in this study using abrasive water jet machining. Various input parameters, such as transverse speed, water pressure, and abrasive flow rate, were considered. The Taguchi method was employed to analyse the process parameters, and ANOVA was conducted to identify the significant factor affecting kerf characteristics. The findings revealed that the transverse speed had a substantial impact on both the kerf taper angle and kerf width (Bandaru, 2015; Boopathy et al., 2022).

Authors focus on conducting experimental investigations to assess the disparities in lignin content among three distinct varieties of kenaf, specifically Everglades (Ananth, 2011). The bast, inner bast, and core samples were prepared from four different positions for analysis The results indicated a decrease in lignin content with an increase in stem height. Furthermore, the disparities in lignin content between the core and bast samples were found to be more pronounced in Everglades and Aokawa varieties compared to Mesta (Azmir & Ahsan, 2011). Several experiments conducted on woven glass fibre reinforced polyester composites in three different orientations to assess their wear and friction coefficient. The characteristics were determined by varying loads and sliding speeds. Scanning electron microscopy was employed to observe changes in the composite laminate. The results revealed a significant improvement in both the friction coefficient and wear loss (Boopathy et al., 2017). Research projects investigated the use of kenaf non-wovens as substrates for laminated products, which have wide applications in wall coverings, upholstery covers, edge banding materials, and other laminates. The objective was to demonstrate the feasibility of utilizing non-woven kenaf fibres in the production of laminated products. The study involved blending kenaf fibres with polypropylene in an 80:20 ratio and preparing batts using a modified cotton card of standard widths. The batts were then processed through calendaring or needle punching and subsequently cured in an oven to create the substrate. These substrates were further laminated with various overlays, including polyester wood grain, phenolic resin, and decorative vinyl (Abbas et al., 2010; Al-Mosawi & Hatif, 2012; Boopathy et al., 2019).

Study focused on the full analysis of hybrid laminates created by mixing non-woven kenaf fibres with Kevlar epoxy, with an emphasis on both quasi-static penetration and ballistic characteristics. The laminates were subjected to high-velocity impacts from strong projectiles at a perpendicular angle, with a thickness spectrum spanning from 3.1 mm to 10.8 mm. The hand layup process was used to create these hybrid composites, which were then cured for 24 h under ambient conditions while carrying a static load. These hybrid composites were made up of three distinct arrangements comprising alternating layers of non-woven kenaf and Kevlar. Furthermore, for the sake of comparison, composite materials were created by combining Kevlar and epoxy, as well as kenaf and epoxy. A mechanical testing device with controlled displacement rates (1.27 mm/min and 2.54 mm/min) was used to perform quasi-static penetration tests. A powder cannon and a 9 mm full metal jacket bullet were used for ballistic testing to submit the laminates to various ballistic scenarios. Parallel to this study, another experiment was carried out to investigate the effects of fibre alignment and alkalization on the mechanical and thermal properties of stretched, randomly oriented kenaf and hemp fibres. The results of this investigation revealed a significant improvement in the mechanical properties of alkalized fibres when compared to their untreated counterparts. Morphological analysis techniques were used to investigate the interior structure of the composite laminate in order to fully appreciate the effects of fibre alignment and alkalization. The outcomes of this study emphasized that hemp polyester composites had enhanced fracture resistance, and treated hemp and kenaf fibres revealed a lack of surface pollutants that were present in untreated fibres (Arsath, 2014; AultrinK, 2016).

2 Material Selection and Manufacturing Process

2.1 Material Selection

Kevlar fibre is a remarkable synthetic material with many benefits over traditional materials, including decreased weight, cost effectiveness, and simplicity of manufacturing. Synthetic materials like Kevlar are essential in today's environment, where we constantly work to address the issues facing society and develop more durable and useful products. When compared by weight, the hard plastic material Kevlar is five times more powerful than steel. It is made using a polymerization method, which combines a large number of synthetic plastics to create long chain molecules. Its remarkable internal structure, in which the molecules are arranged in orderly, parallel lines, is responsible for its extraordinary properties. A strong material is made by firmly knitting these fibres together.

The Dupont Company is credited with developing and introducing Kevlar, a proprietary material with major technical significance. Stephen Kwolek is the man behind this breakthrough. Kevlar belongs to the class of synthetic aromatic polyamides, often known as Aramids. Kevlar and Nomex are two well-known examples of Aramids. Kevlar comes in a variety of compositions, including Kevlar 29 and Kevlar 49, each customized to a specific application.

Kevlar's unusual and durable molecular structure includes benzene-like rings, which are referred to as "aromatic" in this context. The term "polyamide" refers to long chains of these ring-shaped aromatic molecules that are coupled to form a bigger chemical framework. Kevlar is manufactured by spinning solid fibres from a liquid mixture. Poly para phenylene terephthalamide, a synthetic organic polymer, is used in this procedure. One of the hydrogen atoms in ammonia is substituted by amides generated from organic acids in this process. This substitution approach results in a super-strong structure with a reduced modulus and increased impact resistance. Kevlar's inherent qualities make it ideal for applications involving ballistic materials. The combination of ammonia and organic acid condensation produces a polymer with excellent strength and endurance. As a result, Kevlar is widely used in situations where its exceptional structural properties are critical.

2.2 Fabrication Method

A common method for creating hybrid composites is the hand layup process. This method is dependent on the operator's expertise, the composite's hardness, cost factors, production pace, and shape. The hand layup approach, which is simple and affordable, is used in this study endeavour to create the hybrid composites. The equipment needed to fabricate the composite is likewise reasonably priced. Twisted Kevlar fibres and glass reinforced polymer fibres are utilized to create the composite. To ensure a solid binding between the reinforcement and the medium, a thermosetting resin that has been combined with a hardener is used. The technique makes use of a frame and roller made of silica rubber. The hand layup technique is best suited for straightforward, low-volume production.

To make the laminate removal process simple, a releasing agent is first applied to the surface during manufacture. Following the application of a thin coat of resin, Aramid fibres are then spread across the surface. To remove any air bubbles, a weight of 5 kg is placed over the glass fibres. The composite is then left alone for about 3 h. The fibres are normally dried before the layup in order to reduce moisture. At the top and bottommost layers of the composite laminate, Aramid fibres are laid to produce a superior finish. Using the hand layup technique, three distinct categories and three samples were created. Every fibre is laid out in its usual direction. Before the fibres are added to the laminate framework, care is taken to keep them dry.

3 Experimental Analysis

3.1 Tensile Test

Tensile testing, a mechanical test, evaluates the tensile strength and characteristics of materials under tension. During the test, a sample is subjected to an increasing axial force until it either fractures or reaches its maximum deformation. Figure 1 illustrates the standard specimen utilized in the tensile test. The obtained tensile results for the aramid specimens, both with and without nanoparticles, are presented in Fig. 2 and Fig. 3, respectively, using a universal testing machine.

The tensile test results indicate a noticeable difference between the material with added nanoparticles and the material without nanoparticles. When nanoparticles were incorporated into the parent material, the ultimate load at which the material failed increased compared to the material without nanoparticles. Additionally, the yield point, which represents the onset of permanent deformation, also increased when nanoparticles were present. However, in the absence of nanoparticles, the ultimate load and yield point remained the same, indicating that the material could withstand the applied stress within the elastic limit without undergoing permanent deformation.

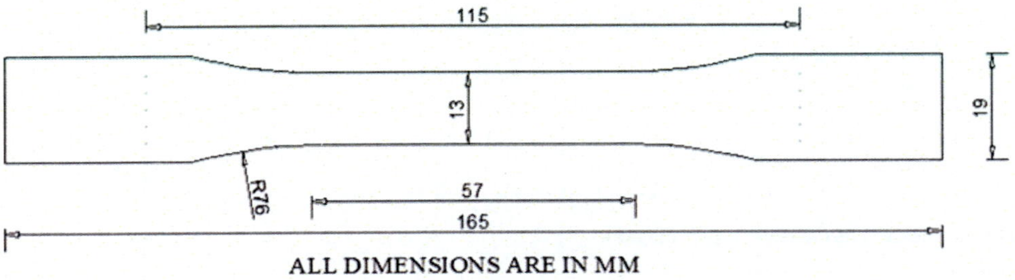

Fig. 1 ASTM D-638 standard for tensile specimen

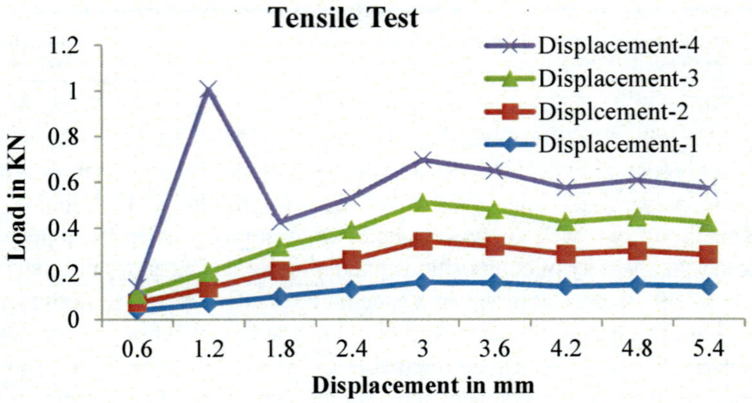

Fig. 2 Aramid fibre without nanoparticles

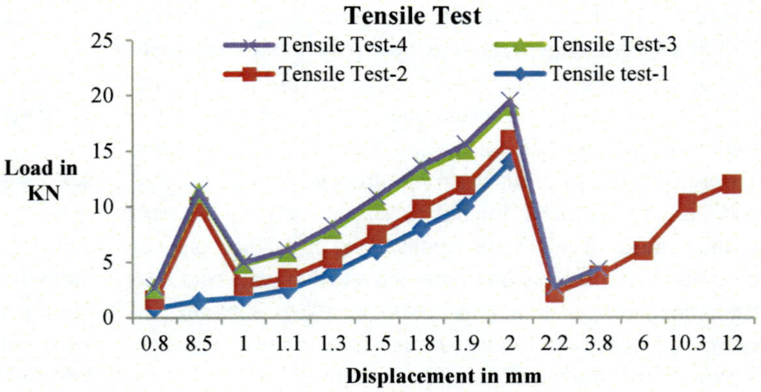

Fig. 3 Aramid fibre with nanoparticles

3.2 Flexural Test

The flexural test, also known as the bending test, is a common mechanical test employed to evaluate the stiffness and bending strength of materials. During this test, a specimen is secured at two points and subjected to a three-point bending configuration. Figure 4 illustrates the standard specimen used for the flexural test. The obtained flexural results for the aramid specimens, both with and without nanoparticles, are presented in Fig. 5 and Fig. 6, respectively, using a universal testing machine.

The flexural test results reveal a distinct difference between the material with added nanoparticles and the material without nanoparticles. When nanoparticles were incorporated into the parent material, the ultimate load, at which the material failed in bending, increased compared to the material without nanoparticles. Additionally, the yield point, which represents the onset of permanent deformation, also increased when nanoparticles were present.

However, in the absence of nanoparticles, the ultimate load and yield point remained the same, indicating that the material could withstand the applied stress within the elastic limit without undergoing permanent deformation.

3.3 Hardness Test

The hardness test is primarily conducted to evaluate the material's ability to absorb sudden shocks or resist indentation. The ASTM standard commonly employed for hardness testing is ASTM D2583.

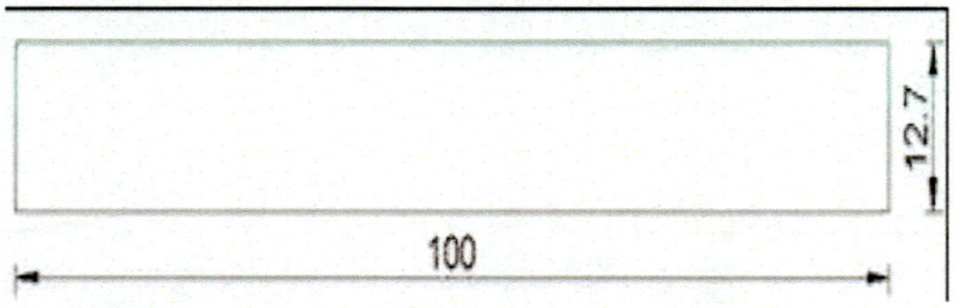

Fig. 4 Flexural test specimen-part diagram

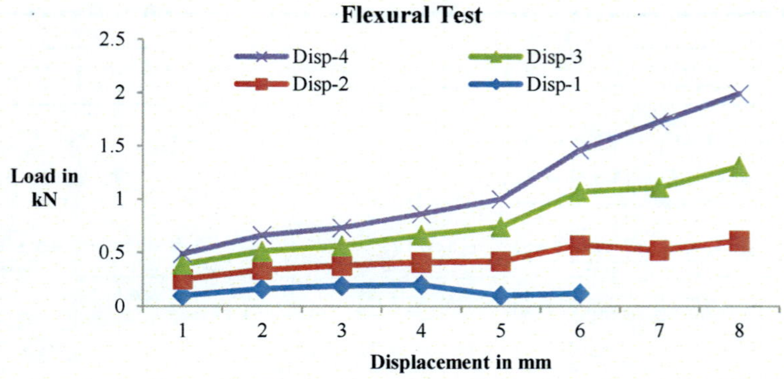

Fig. 5 .Aramid fibre without nanoparticles

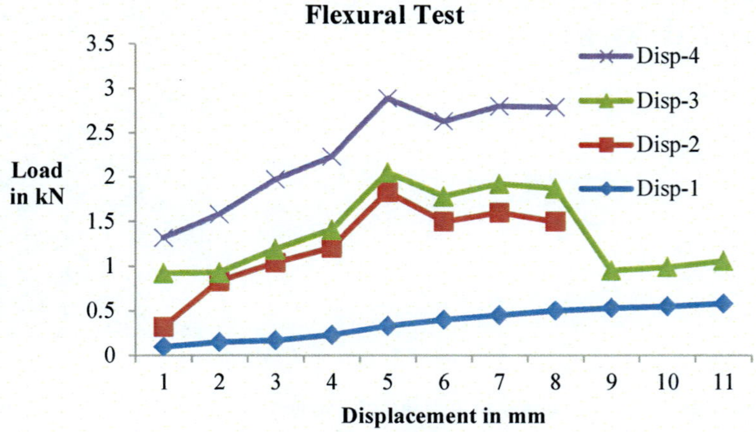

Fig. 6 Aramid fibre with nanoparticles

From Fig. 7, hardness test results indicate a noticeable difference between the material with added nanoparticles and the material without nanoparticles. When nanoparticles were introduced into the parent material, the hardness value increased compared to the material without nanoparticles. This suggests that the incorporation of nanoparticles enhanced the material's resistance to indentation or deformation.

4 Morphological Analysis Using SEM

A scanning electron microscope (SEM) is a type of electron microscope that produces high-resolution images of a sample by scanning it with a focused beam of electrons. This technique allows for the examination of the specimen's topography and morphology. When the electron beam

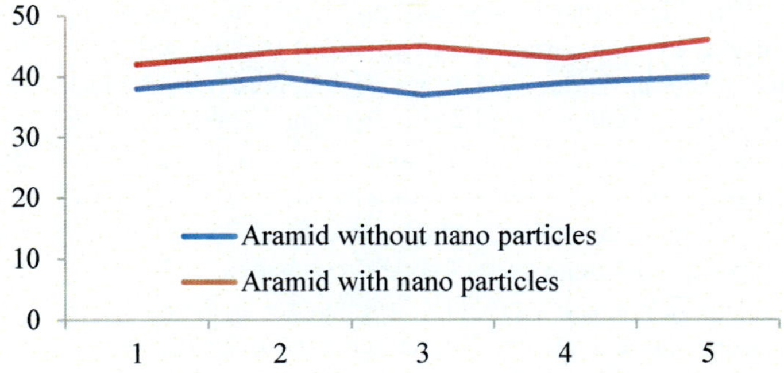

Fig. 7 Aramid particles without and with nanoparticles

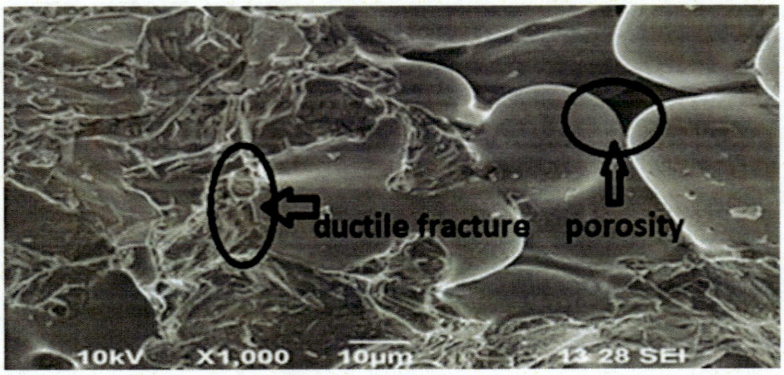

Fig. 8 Fracture surface of SEM images with nanoparticles

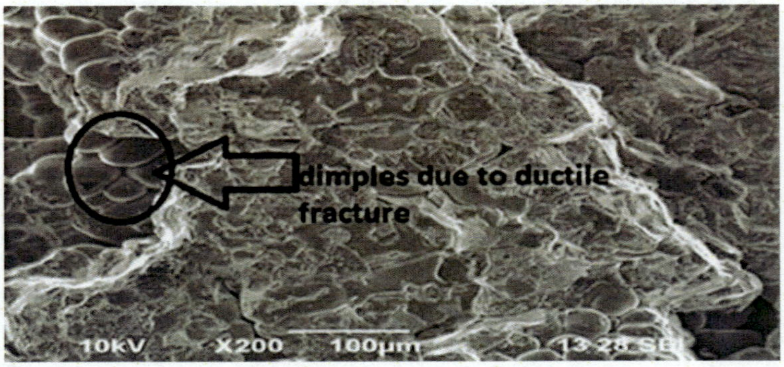

Fig. 9 Fracture Surface of SEM Images with Nanoparticles

interacts with the atoms in the sample, it generates signals that can be detected and provide valuable information about the surface's topography and composition.

SEM analysis is commonly employed to study fractured surfaces of impact test specimens. These samples are selected due to their significant variations in mechanical test results, as well as differences in composition and melting temperature. By using SEM, researchers can examine the microstructure and surface characteristics of these specimens, gaining insights into the nature of the fracture and understanding the material's response to the applied force. Fracture Surface of SEM Images with Nanoparticles is shown in Fig. 8.

The SEM images depicted in Fig. 9 offer significant insights into the particle distribution and fracture surface of the material. These images reveal the presence of black spots, representing the SiC particles, and provide valuable information about their distribution within the material. The presence of these particles indicates that the material has been reinforced with SiC, which can enhance its mechanical properties and improve its fracture resistance.

The SEM analysis provides insights into the nature of the fracture surface. If dimples are present, indicating ductile fracture, it suggests that the material can undergo plastic deformation prior to fracture, indicating high load-bearing capacity. This characteristic is advantageous for applications requiring strength and toughness. Additionally, the absence of porosity, as observed in the SEM image, signifies a sound and compact structure. Porosity can weaken the material and diminish its mechanical properties, making the lack of porosity a favourable attribute.

5 Conclusion

Numerous improvements in the material properties were found in the investigation on aramid matrix composites with and without nanoparticles. The tensile parameters of the aramid matrix composite, such as tensile strength and modulus, improved with the addition of nanoparticles. This shows that the nanoparticle addition improved the composite's total mechanical strength. The study also discovered that the composite's improved flexural properties were positively impacted by the composite's improved tensile capabilities. The aramid matrix's flexural strength and modulus gradually increased after the addition of nanoparticles. This implies that the composite displayed greater stiffness and bending resistance.

With the addition of nanoparticles, the aramid matrix composite's mechanical properties as well as hardness value improved. This suggests that the composite hardened and grew more indentation-resistant, which is advantageous for applications needing high wear resistance. Porosity and other defects were not present in the composite with nanoparticles, according to the microstructure examination. This shows that adding nanoparticles effectively filled any gaps and enhanced the composite's overall density and integrity. These results lead to the conclusion that the aramid matrix composite with nanoparticles is ideally suited for mechanical applications where enhanced hardness, high tensile and flexural strength, and a microstructure devoid of defects are needed.

Conflicts of Interest The authors declare that there are no conflicts of interest regarding the publication of this paper.

Funding Statement Authors declared that no funding was received for this Research and Publication.

Data Availability The data used to support the findings of this study are included in the article. Should further data or information be required, these are available from the corresponding author upon request.

References

Abbas, A., Al-Jeebory, A., & Al-Mosawi, A. I. (2010). Mechanical properties of araldite matrix composites reinforced with hybrid carbon - kevlar fibers. *Al-Qadisiya Journal For Engineering Sciences, 3*(1).

Akil, H. M., Omar, M. F., Mazuki, A. A. M., Safiee, S., Ishak, Z. A. M., & Abu Bakar, A. (2011). Kenaf fiber reinforced composites: a review. *Materials and Design, 32*, 4107–4121

Anuar, H., & Zuraida, A. (2011). Improvement in mechanical properties of reinforced thermoplastic elastomer composite with kenafbastfibre. *Composites: Part B, 42*, 462–465.

Ananth, R. (2011). MahanthKasavajhala, LinxiaGu, 'Fracture analysis of Kevlar-49/epoxy and e-glass/epoxy doublers for reinforcement of cracked aluminum plates.' *Composite Structures, 93*, 2090–2095.

Al-Mosawi, A,. & Ahmed, H. H. (2012). Reinforcing by Palms-Kevlar hybrid fibers and effected on mechanical properties of polymeric 159 composite material. *Journal of Babylon University/Engineering Sciences, 20*(1).

Arsath, R., Suresh Kumar, S., Prithivirajan, R., & GowriShankar, S. (2014). Dry sliding wear behavior of glass and jute fiber hybrid reinforced epoxy composites. *International Journal of Engineering Research and Development, 10*(11), 46–50.

Asumani, O. M. L., Reid, R. G., Paskara, & M. R. (2012). The effects of alkali–silane treatment on the tensile and flexural properties of short fibre non-woven kenaf reinforced polypropylene composites. *Composites: Part A, 43*, 1431–1440

Aultrin, K. S. J., & Dev Anand, M. (2016). Development of an ANN model to predict MRR and SR during AWJM operation for lead tin alloy. *Indian Journal of Science and Technology, 9*(13), 1–8

Azmir, M. A., & Ahsan, A. K. (2008). Investigation on glass/epoxy composite surfaces machined by abrasive water jet machining. *Journal of Materials Processing Technology, 198*, 122–128.

Azmir, M. A., & Ahsan, A. K. (2009). A study of abrasive water jet machining process on glass/epoxy composite laminate. *Journal of Materials Processing Technology, 209*, 6168–6173.

Azmir, M. A., & Ahsan, A. K. (2011). A study of abrasive water jet machining process on glass/epoxy composite laminate', International Journal of Engineering Science and Technology, *3*.

Bandaru, A. K., Vetiyatil, L., & Ahmad, S. (2015). The effect of hybridization on the ballistic impact behavior of hybrid composite armors. *Composites Part B, 76*, 300–319.

Boopathy, G., Gurusami, K., Chinnapandian, M., & Vijayakumar, K. R. (2022). Optimization of process parameters for injection moulding of Nylon6/SiC and Nylon6/B4C polymer matrix composites. *Fluid Dynamics & Materials Processing, 18*(2), 223–232.

Boopathy, G., Vijayakumar, K. R., & Chinnapandian, M. (2017). Fabrication and fatigue analysis of laminated composite plates. *International Journal of Mechanical Engineering and Technology, 8*(7), 388–396.

Boopathy, G., Vijayakumar, K. R., Chinnapandian, M., & Gurusami, K. (2019). Development and experimental characterization of fibre metal laminates to predict the fatigue life. *International Journal of Innovative Technology and Exploring Engineering, 8*(10), 2815–2819.

Design and Development of Multipurpose Quadcopter Using Fused Deposition Modeling

Yasin Sayyed, Nayan Patil, Nikhil Solanki, Bhargav Patil, Saurabh Kale, and Niranjan Padawale

Abstract

The manufacturing of products using rapid prototyping is centered on production technology in this era. Concept designs of drone bodies and mounting structures can be manufactured by using this rapid prototyping and 3D printing technology. In the presented paper, a bio-degradable quad-configuration drone has been designed, manufactured, and tested. The complete drone has been 3D printed with the fused deposition modeling technique. This work mainly focuses on the design and development of structurally optimized drones and object-picking mechanisms. The flexible mountings are given to the base plate of the drone to fit any type of application-based working mechanism. By performing topological optimizations, the weight of the drone body is greatly reduced due to a thirty percent mass reduction at the drone arm as compared to commercial drone bodies available on the market.

Keywords

Fused deposition modeling • Rocker • Shoe rod • Topological optimization • Mini hub wheel • Finite element analysis

Y. Sayyed (✉) · N. Patil · N. Solanki · B. Patil · S. Kale · N. Padawale
Government College of Engineering and Research, Avasari (Kh), Pune, India
e-mail: yasinsayyed78769@gmail.com

N. Patil
e-mail: patilnayan211001@gmail.com

N. Solanki
e-mail: nikhilpsolanki1@gmail.com

B. Patil
e-mail: bhargavpatil@gmail.com

N. Padawale
e-mail: ndpadawale.mech@gcoeara.ac.in

1 Introduction

Industry 4.0 has the emerging technology of rapid prototyping for the flexible production of concept products. In the last few years, the limitations of 3D printing technology have been reduced through advancements in research. Complex engineering geometries can be produced using this non-conventional manufacturing technique. Nowadays, a vast range of printing materials can be printed with high precision and accuracy.

However, designing and producing multirotor and multipurpose UAVs for better structural strength and stability is a crucial part of the research, considering the growth of the drone industry over the last decade. In this paperwork, a four-arm configuration drone called a quadcopter is designed and manufactured due to its better dynamics and control. One of the famous rapid prototyping techniques fused deposition modeling with 0.6-mm deposition of PLA Pro material. Considering the application-based concept, the three-finger claw is designed for picking objects. The drone arm has been topologically optimized for possible iterations for weight reduction and provided with a stiffener for structural rigidity and vibration reduction. The computer-integrated manufacturing cycle has been applied to the whole working process. Flight stability has been achieved by using the Pixhawk flight controller board for greater flying operations and dynamic task performance.

The research offers a roadmap for comprehending the sequential evolution of advanced drones and UAVs in terms of their dimensional structure, aerial mechanism, sensing and visual capabilities, aviation quality, path planning, intelligent behavior, and adaptability (Ahmed et al., 2022). Using the 3D printing-fused deposition modeling method, design and replication procedures are provided for polylactide (PLA)/polyhydroxyalkanoate (PHA) scaffolds with a particular porosity. This study investigated the mechanical characteristics of lattices based on triple periodic minimum surfaces. It examines TPMS-based lattices and composites'

M. Sumesh et al. (eds.), *2nd International Conference on Smart Sustainable Materials and Technologies (ICSSMT 2023)*, Advances in Science, Technology & Innovation, https://doi.org/10.1007/978-3-031-49826-8_12

uses and potential for further study (Cecen, 2023). Weeks et al., "Additive Manufacturing Drone Design Challenge." Presented the basic approach to design and implement additively manufactured small-size drones for student competition purposes (Weeks et al., 2020). Brischetto, Ciano, and Ferro represented the novel multirotor Unmanned Aerial Vehicle (UAV) in this research can alter its configuration rapidly and easily. The main component is built of a universal plate linked with a circular ring to create a rail guide capable of hosting the arms, which can be anywhere between three and eight in number, and the legs. The movable arms are equipped with all the avionic and motor drivers required to connect the main structure to each electric motor. The use of 3D printing technology to create all the structural components is another novelty. In this way, each component is created to be generated using desktop 3D printers and Fused Deposition Modelling (FDM) technology (Brischetto et al., 2016). The purpose of this study is to develop Quadcopter Spreader Drone technology as a new kind of thrower for crystal fertilizer. Putra et al. explore the variation of weight lifting force from this new type of thrower for crystal fertilizer. The method in this study using the research and development method includes five stages, namely: (a) material analysis; (b) design; (c) manufacture and assembly; (d) strength testing; and (e) test results and analysis (Putra et al., 2021). The construction of a dodecacopter system for an unmanned aerial vehicle (UAV) was the main emphasis of this study conducted by Liwauddin et al. A flight controller with Pixhawk 2.4.8 firmware is used to control the dodecacopter. A wireless communication system connects the dodecacopter and flight controller for communication. For an effective and seamless controlling system, the dodecacopter is well-equipped with the most recent technologies from Pixhawk flying controller (Liwauddin et al., 2022). In this paper, through the kinematics analysis of the quadrotor UAV, combined with PID control and simulation, finally realized the modification of the UAV's flight attitude with moving mass technology (Xu, 2489). Yu et al., a UAV flight control algorithm was proposed based on the PID control algorithm and the adaptive genetic algorithm (GA). On the basis of the traditional algorithm, the settling time was considered the influencing factor of the objective function (Yu et al., 2022). PIDs (Proportional, Integral, Derivative) are the most accurate controllers, frequently lauded for their stability, and have won broad appreciation in the industrial and academic sectors for their control applications ranging from temperature, flow, and pressure management to being the most precise controllers. This study focuses on the efficiency of PID controllers in maintaining drone stability as well as their simplicity and speed compared to controllers with intricate workings (Sree Ezhil et al., 2022). An attitude control method for unmanned aerial vehicles addresses the issues with manual rule setting and low control precision of the fuzzy PID control method (Song et al., 2022).

Abbrev. Used

- UAV—Unmanned Aerial Vehicle
- PLA Pro—Polylactic Acid Pro
- OEM—Original Equipment Manufacturer
- RPM—Revolution per minute
- BLDC motor—Brushless direct current motor

2 Comprehensive Quadcopter Design

In the UAV industry, many aviation engineers and researchers carried out the possible configuration of drone bodies and arm arrangements and positions for better flight dynamics. Some famous and well-known configurations for arm positions are shown in Fig. 1.

In Table 1, the benefits and limitations of different arm arrangements of drones have been discussed.

2.1 Parametric CAD Model

The parametric CAD model of the Quadcopter 380 body with electronic equipment is shown in Fig. 2. The whole model is designed in Solid Works licensed geometric modeling software. The drone body is ergonomically designed by mountings for OEM-based motors and other electronic modules. To protect the sensitive electronic modules during a crash landing the protective casing is provided at the top of the drone body. The quadcopter has a standard frame dimension of 380 mm.

Some of the physical parameters of the rover are shown in Table 2.

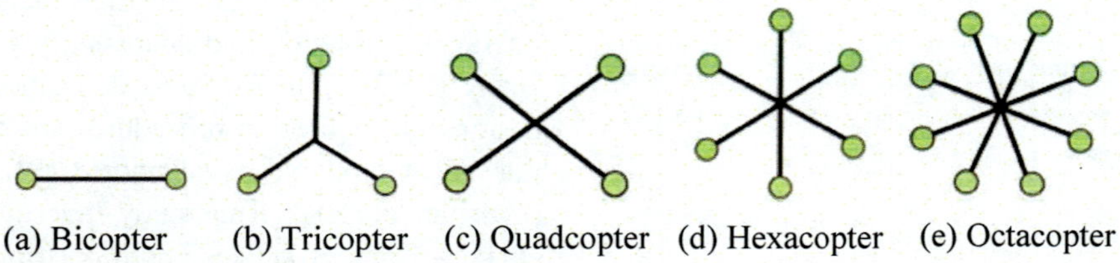

Fig. 1 Different positions in which the drone arm can be arranged

Table 1 Physical parameters

No.	Arm arrangement	Benefits	Limitations
1.	Bicopter	Maximum flight time due to two motors only	Difficult to maintain the aero stability
2.	Tricopter	Able to perform acrobatics	High cost due to advanced control algorithm
3.	Quadcopter	Excellent aero stability	Single motor failure causes instability
4.	Hexacopter	Better in agricultural applications	High cost and maintenance
5.	Octacopter	High payload lifting and capacity with industrial applications	Requires expert-level pilot

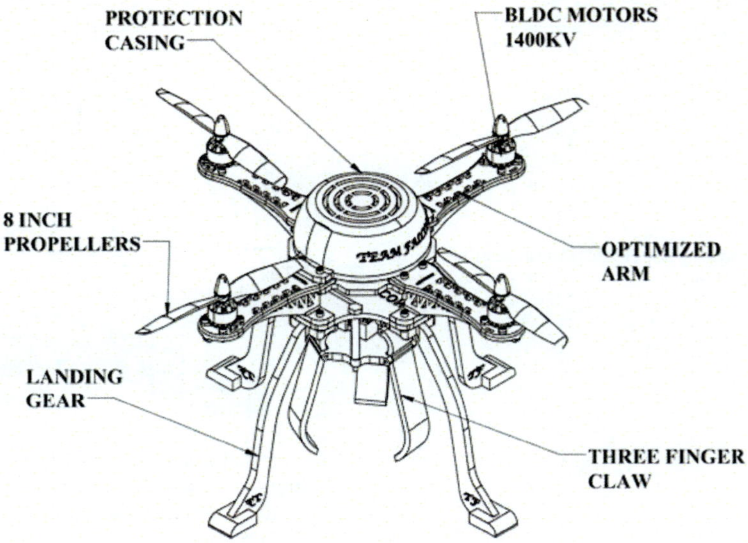

Fig. 2 Parametric CAD model of quadcopter 380

Table 2 Physical parameters

No.	Parameters	Value	Units
1.	Overall mass with payload	1.8	Kilogram
2.	Overall height	0.29	Meters
3.	Width	0.39	Meters
4.	Maximum Thrust	32	Newtons
5.	Thrust-to-weight ratio	2:1	–
6.	Flight Time	10	Minutes

2.2 Drone Arm Design

The arm has special features in terms of structural strength and vibration. Generally, in a drone body, the arm has attached to the endpoint according to the configuration which causes bending stresses and vibrations due to another end where motors are mounted hanging in space. The quad-configuration drone with four motors has been designed.

Topological optimization has been carried out in iteration to reduce the weight of arms by thirty percent. The removal of material from an arm is shown in Fig. 3.

Topological optimization is a technique to develop a structurally optimized product by taking parameters like boundary conditions, material mass, and cost-effectiveness into consideration. The creation of a design with minimum material mass and maximum stiffness is achieved. Due to the quadcopter's arm's self-weight, the weight of the mounting

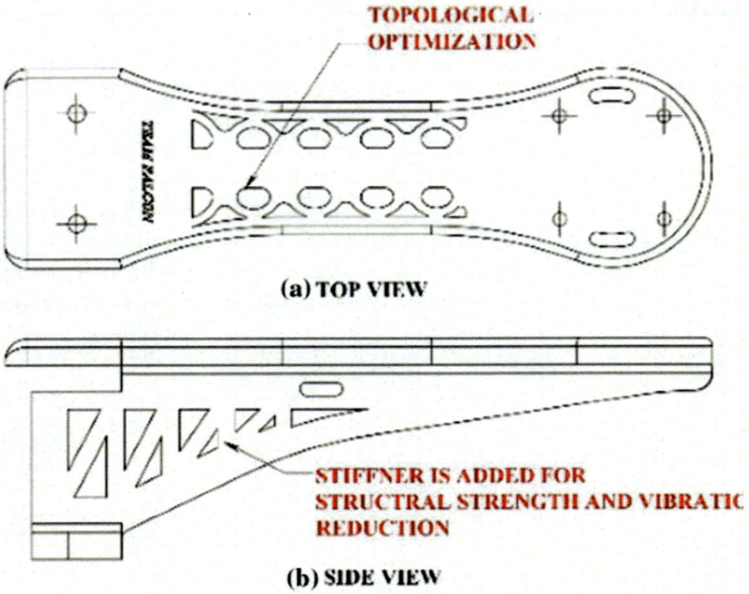

Fig. 3 Topologically optimized arm. **a** Top view **b** side view

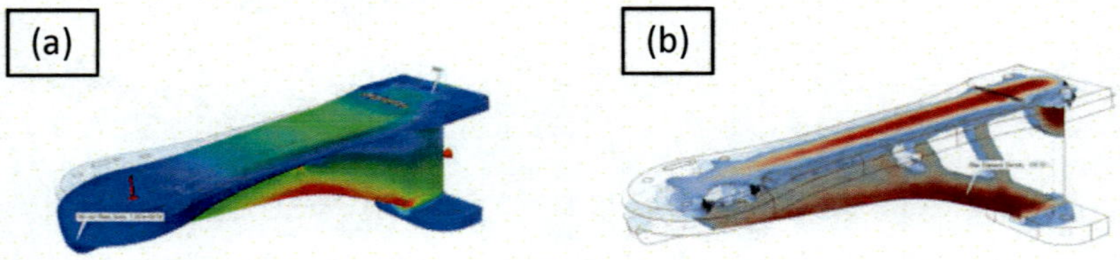

(a) Initial design under FEA for 12.5 N vertically downward force which gives maximum stress of 1.34 Mpa at arm mount location
(b) Mass reduction of 30 percent in the initial design for maximum stiffness by performing six iterations

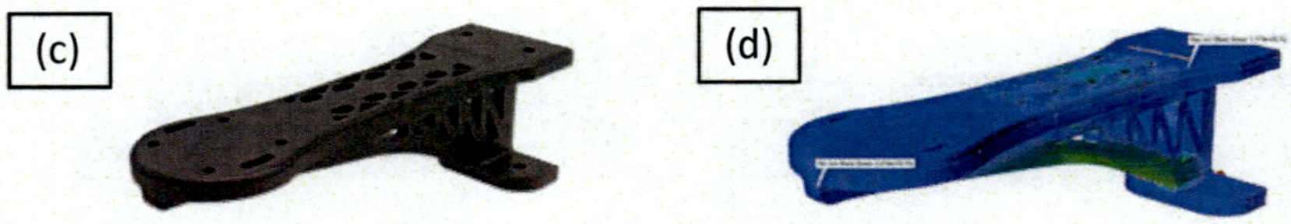

(c) Final concept based on six iterations of optimization considering the aesthetics of the drone arm
(d) Final concept design Under FEA for dame 12.5N vertically downward force which gives maximum stress of 1.36 Mpa at arm mount location

Fig. 4 Topological optimization process

structure, and the downward drag by the motor, the arm comes under stress. Sim scale and inspire, a finite element analysis software, were used to examine the identical scenario virtually. The process of analysis and optimization is shown in Fig. 4.

The PLA Pro's yield strength is around 38.7 megapascals, and the arm's maximum stress observed in the initial design was 1.34 megapascals under 12.5 N vertically downward force. After thirty percent reductions in mass by following six iterations, the final concept design was tested under the same downward force value which gives 1.36 megapascals of stress at the mount location as indicated in the figure. Overall safe design is accomplished.

2.3 Pick and Place Claw

The quadcopter is equipped with a special kind of uniquely designed three-finger pick and place claw mechanism as shown in Fig. 5. Very simple mechanism design which operates on a single mini metallic servo.

Fig. 5 Three-finger claw

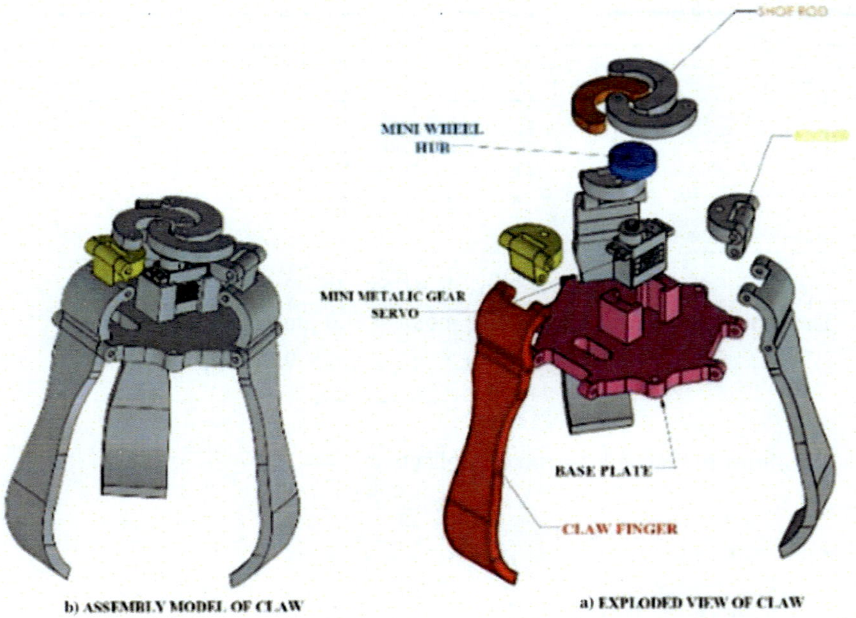

The claw has three fingers connected to the shoe rod using a rocker. Further shoe rod is connected to the mini wheel hub through which the toque produced by servo motors is transmitted to the finger of a claw.

2.4 Material Selection

The whole structural body of the drone is made up of PLA pro material. The life cycle has shown in Fig. 6.

Using the 3D printing technique and high-strength PLA pro fast prototyping material, additional mechanical subsystems and linkages are also produced. This material can be processed more easily. Through breakdown, it may become a non-toxic byproduct. The amount of plastic trash is decreased by being biocompatible and degradable.

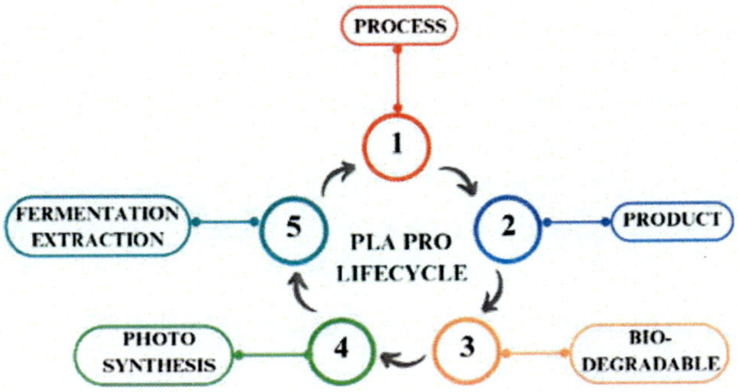

Fig. 6 PLA pro lifecycle

3 Loading Calculations and Maneuverability

3.1 Arm Loading Calculations

Drone Initial Sizing and Dimension Restrictions:

Propeller Length = 8 inches = 203.2 mm.

Diagonal frame length (*x*)/Propeller Length = 1.85 < 2 … [But near 2]

$$1.85 = x/203.2$$
$$x = 375.92$$
$$\therefore x = 380 \text{ mm}$$

By Pythagoras's theorem,

$$2(\text{Side})^2 = (380)^2$$
$$\text{Side} = 380\big/\sqrt{2}$$

According to the data listed in Table 3, a 1400 kV BLDC motor is selected.

$$\text{Weight of the drone (excluding payload)} = 1.5 \text{ kg}$$
$$\text{Maximum payload carrying capacity} = 0.3 \text{ kg}$$
$$\text{Total weight of the drone (N)} = 1.8 \times 9.81$$
$$= 17.65$$
$$= 18 \text{ N}$$

Table 3 Physical parameters

No.	KV (rpm/V)	Voltage (V)	Propeller	Lifting Capacity
1.	930	11.1	1060	660
2.	1000	11.1	1047	885
3.	1400	11.1	0845	860
4.	1800	11.1	0860	805
5.	2200	11.1	0630	239

The thrust produced by one single motor = 800 gm

$$= 0.8 \times 9.81$$
$$= 8 \text{ N}$$

The thrust produced by 4 motors collectively $= 4 \times 8$

$$= 32 \text{ N}$$

For a condition to hovel,

$$\text{Thrust/Weight} = 18/18$$
$$= 1\ldots[\text{Drone will hover}]$$

For a condition to fully controllable drone,

$$\text{Thrust/Weight} = 32/18 = 1.777 = 1.8 > 1$$

Now for analyzing the deflection in the beam under static conditions, consider the arm of the drone as a cantilever beam and carry out further calculations. Drone arm as a cantilever beam is shown in Fig. 7.

M/EI Diagram—Fig. 8.

$$\text{Moment at B} = 0.54 \times 0.00785$$
$$= -0.00042 \text{ N m}$$

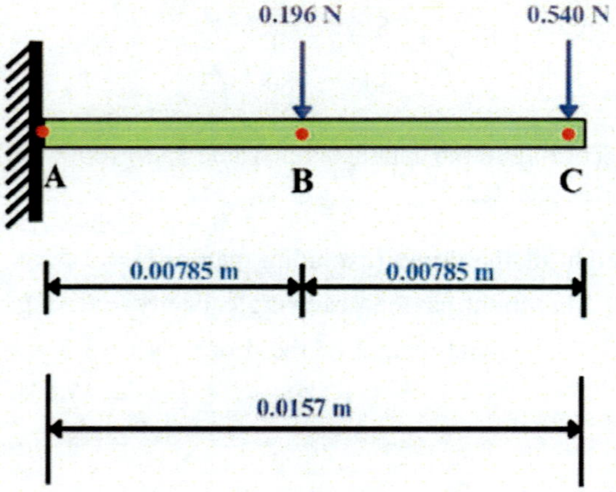

Fig. 7 Drone arm as a cantilever beam

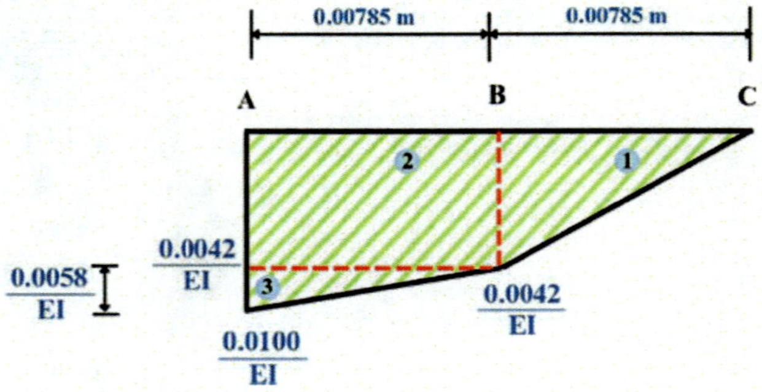

Fig. 8 (M/EI) Diagram of arm

Rhs:
Clockwise: Negative.
Anticlockwise: Positive.
Lhs:
Clockwise: Positive.
Anticlockwise: Negative.

$$\text{Moment at A} = [-(0.196 \times 0.00785)] + [-(0.54 \times 0.00785)]$$
$$= -0.0100 \text{ N}$$

Slope:

$$\theta = \text{Area of M/EI Diagram}$$
$$= \{1/2 \times (0.00785) \times (-0.0042/\text{EI})\}$$
$$+ \{0.00785 \times (-0.0043/\text{EI})\}$$
$$+ \{1/2 \times (-0.0058/\text{EI}) \times 0.00785)\}$$
$$= 4.689 \times 10 - 3 \text{ m2}$$

Deflection:

$$\Delta C = \text{Moment of M/EI Diagramabout C}$$
$$= \text{Area} \times \text{Centroidal Distance}$$
$$= \{[1/2 \times (0.00785) \times (-0.0042/\text{EI})] \times [2/3 \times 0.00785]\}$$
$$+ \{[0.00785 \times (-0.0043/\text{EI})] \times [(0.00785) + (0.00785/2)\}$$
$$+ \left\{ \begin{array}{c} [1/2 \times (-0.0058/\text{EI}) \times (0.00785)] \\ \times [(0.00785) + (2/3 \times 0.00785)] \end{array} \right\}$$
$$= 0.05 \text{ mm}$$

3.2 Dynamics of Quadcopter

Consider the force vector Fig. 9 shown,

Drone dynamics has been illustrated by Newton Euler's Equation,

$$mr = \begin{pmatrix} 0 \\ 0 \\ -mg \end{pmatrix} \begin{pmatrix} 0 \\ 0 \\ F1 + F2 + F3 + F4 \end{pmatrix}$$

$$j\begin{pmatrix} p \\ q \\ r \end{pmatrix} = \begin{pmatrix} L(F2 - F4) \\ L(F3 - F1) \\ M1 - M2 + M3 - M4 \end{pmatrix} - \begin{pmatrix} p \\ q \\ r \end{pmatrix} .1 \begin{pmatrix} p \\ q \\ r \end{pmatrix}$$

Here,

R = Rotation of thrust from B to A.
Components in the inertial frame along a1, a2 & a3.
Components in the body frame along b1, b2 & b3.
Angular velocity components in B.
Angular velocity components p, q, r

$$\begin{pmatrix} p \\ q \\ r \end{pmatrix} = \begin{pmatrix} c\theta & 0 & -\cos\varnothing \\ 0 & 1 & s\varnothing \\ s\theta & 0 & \cos\varnothing \end{pmatrix} \begin{pmatrix} \theta \\ \varnothing \\ \gamma \end{pmatrix}$$

Planar Quad Robot Model,

$$\begin{pmatrix} y \\ z \\ \varnothing \end{pmatrix} = \begin{pmatrix} 0 \\ -g \\ 0 \end{pmatrix} + \begin{pmatrix} (\frac{-1}{m}) \sin\varnothing & 0 \\ (\frac{1}{m}) \sin\varnothing & 0 \\ 0 & 1/Ixx \end{pmatrix} \begin{pmatrix} u1 \\ u2 \end{pmatrix}$$

Travel functions, Static vector,

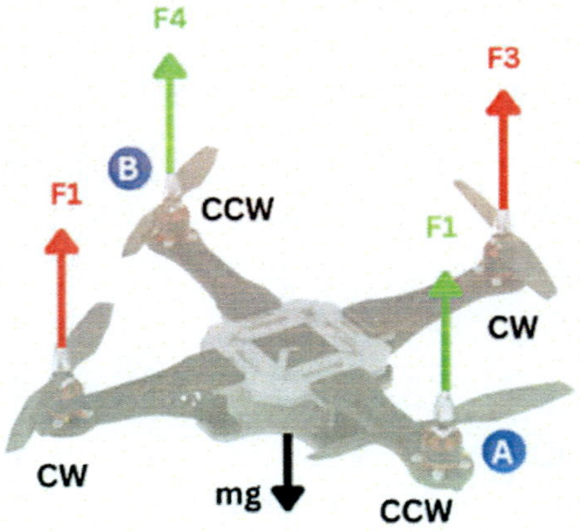

Fig. 9 Force diagram of the quadcopter

$$q = \begin{pmatrix} x \\ y \\ z \\ \theta \\ \varnothing \end{pmatrix} X = \begin{pmatrix} u \\ q \end{pmatrix} q = \begin{pmatrix} y \\ z \\ \varnothing \end{pmatrix}$$

Hovering Equilibrium State,

$$Qo = \begin{pmatrix} xo \\ yo \\ zo \\ \theta \\ \varnothing \end{pmatrix} x = \begin{pmatrix} qc \\ o \end{pmatrix} qe = \begin{pmatrix} yo \\ zo \\ o \end{pmatrix}$$

Motor speed:

$$Kf\omega 2 = (1/4)\ \text{mg}$$
$$\text{Tf} = \text{Km}\ x\omega^2$$
$$F = F1 + F2 + F3 + F4 - \text{mga3}$$

Resultant Movement

$$M = r1 \times F1 + r2 \times F2 + r3 \times F3 + r4 \times F4 + M1 + M2 + M3 + M4$$

$$\sum_{i=2}^{4} Kf\omega 2 + mg = ma$$

$$\omega = Kv \times Voltage$$
$$= 1400 \times 11.1$$
$$= 15,540$$
$$Kf = 1.19g/L$$

$$F1 = 15540 \times 1.19$$
$$F2 = -15540 \times 1.19$$
$$F3 = 15540 \times 1.19$$
$$F4 = -15540 \times 1.19$$

(Due to clockwise and counterclockwise)

$$F1 + F2 + F3 + F4 + 1120\ \text{gm} \times 9.8 = 1120\ \text{gm} \times \text{acceleration}$$

Force components developed by the motors are to be included:

910 N thrust
1400 kV of the motor with an 8 - inch propeller
450 N thrust for 50% throttle

At 50% throttle:

$$\begin{aligned} & mg + 450 \times 4 = ma \\ & 9.8 \times 1120 + 1800 + 1120 \times \text{acceleration} \\ & \text{Acceleration} = 11.407\ \text{m/s} \end{aligned}$$

At 100% throttle:

$$\begin{aligned} & mg + 910 \times 4 = 1120 \times \text{acceleration} \\ & 9.8 \times 1120 + 3640 = 1120 \times \text{acceleration} \\ & \text{Acceleration} = 13.05\ \text{m/s} \end{aligned}$$

3.3 Loading Calculations of Claw

The gripping force calculations are given as,

$$\begin{aligned} \text{Stall Torque} &= 1.8\ \text{kg} - \text{cm} = 0.1765\ \text{N} \\ \text{Operating Voltage} &= 4.8\ \text{v} \\ r = \text{Radius} &= 0.030\text{m} \end{aligned}$$

where,

$$\begin{aligned} F &= \text{Gripping force (N)} \\ m &= \text{Weight of an object (kg)} \\ g &= \text{acceleration due to gravity (m/s2)} \end{aligned}$$

From the stall Torque of the servo, we can calculate the gripping force:

$$\begin{aligned} T &= r.F \\ F &= T/r \\ F &= 0.1765/0.030 \\ F &= 5.883\ \text{N} \end{aligned}$$

Condition in which object does not drop when the object is gripped:

$$\begin{aligned} F\mu &> W \\ F &= mg/\mu \\ m &= F.\mu/g \\ m &= 5.883 \times 0.5/9.81 \\ m &= 0.299\ \text{kg} \end{aligned}$$

Maximum weight lift by the claw is 0.299 kg.

The summary of calculated data is shown in Table 4.

3.4 Quadcopter Flight Modes

The basic flight modes available in the quadcopter are given as follows,

1. Hover Mode: In this mode, the quadcopter remains in a stable position without requiring any external inputs from pilot. Hover mode can be enabled by a flight controller, which combines data from various sensors, such as accelerometers and gyroscopes to stabilize the quadcopter and maintain the position.
2. Manual Mode: In this mode, the quadcopter is directly controlled by pilot. Additionally, a remote controller is used to adjust the throttle, yaw, roll, and pitch of the quadcopter in order to maintain a precise control over its movements.
3. Altitude Hold Mode: In this mode, the quadcopter is allowed to maintain in a specific altitude without receiving any constant input from the pilot. Additionally, the flight controller uses an onboard barometer or altimeter sensor to measure the changes in air pressure and automatically adjust the throttle to hold at a desired altitude.
4. GPS Mode: In this mode, the GPS receiver of the quadcopter is utilized to enhance the flight capabilities and enable autonomous features. With the GPS mode, the quadcopter can perform different functions such as

Table 4 Physical parameters

No.	Parameters	Value
1.	Thrust-to-weight ratio	1.8:1
2.	Thrust is produced by four BLDC motors	32 N
3.	The maximum displacement of the arm due to thrust force	0.05 mm
4.	Maximum RPM of motor	15,540
5.	Maximum lifting capacity of claw	300 gm
6.	Gripping force of the claw	5.88 N
7.	Peak acceleration	13.05 m/s^2

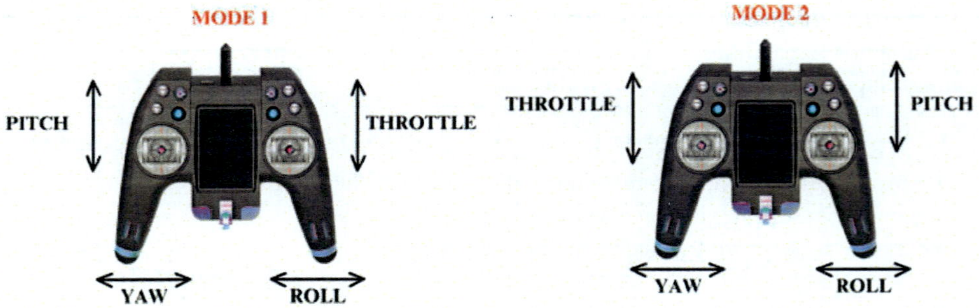

Fig. 10 Transmitter Control Configurations Modes (mode 1 and mode 2)

position hold, Return To Home (RTH), and waypoint navigation. In particular, the position hold function uses GPS data and other sensors to maintain the quadcopter in a stationary position in a specific location.

3.5 Different Transmitter Control Configurations Modes

Mode 1 and Mode 2 refers to the different transmitter control configurations to control the quadcopter's movement. These modes determine the functions assigned to the control sticks as shown in Fig. 10.

(a) Mode 1: In this mode, the control configuration has the throttle (up and down movement) and the yaw (rotation) controls on the left stick, while the pitch (forward and backward movement) and roll (sideways movement) controls are given on the right stick. This mode is widely used in countries like the United Kingdom and Australia.
(b) Mode 2: This is the most widely used configuration. In this mode, the throttle and yaw controls are given on the left stick, while the pitch and roll controls on the right stick.

3.6 Stability Equation

The Proportional Integral Derivative (PID) controller is commonly used to control the stability and performance of quadcopters. The PID control equation for a quadcopter can be expressed as follows:

$$u(t) = Kp * e(t) + Ki * \int e(t)dt + Kd * de(t)/dt$$

where:

– u(t) is the control signal or output at time t, which is applied to control the quadcopter's motors or actuators.

– Kp, Ki, and Kd are the proportional, integral, and derivative gains respectively. These gains are determined through tuning and are used to adjust the contribution of each term in the control equation.

– e(t) is the error at time t, which is the difference between the desired setpoint and the current state of the quadcopter.

– ∫e(t) dt is the integral term, which integrates the error over time and helps in reducing steady-state errors.

– de(t)/dt is the derivative term, which represents the rate of change of the error and helps in reducing overshoot and stabilizing the system.

The gains (Kp, Ki, Kd) in the PID equation need to be properly tuned to achieve stable and responsive flight performance.

3.7 Component Selection and Circuit Diagram

1. Quadcopter drone frame that was self-designed and extremely effective. Mounts for motors, batteries, and other parts make it more stable. Special safety precautions were taken for the flight controller, the material used in the 3D printing of the drone frame from PLA Pro material. Dimensions: 380 mm X 380 mm.
2. Choosing the proper propeller provides optimal stability and produces a good quantity of push. An 8-inch propeller fits it exactly based on the drone's frame's dimensions. Its pitch is 4.5 and it is 8 inches in length. 20 gm is the weight of a single propeller.
3. The 1400 kV BLDC motor used for drones has a current capacity of 12 amperes.
4. Pixhawk 2.4.8, which has the best stability and is a 32-bit flight controller good at detecting drone movement and responding quickly to commands. It can use autopilot and is lightweight.

Factors that contribute to the stability of the Pixhawk flight controller—

(a) Sensor Integration (b) Advanced Control Algorithms (c) Autopilot Capabilities (d) Firmware Updates and Community Support

5. A BLhelli 30A ESC is used, which is performance-oriented, heat-resistant, and allows for the management of the motor's voltage and amps. It is necessary to run and regulate the motors' speed.

Working to control the motors:

Motor Speed Control: ESCs are connected to each motor of the drone. They receive control signals from the flight controller or receiver, which instruct them to regulate the speed of the corresponding motor. These control signals are typically in the form of pulse width modulation (PWM) signals.

6. The circuit diagram including all the quadcopter components is shown in Fig. 11.

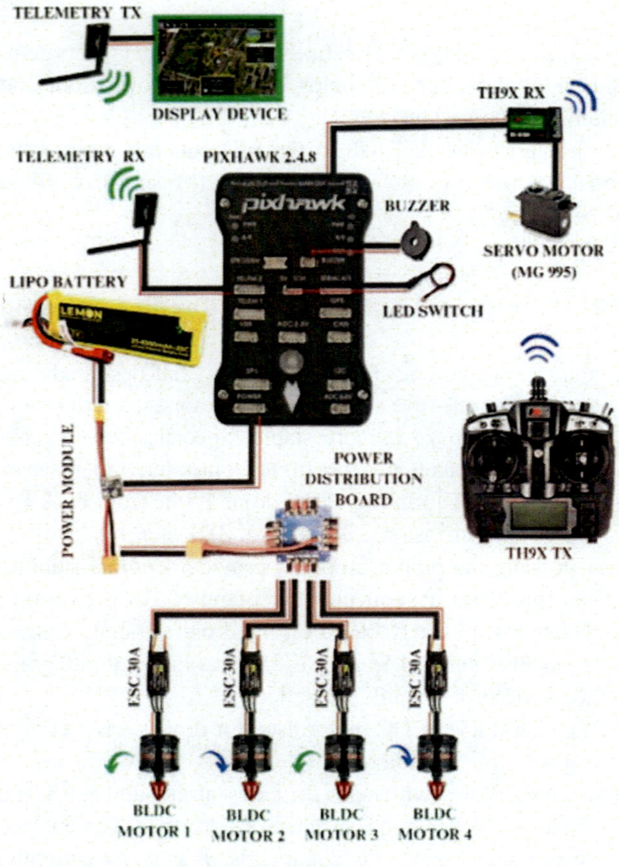

Fig. 11 Circuit diagram for quadcopter

The Pixhawk flight controller is centrally placed and is connected by the different electronic components. The four bldc motors have been connected to the respective ESCs which are further connected to the flight controller via the power distribution board. The power distribution board is powered by the Lipo battery.

3.8 Fused Deposition Modeling

The Fused Deposition Modeling starts from the software process of slicing operation. FDM takes stereolithographic CAD format or part program in the form of geometric code as input for model generation. It requires two kinds of material; one is the actual model material and another is supporting structure material. The typical construction of the FDM setup is shown in Fig. 12.

The thermoplastic material is used for the FDM process which is supplied as a filament. The material enters the extruder from one end which undergoes melting due to the heating effect caused by heating coils. Following the additive principle, the melted material from the nozzle gets deposited layer by layer on the build table to form a three-dimensional physical model.

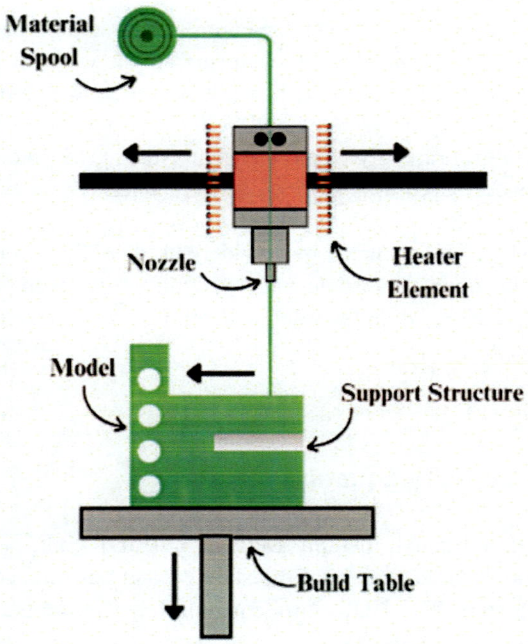

Fig. 12 Setup of fused deposition modeling

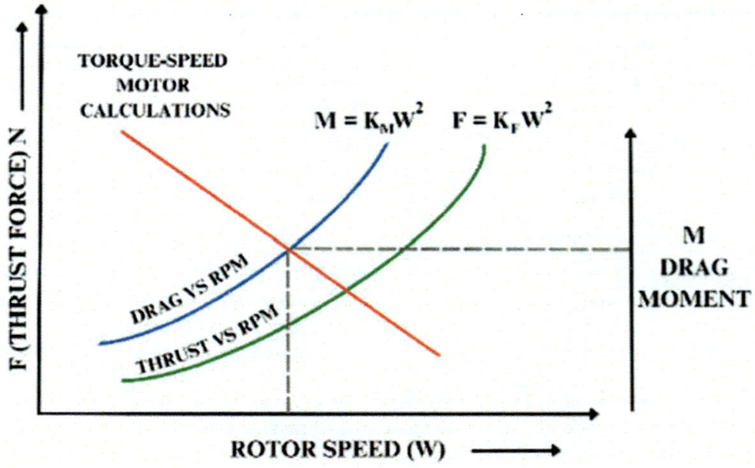

Fig. 13 Performance characteristics of the quadcopter

Applications:

1. Low Volume productions of conceptual and intricated components.
2. Tool and jigs fixtures.
3. Bionic parts, for example—prosthetic arms.

4 Results

The present quadcopter prototype can easily fly by analyzing the data gathered from calculations, design, and FEA analysis with topological optimization. The complete drone has been 3D printed with the fused deposition modeling technique and performed very well structurally and ergonomically. Through design and topology iterations the overall body weight of the drone is greatly reduced as compared to the existing commercial molded drone frames available in the market. With the new frame dimension idea, our quadcopter frame performed in flight dynamics and stability. The unique design of three-finger claw can carry small objects very easily having a mass of up to 0.3 kg. Mounting structure at the base of the drone gives provision to perform multiple tasks as per requirement.

Using the high-end flight controller for flying operations would help to maintain stability while performing various tasks. Better ergonomic design easily gives space to mount the electronic components on the quadcopter body. The performance characteristics of drones are shown in Fig. 13 which shows that with an increase in rotor speed, the thrust generation gets increased, and the same observation can be seen in the case of drag vs revolution per minute. Reduction in torque takes place as the speed of the rotor increases. The final model of the Physical quadcopter is shown in Fig. 14.

Fig. 14 Physical quadcopter model

5 Conclusion

By performing the design iterations and finite elemental analysis, the structurally stable and durable drone can be manufactured using 3D printing technology. Using some advanced 3D printing material can enhance the structural capabilities of UAVs. Flexibility in drone designs and ergonomics can open the vast scope to add more application-based mechanisms to drones.

The quadcopter which has been designed can further go for automation purposes by adding some microprocessors, sensors, cameras, and actuators to perform autonomous tasks and can solve the problem statements given by the Society of Automotive Engineers for the UAV industry.

Acknowledgements For this work, the project is provided with sponsored assets from renowned market players like MG Seating Systems Pvt Ltd for the monetary fund, 3d Spectra Tech for complete drone 3D printing using FDM, MTS Electronics for providing brushless DC motors, NXG Robotics for providing propeller blades, Robotics pare for providing electronic system controllers, AM electronics for providing servo motors, XPS India for providing fasteners. We are intrigued by the golden opportunity to officially connected with all of them. We would like to express our appreciation to our Head of Department, Dr. S. V. Kshirsagar for his valuable support to the team. Also, we are thankful to Prof. N. D. Padawale, Assistant Professor for his valuable guidance throughout the journey and thanks to Mr. Gopal Sah, Managing Director of MG Seating Systems Pvt. Ltd. for economic assistance and industrial guidance. We are appreciative to everyone who helped us in the design and development of a multipurpose quadcopter.

References

Ahmed, F., Mohanta, J. C., Keshari, A., et al. (2022). Recent advances in unmanned aerial vehicles: A review. *Arabian Journal for Science and Engineering, 47*, 7963–7984.

Brischetto, S., Ciano, A., & Ferro, C. G. A. (2016). Multipurpose modular drone with adjustable arms produced via the FDM additive manufacturing process. *Curved and Layered Structures, 3*(1).

Cecen, B. (2023). FDM-based 3D printing of PLA/PHA composite polymers. *Chemical Papers, 77*, 4379–4386.

Liwauddin, M. L., Ayob, M. A., Zakaria, M. F., Shamsudin, A. U., & Rohaziat, N. (2022). Development of a Dodecacopter using Pixhawk 2.4.8 autopilot flight controller. In: *2022 IEEE 5th International Symposium in Robotics and Manufacturing Automation (ROMA), August 6, 2022*.

Putra, A. B. N. R., Mizar, M. A., Masek, A. B., Nagari, P. M., Habibi, M. A., & Subandi, M. S. (2021). The hybrid technology of quadcopter spreader drone with crystal fertilizer as an unmanned aerial vehicles innovation. In: *2021 7th International Conference on Electrical, Electronics and Information Engineering (ICEEIE)*.

Song, H., Hu, S., Jiang, W., & Guo, Q. (2022). EKF-AF PID-based attitude control algorithm for UAVs. Edited by Liping Zhang. Mobile Information Systems 2022 (June 20, 2022), pp. 1–11

Sree Ezhil, R., Rangesh Sriram, V. B. S., Christopher Vijay, R., Yeshwant, S., Sabareesh, R. K., Dakkshesh, G., & Raffik, R. (2022). Investigation on PID controller usage on unmanned aerial vehicle for stability control. *Materials Today: Proceedings, 66*, 1313–18.

Weeks, S., Osorno, R. M., Prestwich, B., Sanford, L., & Amin, A. (2020). Additive manufacturing drone design challenge. In: *2020 intermountain engineering, technology and computing (IETC)*.

Xu, T. (2023). PID control and simulation of moving mass quadcopter UAV. *Journal of Physics: Conference Series, 2489*(1), 012015.

Yu, X., Yan, L., Guan, Z., Wu, Y., Peng, F., & Yan, F. (2022). Control of fixed-wing UAV using optimized PID controller with the adaptive genetic algorithm. In: *2022 IEEE International Conference on Real-Time Computing and Robotics*.

Thermodynamic Characterization of Building Thermal Insulating Materials Using Cardboard Wastes and Zea Mays Everta

Houda Ajabli, Amine Zoubir, Rabie Elotmani, Mohamed Louzazni, Khalid Kandoussi, and Abdelmajid Daya

Abstract

Building envelope factors have a significant impact on the amount of energy they consume. The thermal performance of the walls is a crucial factor in increasing the construction industry's energy efficiency and lowering greenhouse gas emissions. Thermal insulation is one of the most effective techniques to raise a building's energy efficiency. The major objective of this study is to establish the degree to which the building envelope must incorporate composite materials in order to maintain the optimum level of thermal comfort. In fact, the goal of this study is to evaluate the insulation's thermal performance made from waste cellulose and reinforced with popcorn (Zea mays Everta). In order to compare the thermal performances of the various materials investigated, our research will focus on experimental aspects under diverse climatic circumstances. The results show that cellulose performs effectively as a thermal insulator and that adding vegetal wastes can improve thermal properties. This work will significantly increase sustainability by recycling waste in thermal insulators.

H. Ajabli (✉) · R. Elotmani · M. Louzazni · K. Kandoussi
Chouaïb Doukkali University, National School of Applied Sciences, Science Engineer Laboratory for Energy of El Jadida, El Jadida, Morocco
e-mail: ajabli.h@ucd.ac.ma

R. Elotmani
e-mail: elotmani.r@ucd.ac.ma

M. Louzazni
e-mail: louzazni.m@ucd.ac.ma

A. Zoubir
Cadi Ayyad University, Faculty of Science and Technology of Marrakech, Marrakech, Morocco

A. Daya
Department of Physics, Laboratory M3ER, Faculty of Sciences and Techniques Errachidia, Moulay Ismail University of Meknes, 509, Meknes, Morocco
e-mail: a.daya@fste.umi.ac.ma

Keywords

Building envelope • Energy efficiency • Thermal comfort • Waste valorization

1 Introduction

Building envelope factors have a significant impact on its energy consumption. In fact, in order to ensure thermal comfort, the building needs a significant quantity of energy for cooling and heating. The proper design of the building envelope to maximize its thermal performances is a requirement of contemporary building energy efficiency strategies (Buratti et al., May 2018).

The construction industry may become more energy efficient and emit less greenhouse gases by improving the thermal performance of the walls. The creation of envelope constructions that can achieve adequate levels of thermal insulation while utilizing materials with a low environmental impact is seen to be the most efficient solution to handle such challenges (Calero et al., 2018). In this context, natural materials stand out for their particularly appealing qualities, notably in terms of thermal comfort, derived from recycled industrial and agricultural waste. Vegetable fibers are used in the construction industry because they are sustainable in terms of cost, energy, and the environment. Numerous research investigations have focused on the design and characterization of novel bio-based insulation intended to improve the thermal insulation of buildings (Ajabli et al., 2023).

Many researchers have approached the study of natural materials, especially investigating their thermal insulating properties like cork, corncob, sugarcane, straw bales, coconut, and hemp (Jelle, 2011; Thorsnes & Bishop, 2013). Paiva et al. (Paiva et al., 2012) calculated the heat conductivity of particleboards made from wood glue and corncobs, the best thermal conductivity value found was around 0.101 W/m. K.

M. Sumesh et al. (eds.), *2nd International Conference on Smart Sustainable Materials and Technologies (ICSSMT 2023)*,
Advances in Science, Technology & Innovation, https://doi.org/10.1007/978-3-031-49826-8_13

This work aims to contribute to this field of research by detailing an experimental examination of new types of ecologically friendly assemblies that contain plant matter and ecological waste, mostly acquired from recycling efforts. Samples composed of popcorn (Zea Mays Everta), corncobs, and cardboard scraps were put together in order to enhance the energy performance of the thermal insulation and provide thermal comfort for the inhabitants.

The study summarizes the findings of a series of measurements of the thermal conductivity of samples with a specific focus on the thermal properties of these novel composites.

2 Materials and Methods

2.1 Cardboard/Paper Wastes (Cellulose)

The majority of cardboard waste is composed of cellulose (Villanueva & Wenzel, 2007; Yakut, 2012). Cardboard waste can be recycled to reduce its environmental impact, proper disposal helps minimize the negative effects on the environment. The cardboard used in our study Fig. 1 originates from packaging and paper wastes.

2.2 Popcorn (Zea Mays Everta) and Corncob Wastes

A form of flint corn known as popcorn has a soft starchy interior that is encircled by an extremely hard external shell. Corncobs Fig. 2 are leftovers of corn fields and the processing industry. Several tabique buildings, (Pinto et al., 2011) which were popular in Portugal in the eighteenth and nineteenth centuries, were built using corncobs as filler.

2.3 Waste Management

Life Cycle Thinking—LCT—is a method that takes into account every stage of a process, product, or activity's life cycle, including the extraction of raw materials, manufacture, use, and disposal. At each stage, it seeks to evaluate and lessen any negative effects on the environment, society, and economy Fig. 3.

The Life Cycle Assessment—LCA—tool evaluates the emissions, resources used, and pressures on health and the environment for any product in order to quantify these effects. As a result of this information, decisions about waste management are made more sustainably and put prospective environmental benefits and drawbacks into context (Curran, 2013).

It has been demonstrated that Life Cycle Thinking—LCT—and Life Cycle Assessment—LCA—are effective tools for improving waste management strategies and making them more sustainable. Because of this, the life cycle thinking is becoming more and more important for decision-makers (Bahramian & Yetilmezsoy, 2020).

In this context, many researches focuses on the insulation and acoustic potential of this some vegetal wastes, that can serve as a solid starting point for creating more effective and secure insulation materials in comparison to other traditional materials, in order to improve the life cycle of these materials.

Our research is part of the same goal of improving the life cycle of some vegetal wastes and demonstrate its significance in reducing the energy consumption in the construction industry.

Fig. 1 Cardboard/Paper wastes

Fig. 2 Popcorn (Zea Mays Everta) and Corncob

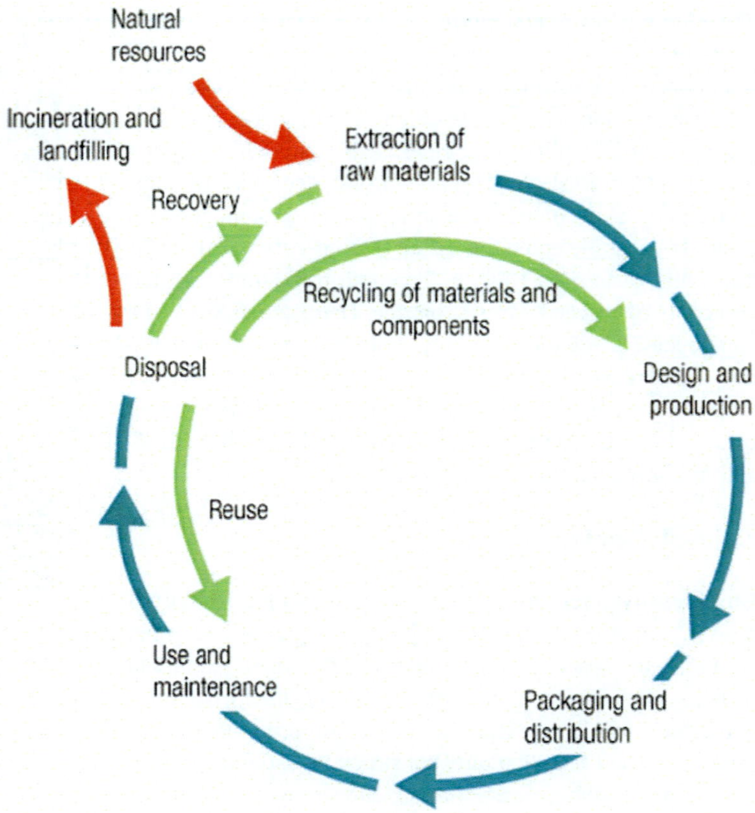

Fig. 3 A typical representation of the product lifecycle

2.4 Thermal Insulation

Thermal insulation is a method used to reduce the transfer of heat from objects and spaces. This method helps to maintain a constant temperature inside a building, conserve energy, and reduce heating and cooling costs. The materials that are commonly used to enable thermal insulation are fiberglass, foam, cellulose, and wool. These materials trap the air and slow down the transfer of heat. Thermal conductivity is a measure used to analyze a particular material's ability to conduct heat and also to quantify how quickly heat can travel through a substance.

The steady-state heat flow through a unit area of a homogeneous material that is 1 m thick caused by a 1 K temperature differential at the faces is known as thermal conductivity. It is stated in W/m K. If the thermal conductivity of a material is less than 0.07 W/m K, it is typically considered to be a good insulator (Florea & Manea, 2019).

2.5 The Experimental Protocol

A. Samples Description

The Popcorn (Zea Mays Everta) and Corncob wastes are combined with cardboard/paper in four mass fractions (10%, 20%, 30%, and 40%) to create the thermal insulation materials under study Fig. 4.

A reference sample Fig. 5 will be made entirely of cardboard/paper to demonstrate the value of cellulose as an insulator and to test the influence of adding vegetal wastes on the properties of our composites.

Fig. 4 Example of the studied sample

B. **Composites Preparation**

The corrugated cardboard collected in small pieces, was crushed and moistened for 30 min in water to have a viscous paste that will be used as a binder. This paste will be mixed with vegetable waste for 5 min to ensure a good distribution of fibers in the composite and to obtain a homogeneous paste. In a parallelepiped mold measuring 250 × 250 × 20 mm^3, this paste is cast Fig. 6. To confirm the impact of the inclusion of vegetal wastes on the characteristics of our composites, a reference sample, will be constructed using only cardboard. In line with NF P 75-101 (AFNOR NF, 1983), the samples will be air-dried for 72 h before being baked at 50 °C. Ambient air will be sampled at 20 ± 2 °C and a relative humidity of 65 ± 5%.

C. **Bulk Density**

Bulk density refers to the mass of a substance (usually a powder, granular material, or aggregate) per unit volume, including the spaces between particles. It is often expressed in units like kilograms per cubic meter (kg/m^3) or pounds per cubic foot (lb/ft^3). Bulk density is important in various industries, such as agriculture, construction, and mining, as it affects the handling, transportation, and storage of materials.

According to ISO/TS 17892-2:2004, which outlines test procedures for calculating bulk and dry material densities, the bulk densities of the materials under study were calculated (Vomocil, 1957).

The dry mass is properly weighed using an electronic scale after drying, and the dimensions are precisely measured using a calibrator.

Fig. 5 Example of the reference sample

Fig. 6 Popcorn (Zea Mays Everta) and Corncob wastes mixed with cardboard wastes

The bulk/apparent density ρ_{app} is calculated by the Eq. (1), $\mathbf{m_{app}}$ is the mass in Kg, and $\mathbf{V_{app}}$ is the sample volume in m^3.

$$\rho_{\text{app}} = \frac{m_{app}}{V_{app}} \tag{1}$$

Equation 1. Equation of bulk density.

D. **Thermal Measurement**

The high insulation housing approach's accuracy is proven by its error uncertainty of less than 10% (Benallel et al., 2021). It features four side walls that each have 210 × 210 mm^2 square hole that is covered by a 5 cm thick polystyrene plate Fig. 7.

To minimize the effects of radiation, a 100 W bulb is used as the heat source, and it is evenly distanced from each wall surface inside a little black box. The bulb is attached to a regulator that is intended to keep the temperature of the interior air at 50 °C.

Air-conditioner is generally used in a closed room in order to maintain a room temperature of about 25 °C. The sample used here has the dimensions of 250 × 250 × 30 mm^3 and is placed on one of the sidewalls. In order to prevent heat loss, the remaining side of the walls are covered with polystyrene panels. To assess the internal temperature as well as the

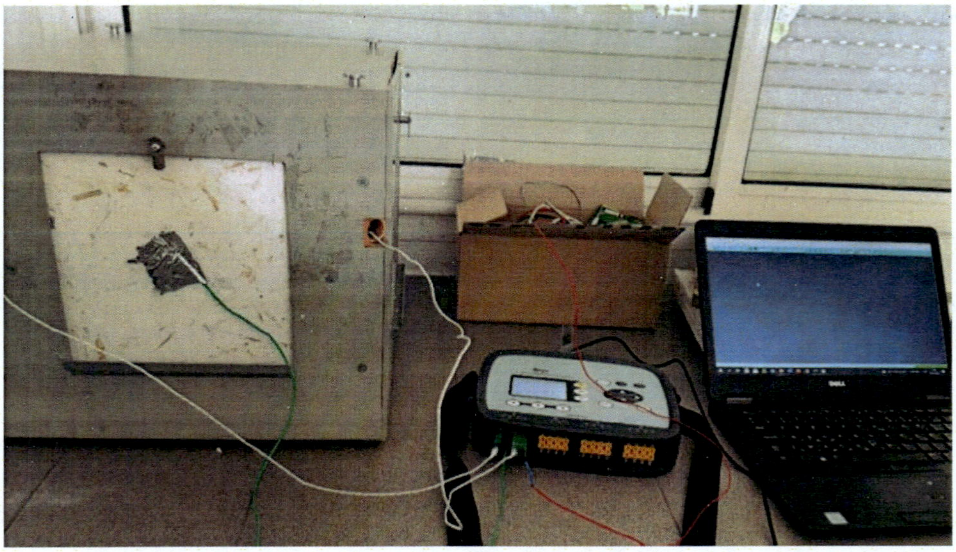

Fig. 7 High Insulation Thermal House

sample's inner and outer surface temperature, thermocouples were integrated in the holes of the corner posts.

E. **Thermal Conductivity Equation**

Equation (2) for the sample's interior wall surface and internal air temperature transfer:

$$\varphi = h_{\text{int}} \bullet S \bullet (T_1 - T_2) \quad (2)$$

Formula (3) is used to determine how much heat is transferred through the sample.

$$\varphi = \lambda \bullet S \bullet \left(\frac{T_2 - T_3}{e}\right) \quad (3)$$

Equation (4) and the principle of heat flux conservation are used to calculate the thermal conductivity:

$$\lambda = h_{\text{int}} \bullet e \bullet \left(\frac{T_1 - T_2}{T_2 - T_3}\right) \quad (4)$$

- φ: The coefficient of heat flux (W)
- λ: Thermal Conductivity (W/m K)
- **S**: Surface of Wall (m^2)
- **e**: Thickness (m)
- $\mathbf{T_1}$: The high insulation thermal house inside temperature
- $\mathbf{T_2}$: The sample's inner wall surface temperature
- $\mathbf{T_3}$: The sample's outside wall surface temperature of the sample
- $\mathbf{h_{int}}$: The coefficient of heat transfer for internal convection. When using natural ventilation in confined spaces, the equipment's manufacturer suggests using the $\mathbf{h_{int}}$ = **8.1 W/K** $\mathbf{m^2}$ value (P2360300 PHYWE Series of Publications).
- $\mathbf{T_1}$, $\mathbf{T_2}$, $\mathbf{T_3}$, and hint are represented in Fig. 8.

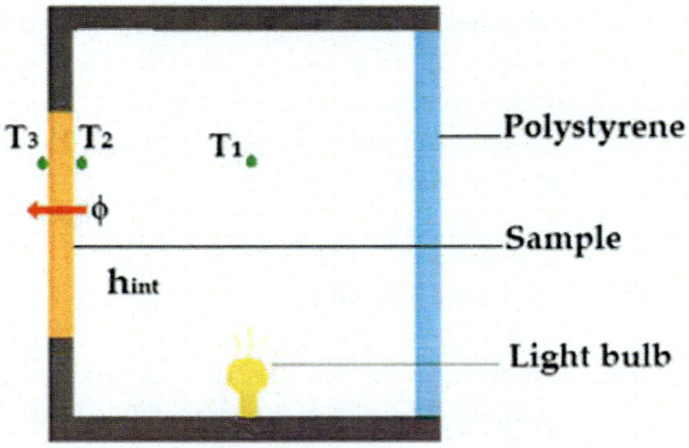

Fig. 8 Representation of the variables used to calculate thermal conductivity

3 Results and Discussion

3.1 Bulk Density

In this investigation, an electronic balance was used to weigh the sizes of the five samples and measure their masses with a high-precision caliper. Table 1. The results of Fig. 9 show that as the percentage of cardboard increases, the apparent density of the insulation materials under consideration

Table 1 The density value

Percentage of vegetal waste	Density (kg/m³)
0%	113.91
10%	165.00
20%	170.08
30%	180.22
40%	215.00

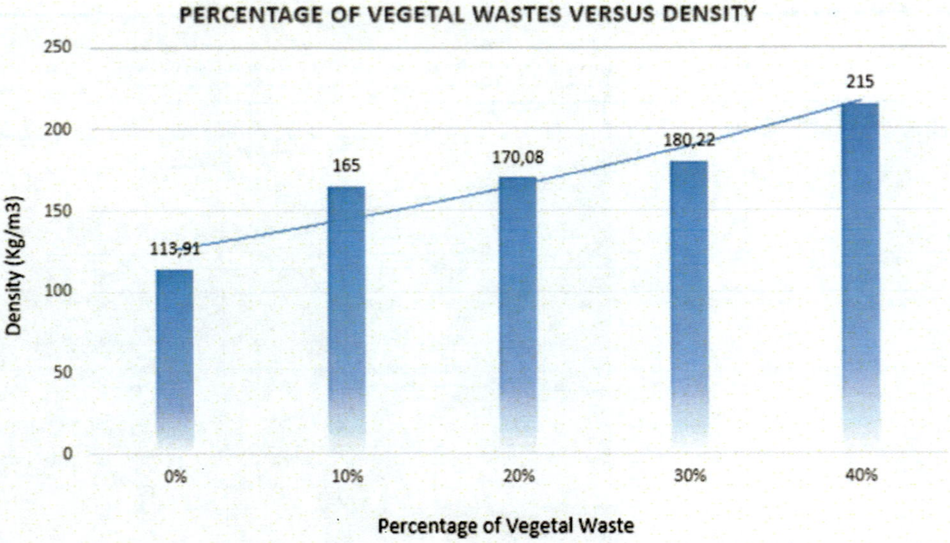

Fig. 9 Percentage of vegetal wastes versus density

decreases. The studied material made of 60% cardboard and 40% vegetal wastes has the highest density, 215 kg/m^3, whereas lowest density was around 175 kg/m^3 corresponding to the insulation panel made of 90% cardboard and 10% corncob and popcorn wastes.

Table 2 Thermal conductivity value

Percentage of vegetal waste	Thermal conductivity (W/m K)
0%	0.0495
10%	0.0680
20%	0.0780
30%	0.0970
40%	0.1230

3.2 Thermal Tests Results

The thermal conductivity of the various test samples was measured using the high insulation home method. To satisfy the requirements of the experiment, each sample was evaluated three times. Table 2's findings show that as the amount of cardboard grew, the heat conductivity decreased. This decrease is attributable to a considerable rise in the number of air-filled holes in the material, which is also a result of cardboard's poor thermal conductivity. Comparing the thermal conductivity values of the various insulator samples under investigation, we found that the sample manufactured with 90% recycled cardboard waste had the best value 0.0680 W/m K in Fig. 10.

Thermal insulation ability of corncob particleboards was characterized by Paiva et al. (2012) the findings indicate a thermal conductivity of 0.101 W/m K. The technique used to make the particleboard out of corncob is still ongoing, but the thermal conductivity value that was attained is still suitable for use in thermally insulating buildings. There are unquestionably strong grounds to think that the alternative

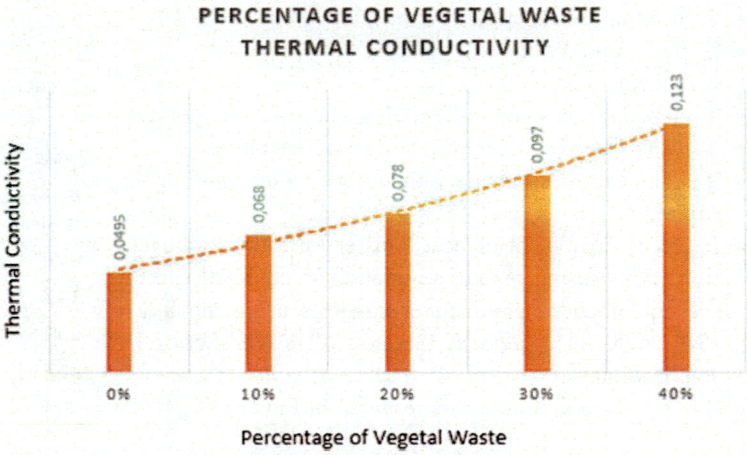

Fig. 10 Percentage of vegetal wastes versus thermal conductivity

Table 3 Density and thermal conductivity values

Percentage of vegetal waste	density (kg/m^3)	Thermal conductivity (W/m K)
0%	113.91	0.0495
10%	165.00	0.0680
20%	170.08	0.0780
30%	180.22	0.0970
40%	215.00	0.1230

product under consideration is promising from the standpoint of thermal insulation.

Given that the respective values of extruded polystyrene (XPS) and expanded polystyrene (EPS), the two most frequently used insulation materials, are roughly 0.037 W/m K and 0.038 W/m K, respectively, this vegetal waste can demonstrate promising results in terms of thermal conductivity by simply adding a percentage of cardboard wastes, as shown in our study (Asdrubali et al., 2015).

3.3 The Effect of Density on Thermal Conductivity

Because the thermal conductivity significantly decreases with apparent density, there is a strong correlation between thermal conductivity and apparent density (Chikhi et al., 2013). However, as the sample's density rises, it loses some of its insulation properties.

The sample is made lighter and more thermally insulated by the cardboard's inclusion Table 3. Due to the cardboard's inclusion, the sample is lighter and has superior thermal insulation. Moreover, the apparent density causes a significant reduction in thermal conductivity, when the sample's density increases, the thermal conductivity decreases.

4 Conclusion

The objective of this research was to evaluate the insulating qualities of green structures produced from a combination of cardboard, popcorn (Zea Mays Everta), and corncob wastes. A High Insulation House was used specifically to assess the thermal conductivity of various substances. Since the process of using corncob wastes as insulation material is still under progress, the thermal conductivity value obtained is still acceptable for thermal insulating building purposes. Although the samples have strong insulation qualities, it would also be interesting to investigate how humidity affects the thermo-physical characteristics of insulation products made of these vegetal wastes.

References

AFNOR NF P 75–101, Isolants thermiques destinés au bâtiment - Définition (October 1983).

Ajabli, H., Lairgi, L., Daya, A., Elotmani, R., Kandoussi, K., & Zoubir, A. (2023). Research on the energy signature of a residential building in an arid climate of Errachid-ia city (South Eastern Morocco). *Materials Science for Energy Technologies, 6*, 1–6. https://doi.org/10.1016/j.mset.2022.10.008

Asdrubali, F., D'Alessandro, F., & Schiavoni, S. (2015). A review of unconventional sustain-able building insulation materials. *Sustainable Materials and Technologies, 4*, 1–17. https://doi.org/10.1016/j.susmat.2015.05.002

Bahramian, M., & Yetilmezsoy, K. (2020). Life cycle assessment of the building industry: An overview of two decades of research (1995–2018). *Energy and Buildings, 219*(15), 109917. https://doi.org/10.1016/j.enbuild.2020.109917

Benallel, A., Tilioua, A., Ettakni, M., Ouakarrouch, M., Garoum, M., & Alaoui Hamdi, M. A. (2021). Design and thermophysical characterization of new thermal insula-tion panels based on cardboard waste and vegetable fibers Scanning elec-tron Microscopy. *Sustainable Energy Technologies and Assessments,, 48*(February), 101639.

Buratti, C., Belloni, E., Lascaro, E., Merli, F., & Ricciardi, P. (2018). Rice husk panels for building applications: Thermal, acoustic and environmental characterization and comparison with other innovative recycled waste materials. *Construction and Building Materials, 171*(20), 338–349. https://doi.org/10.1016/j.conbuildmat.2018.03.089

Calero, M., Alameda-Hernandez, E., Fernández-Serrano, M., Ronda, A., & Martín-Lara, M. Á. (2018). Energy consumption reduction proposals for thermal systems in residential buildings. *Energy Building, 175*, 121–130. https://doi.org/10.1016/j.enbuild.2018.07.028

Chikhi, M., Agoudjil, B., Boudenne, A., & Gherabli, A. (2013). Experimental investigation of new biocomposite with low cost for thermal insulation. *Energy and Buildings, 66*, 267–273. https://doi.org/10.1016/j.enbuild.2013.07.019

Curran, M. A. (2013). Life cycle assessment: A review of the methodology and its application to sustainability. *Current Opinion in Chemical Engineering, 2*(3), 273–277. https://doi.org/10.1016/j.coche.2013.02.002

Florea, I., & Manea, D. L. (2019). Analysis of Thermal Insulation Building Materials Based on Natural Fibers. *Procedia Manufacturing, 32*, 230–235. https://doi.org/10.1016/j.promfg.2019.02.207

Jelle, B. P. (2011). Traditional, state-of-the-art and future thermal building insulation materials and solutions–Properties, requirements and possibilities. *Energy Building, 43*(2011), 2549–2563. https://doi.org/10.1016/j.enbuild.2011.05.015

P2360300 PHYWE Series of Publications, Laboratory Experiments, Physics, PHYWE SYSTEME GMBH & Co. KG, Göttingen.

Paiva, A., Pereira, S., Ana Sá, A., Cruz, D., Varum, H., Pinto, J. (2012). A contribution to the thermal insulation performance characterization of corn cob parti-cleboards. *Energy and Buildings, 45* (2012), 274–279

Pinto, J., Paiva, A., Varum, H., Costa, A., Cruz, D., Pereira, S., Fernandes, L., Tavares, P., & Agarwal, J. (2011). Corn's cob as a potential ecological thermal insulation material. *Energy and Buildings, 43*, 1985–1990.

Thorsnes, P., & Bishop, T. (2013). The value of basic building code insulation. *Energy Economics, 37*, 68–81. https://doi.org/10.1016/j.eneco.2013.02.001

Villanueva, A., & Wenzel, H. (2007). Paper waste – Recycling, incineration or landfilling? A review of existing life cycle assessments. *Waste Management, 27*(8), S29–S46. https://doi.org/10.1016/j.wasman.2007.02.019

Vomocil, J. A. (1957). Measurement of Soil Bulk Density and Penetrability: A Review of Methods. *Advances in Agronomy, 9*, 159–175. https://doi.org/10.1016/S0065-2113(08)60112-1

Yakut, A. (2012). Investigation of the new paper production with transformable used paper. *Plant Engineering, 127*, 69–75.

Theoretical and Numerical Analysis of MHD Flows Through a Porous Material

S. Krishnakumar, K. Lakshmi Narayanan, S. Thamizh Suganya, and L. Rajendran

Abstract

This paper studies the transient MHD flow around a porous material passing through an ascending perpendicular plate. The medium appears as a grey fluid that absorbs radiation but does not scatter. This model is based on nonlinear differential equations. The most important aspect of this research is finding a closed-form solution to a nonlinear coupled partial differential equation system. Numerical techniques and homotopy perturbations are employed to solve a set of nonlinear equations. The influence of flux parameters on velocity and temperature profiles is discussed. Analytical results are confirmed only by simulation results. A satisfactory agreement is recorded.

Keywords

Mathematical modelling • Nonlinear equation • HPM • Magnetohydrodynamic flow • Porous material • Unsteady flow

1 Introduction

The inspection of the magnetic characteristics and behaviours of electrically conductive fluids is known as magnetohydrodynamics. Plasmas, liquid metals, salt water, and electrolytes are examples of magneto fluids (Hunt, 2016). MHD flows are helpful in various contexts and measurements, such as MHD power generation, geophysics, and other fields (Shankar Goud et al., 2023). MHD streams are employed in a number of industries.

One of these numerous industrial uses is explained in Moreau et al. (Moreau, 1990). Das et al. (Das et al., 1999) studied free convective flow through an infinite perpendicular plate where the temperature varies periodically. Non-Darcy transpiration in a porous multilayered medium was studied for the impact of various viscosities (Afify, 2007). Stationary convection flows over infinite vertical plates of porous materials characterized by non-stationary MHD radiation with the effects of fluctuating and oscillating surface temperatures, viscosity, and heating (Kishore et al., 2010). Reddy and co-workers (Machireddy & Polarapu, 2015) investigated the MHD flux's heat source and turbulent dissipation in a stretched sheet.

Ogulu and Makinde (Ogulu & Makinde, 2008) used the numerical method to study the continuous and instantaneous heat transfer problem in the vertical plane. The thermal stratification fluid is delivered to the absorption medium through an accelerating vertical plate (Shankar Goud et al., 2020). The phases of the MHD fluid flow have been studied by numerous scientists (Akram et al., 2022; Asogwa et al., 2022; Kanayo et al., 2022; Pramod et al., 2020; Shankar, 2020; Shankar & Dharmendar, 2021; Shankar & Nandeppanavar, 2021; Shankar et al., 2020, 2022; Shankar Goud, 2020). Raju et al. (Raju, 2016; Ramya et al., 2016; Srinivasa Raju et al., 2016a, 2016b) analyse the influences of heat diffusion in MHD flow using numerical techniques. On constant natural convection flow through an inclined wavy surface with heat absorption, Krishna et al. (Ahammad & Krishna, 2021; Krishna, 2021, 2022a, 2022b) applied the numerical technique. In Shankar et al. (Hymavathi & Shankar, 2009), the quasi-linearization problem of the heat transfer of viscous and compressive fluids on the stretched sheet is solved numerically. The issue of unstable laminar boundary-layer flow across a stretching sheet was

S. Krishnakumar · K. Lakshmi Narayanan
Department of Mathematics, Sethu Institute of Technology, Pullor, Kariapatti, 626115, India

S. Thamizh Suganya
Department of Mathematics, Vel Tech Multi Tech Dr. Rangarajan Dr. Sakunthala Engineering College, Avadi, Chennai, 631502, India

L. Rajendran (✉)
Department of Mathematics, AMET Deemed to be University, Kanathur, Chennai, 603112, India
e-mail: raj_sms@rediffmail.com

M. Sumesh et al. (eds.), *2nd International Conference on Smart Sustainable Materials and Technologies (ICSSMT 2023)*, Advances in Science, Technology & Innovation, https://doi.org/10.1007/978-3-031-49826-8_14

Table 1 Nomenclature

Symbols	Meaning
u	Non-dimensional velocity
θ	Non-dimensional temperature
Gr	Grashof number
M	Magnetic parameter
K	Porosity parameter
R	Radiation parameter
Pr	Prandtl number
S	Non-dimensional stratification parameter
Ec	Eckert number
t	Non-dimensional time
y	Axis vertical to the plate

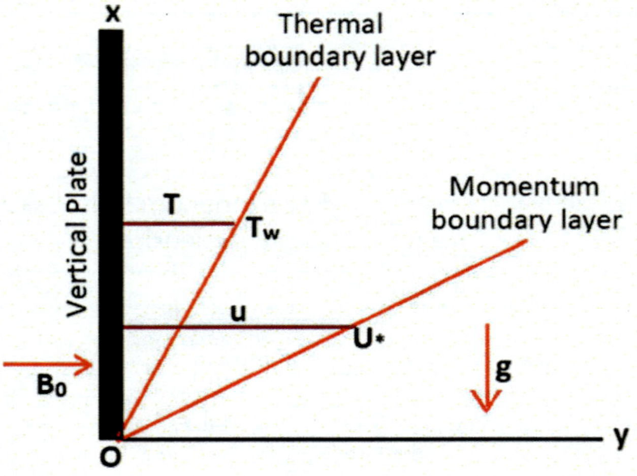

Fig. 1 Flow geometry of the problem

investigated statistically by Ibrahim et al. (Ibrahim & Shankar, 2011). Thamizh Suganya et al. (Thamizh Suganya et al., 2020) obtained an analytical result for the velocity and temperature profiles and the sensitivity analysis of kinetic variables in MHD flow.

Shankar Goud et al. (Shankar Goud et al., 2023) consider unsteady MHD flow through an accelerating perpendicular sheet around a porous material in the flow of a thermally stratified fluid with viscous dissipation. Also, Shankar Goud et al. (Shankar Goud et al., 2023) obtained the temperature and velocity by solving a nonlinear differential equation using the Crank & Nicolson method. There are no analytical representations of the velocity and temperature of the unsteady MHD flow. In this paper, the homotopy perturbation technique procures an analytical expression for velocity and temperature. The above-quoted analytical results are also validated and evaluated using numerical methods. Nomenclature is tabulated in Table 1.

2 Theory

Consider the random flow of a viscous, electrically conductive, inert fluid. The y-axis is chosen for parallel to the plate, and the x-axis is used for perpendicular. The fluid and plate are static and have a constant temperature rise or fall, and the concentration phase near the plate is constant. Figure 1 depicts the flow pattern of this problem.

The dimensionless nonlinear mass balance equations of velocity and temperature MHD flows are (Shankar Goud et al., 2023),

$$u_t(y,t) = u_{yy}(y,t) + Gr\theta(y,t) - Au(y,t) \quad (1)$$

$$B\theta_t(y,t) = \theta_{yy}(y,t) + BSu(y,t) + BEc\theta_y(y,t)^2 \quad (2)$$

where $A = M + \frac{1}{K}, \frac{1}{B} = \frac{1}{Pr}\left[1 + \frac{4}{3R}\right]$. The appropriate non-dimensional initial and boundary conditions are given as follows:

$$u(y,0) = 0, \theta(y,0) = 0. \quad (3)$$

$$u(0,t) = t, \theta(0,t) = 1. \quad (4)$$

$$u(\infty,t) \to 0, \theta(\infty,t) \to 0. \quad (5)$$

where u is non-dimensional velocity, θ is non-dimensional temperature. As in literature (Shankar Goud et al., 2023), the other symbols have their usual meanings.

3 Result and Discussion

The significant interdependence of the system variables makes nonlinear systems complex. Recently many nonlinear equations is solved using HPM (Thamizh Suganya et al., 2020). This technique combines topology's homotopy with the conventional perturbation method. This approach, which doesn't call for a small equation parameter, offers an analytical approximation to an extensive range of nonlinear problems associated with applied sciences. The basic concept of HPM is given in Thamizh Suganya et al. (2020). We obtain the analytic expression for velocity and temperature by solving the Equations (1) and (2) by using HPM and the Laplace transform method (Appendix 1) as follows:

$$
\begin{aligned}
u(y,t) = {} & \left(t+\frac{y^2}{2}\right) erfc\left(\frac{y}{2\sqrt{t}}\right) - \frac{y\sqrt{t}.e^{\frac{-y^2}{4t}}}{\sqrt{\pi}} \\
& - \frac{yA}{2\sqrt{\pi t}}\left[\left(t+\frac{y^2}{2}\right) erfc\left(\frac{y}{2\sqrt{t}}\right) - \frac{y\sqrt{t}.e^{\frac{-y^2}{4t}}}{\sqrt{\pi}}\right] \\
& - \frac{Gr}{(B-1)}\left[\left(t+\frac{y^2}{2}\right) erfc\left(\frac{y}{2\sqrt{t}}\right) - \frac{y\sqrt{t}.e^{\frac{-y^2}{4t}}}{\sqrt{\pi}}\right. \\
& \left. - \left(t+\frac{By^2}{2}\right) erfc\left(\frac{\sqrt{B}y}{2\sqrt{t}}\right) + \frac{y\sqrt{Bt}.e^{\frac{-By^2}{4t}}}{\sqrt{\pi}}\right]
\end{aligned} \tag{6}
$$

$$
\begin{aligned}
\theta(y,t) = {} & erfc\left(\frac{\sqrt{B}y}{2\sqrt{t}}\right) + \frac{B.Ec.t^2}{2(B-4)}\left[erfc\left(\frac{y}{\sqrt{t}}\right) - erfc\left(\frac{\sqrt{B}y}{2\sqrt{t}}\right)\right] \\
& + \frac{BS}{24(B-1)}\left[\left(12t^2+12ty^2+y^4\right) erfc\left(\frac{y}{2\sqrt{t}}\right)\right. \\
& - \left(12t^2+12Bty^2+B^2y^4\right) erfc\left(\frac{\sqrt{B}y}{2\sqrt{t}}\right) \\
& \left. - \frac{2y\sqrt{t}}{\sqrt{\pi}}\left(e^{\frac{-y^2}{t}}\left(10t+y^2\right) - \sqrt{B}e^{\frac{-By^2}{t}}\left(10t+By^2\right)\right)\right]
\end{aligned} \tag{7}
$$

3.1 Validation of Analytical Result with Numerical Simulation

The nonlinear differential Equations (1) and (2) can also be solved numerically using Maple software. Its numerical solution is compared to the analytical result in Figs. 2 and 3 and gives a satisfactory agreement.

3.2 Consequences of the Variables

Equations (6) and (7) denote the new approximate analytical expression of velocity and temperature. The impact of parameters such as magnetic field (*M*), heat source (*S*), Prandtl (*Pr*), permeability (*K*), Eckert (*Ec*), and Grash of (*Gr*) on velocity and temperature is discussed using Figs. 4 and 5.

The impact of various parameters on velocity is plotted in Figs. 4(a–f). The transverse magnetic field generates Lorentz forces, which cause the flow velocity to decrease as the magnetic field intensity rises. As *Gr* increases, the velocity field decreases. This is the fluid velocity profile that increases over time. Declining values signify the cooling of the plate in this instance. When values rise, the temperature drops while the fluid's velocity raises. It is the fluid's velocity profiles that enhance as time increases. Figure 4d illustrates how a drop in the velocity profile results from increased Prandtl numbers. Figure 4e shows how radiation parameters affect the velocity distribution. This is shown in the graph as the velocity curve reduces as the radiation variables rise.

Figures 5(a) and 5(b) show how *Ec* affects temperature. Friction heating results in heat production, which helps raise the wall's temperature due to *Ec*. . These data demonstrate the temperature rise caused by an increase in the Non-dimensional stratification parameter (Eckert number). Figures 5(c-e) illustrate the influence of various parameters on temperature. The Prandtl number affects the temperature distributions, as seen in Fig. 5c. The fluid temperature decreases as *Pr* rises, as shown in this graph. Kinematic viscosity and thermal diffusivity are commonly used to

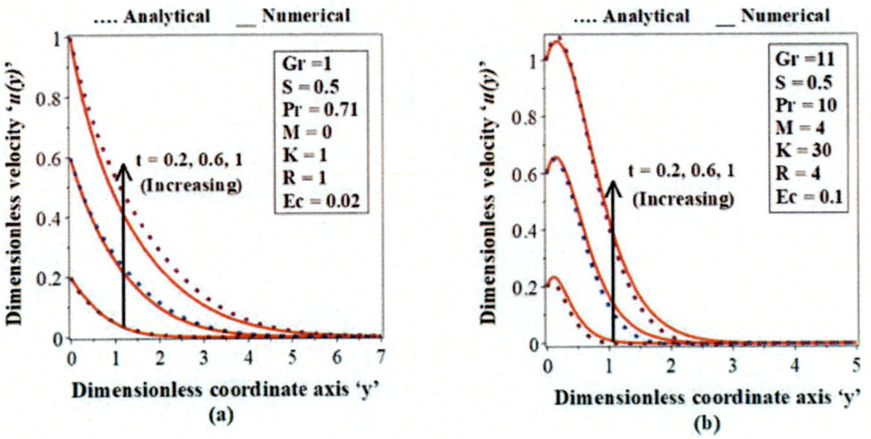

Fig. 2 Comparison of dimensionless velocity with numerical results for various values of parameters

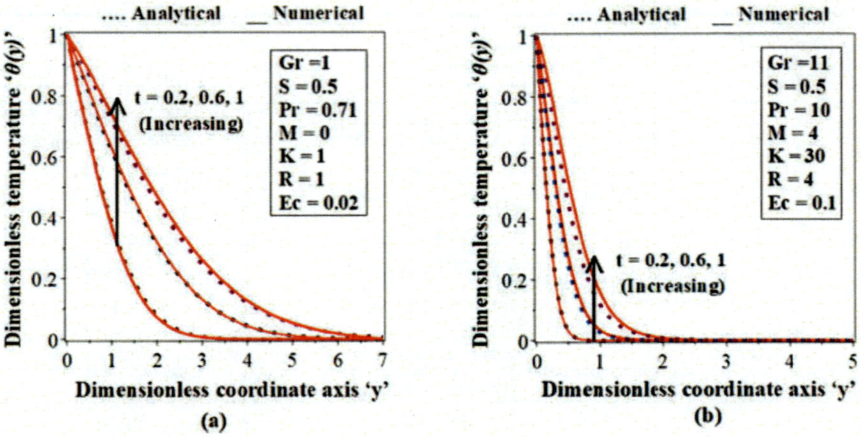

Fig. 3 Comparison of dimensionless temperature with numerical results for various values of parameters

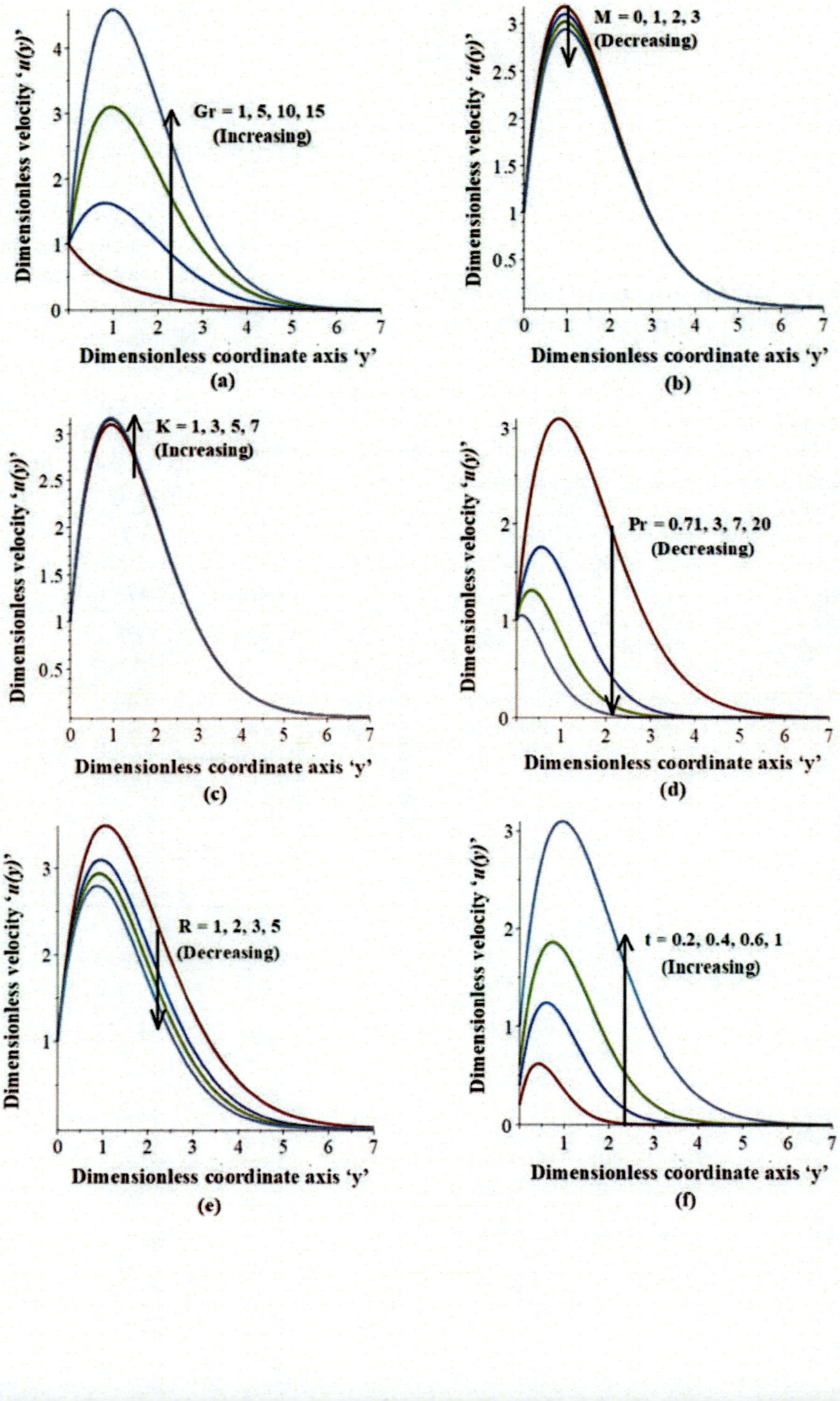

Fig. 4 Impact of the parameters on velocity. The fixed experimental values of other parameters are $Gr = 10$, $S = 0.5$, $Pr = 0.71$, $\mathrm{M} = 1$, $K = 1$, $R = 2$, $Ec = 0.02$, $t = 1$

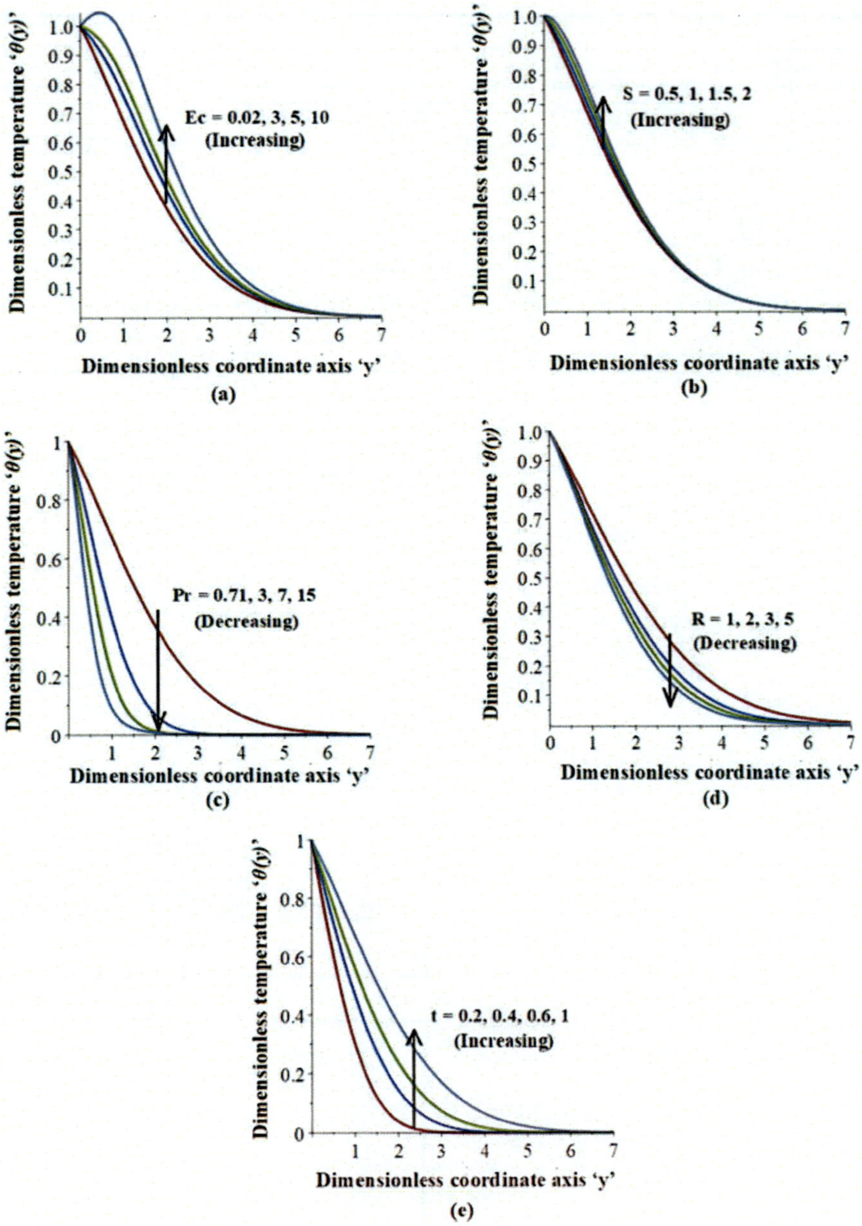

Fig. 5 Effect of parameters on temperature. The fixed experimental values of other parameters are $Gr = 10$, $S = 0.5$, $Pr = 0.71$, $M = 1$, $K = 1$, $R = 2$, $Ec = 0.02$, $t = 1$

define the *Pr* values. The effect of the radiation parameter on the temperature distribution is described in Fig. 5d. Temperature curves become flatter as the parameter R rises, as shown in the graph. From Fig. 5e, it is inferred that the velocity increases as time rises.

3.3 Sensitive Analysis of Parameters on Skin Friction and Nusselt Number

The analytical expression of the skin friction coefficient (τ) and Nusselt number (Nu) at the wall is obtained as follows:

Skin Friction Coefficient: It is practicable to choose which insert design in a vertical plate will improve heat transfer most effectively while requiring the least amount of energy to allow fluid to flow through the medium by determining the coefficient of skin friction. From Equation (6), an analytic expression for the dimensionless coefficient of skin friction (τ) is given by

$$\tau = -\left(\frac{\partial u}{\partial y}\right)_{y=0} = \frac{\sqrt{t}}{\sqrt{\pi}}\left[\frac{K(M+4)+1}{2} - \frac{2Gr\sqrt{3R+4}}{\left(\sqrt{3R+4}+\sqrt{3R.Pr}\right)}\right] \tag{8}$$

Nusselt Number: A non-dimensional heat transfer coefficient is the Nusselt number. It is employed to establish the type of heat transfer occurring through conduction or convection. Using Equation (7), the dimensionless Nusselt number (*Nu*) is analytically expressed as follows:

$$Nu = -\left(\frac{\partial\theta}{\partial y}\right)_{y=0} = \frac{3RPr}{\sqrt{t.\pi}}\left[1 - \frac{\sqrt{3R+4}.\sqrt{3R.Pr}.t^2(4S+Ec)}{18\left[R(Pr+2)+8+\sqrt{3R+4}.\sqrt{3R.Pr}\right]}\right] \quad (9)$$

Equations (8) and (9) present the Nusselt number and skin friction coefficient, respectively. The values of skin friction rise whenever M, S, Pr, R, and t increase, whereas Gr, K, and Ec numbers rise and have the opposite effect. However, the Nusselt number decreases when M and Ec values increase. The opposite result was discovered simultaneously with an enhancement of Gr, K, S, Pr, R, and t.

The influence of the kinetics parameters is summarized in the sensitivity analysis diagram shown in Fig. 5. The experimental values used in this figure are $M = 0.5, K = 0.4, Ec = 0.4, Gr = 10, \mathrm{Pr} = 1.4, S = 0.75, R = 5, \textit{and}\, t = 0.5$. Figs. 6a, b show that the time domain was found to have

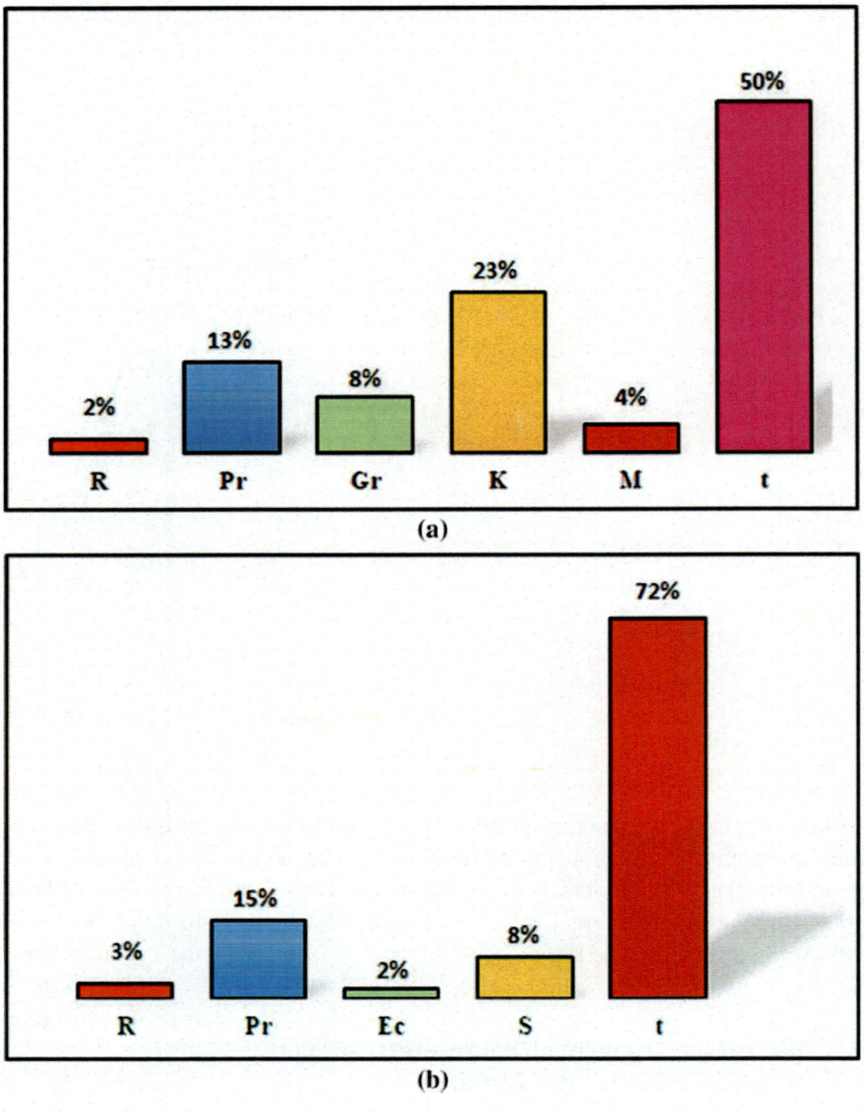

Fig. 6 **a** Sensitive analysis of dimensionless parameters on Skin friction coefficient (τ). **b** Sensitive analysis of dimensionless parameters on Nusselt number (***Nu***)

the most significant impact on the Skin friction and Nusselt number. It is shown that 't' has the most impact on the rate of change of both skin friction coefficient (τ) 50% and Nusselt number (Nu) 72%. Figure 6(a, b) displays the percentage of the impact of other parameters.

4 Conclusion

This article presents an inclined plate containing viscous and incompressible fluids in free convection and mass transfer magnetic fields. The HPM method was used to obtain theoretical results for fluid velocity, temperature, and species concentration. These mathematical formulas employ and investigate how system characteristics affect the temperature and velocity profiles. Analytical expressions of skin friction, Nusselt, exist in solid agreement with the numerical simulations created by MATLAB. It demonstrates the accuracy of the analytical solutions.

Thus, the following concluding remarks can be emphasized: Rising radiation levels, Grashof number, porosity parameter, heat generation/absorption, and time increase the fluid velocity. The increase in the radiation parameter and time during the heat absorption process causes the surface temperature of the fluid to rise. But when the Prandtl number falls, the temperature rises. This HPM method can be extended for the nonlinear problem in MHD radiating convective nanofluid flow. This technique can also applied to future technological and scientific nonlinear problems.

Appendix 1: Derivation of Analytical Expression of Velocity and Temperature Using HPM and Laplace Transform Method

This Appendix analytically solves the partial differential equations using the homotopy perturbation method and the Laplace transform. The Eqs. (1) and (2) in the Laplace plane become as follows:

$$\frac{d^2\overline{u}(y,s)}{dy^2} + Gr.\overline{\theta}(y,s) - A\overline{u}(y,s) - s\overline{u}(y,s) = 0 \quad (A1)$$

$$\frac{d^2\overline{\theta}(y,s)}{dy^2} + BS.\overline{u}(y,s) + BEc\left(\frac{d\overline{u}(y,s)}{dy}\right)^2 - B.s.\overline{\theta}(y,s) = 0 \quad (A2)$$

The corresponding boundary conditions are.

$$\begin{aligned} &\overline{u}(y,0) = 0,\ \overline{\theta}(y,0) = 0,\ \overline{u}(0,s) = \frac{1}{s^2}, \\ &\overline{\theta}(0,s) = \frac{1}{s},\ \overline{u}(\infty,s) \to 0, \overline{\theta}(\infty,s) \to 0. \end{aligned} \quad (A3)$$

Compute the homotopy for the equations (A1) and (A2) as follows:

$$\begin{aligned} &(1-p)\left[\frac{d^2\overline{u}(y,s)}{dy^2} - s.\overline{u}(y,s)\right] \\ &+p\left[\frac{d^2\overline{u}(y,s)}{dy^2} + Gr.\overline{\theta}(y,s) - A\overline{u}(y,s) - s\overline{u}(y,s)\right] = 0 \end{aligned} \quad (A4)$$

$$\begin{aligned} &(1-p)\left[\frac{d^2\overline{\theta}(y,s)}{dy^2} - B.s.\overline{\theta}(y,s)\right] \\ &+p\left[\frac{d^2\overline{\theta}(y,s)}{dy^2} + BS.\overline{u}(y,s) + BEc\left(\frac{d\overline{u}(y,s)}{dy}\right)^2 - B.s.\overline{\theta}(y,s)\right] = 0 \end{aligned} \quad (A5)$$

The approximate solution of equations (A4) and (A5) are

$$\overline{u}(y,s) = \overline{u}_0(y,s) + p\overline{u}_1(y,s) + p^2\overline{u}_2(y,s) + \ldots \quad (A6)$$

$$\overline{\theta}(y,s) = \overline{\theta}_0(y,s) + p\overline{\theta}_1(y,s) + p^2\overline{\theta}_2(y,s) + \ldots \quad (A7)$$

Applying the equations (A6) and (A7) in (A4) and (A5) and we get

$$p^0 : \frac{d^2\overline{u}_0(y,s)}{dy^2} - s.\overline{u}_0(y,s) = 0 \quad (A8)$$

$$p^1 : \frac{d^2\overline{u}_1(y,s)}{dy^2} - s\overline{u}_1(y,s) + Gr.\overline{\theta}_0(y,s) - A\overline{u}_0(y,s) = 0 \quad (A9)$$

$$p^0 : \frac{d^2\overline{\theta}_0(y,s)}{dy^2} - Bs.\overline{\theta}_0(y,s) = 0 \quad (A10)$$

$$\begin{aligned} p^1 :& \frac{d^2\overline{\theta}_1(y,s)}{dy^2} \\ &- B.s.\overline{\theta}_1(y,s) + BS.\overline{u}_0(y,s) + B.Ec\left(\frac{d\overline{u}_0(y,s)}{dy}\right)^2 \\ =& 0 \end{aligned} \quad (A11)$$

The solution of the differential equations (A8) and (A10) with corresponding boundary conditions we get,

$$\overline{u}_0(y,s) = \frac{e^{-\sqrt{s}y}}{s^2} \quad (A12)$$

$$\overline{\theta}_0(y,s) = \frac{e^{-\sqrt{Bs}y}}{s} \quad (A13)$$

Substituting equation (A12) and (A13) in equations (A9) and (A11) and solving this differential equations with applying boundary conditions, we get

$$\overline{u}_1(y,s) = \frac{Gr}{s^2(B-1)}\left(e^{-\sqrt{s}y} - e^{-\sqrt{Bs}y}\right) - \frac{yAe^{-\sqrt{s}y}}{2s^2\sqrt{s}} \quad \text{(A14)}$$

$$\overline{\theta}_1(y,s) = \frac{BS}{(B-1)}\left(\frac{e^{-\sqrt{s}y}}{s^3} - \frac{e^{-\sqrt{Bs}y}}{s^3}\right) + \frac{BEc}{(B-4)}\left(\frac{e^{-2\sqrt{s}y}}{s^4} - \frac{e^{-\sqrt{Bs}y}}{s^4}\right) \quad \text{(A15)}$$

The analytical solution of dimensionless velocity and dimensionless temperature is,

$$\overline{u}(y,s) \approx \overline{u}_0(y,s) + \overline{u}_1(y,s) \quad \text{(A16)}$$

$$\overline{\theta}(y,s) \approx \overline{\theta}_0(y,s) + \overline{\theta}_1(y,s) \quad \text{(A17)}$$

We may obtain the velocity and temperature (Eqs. (6) and (7) in the text) by applying the inverse Laplace transform in equations (A16) and (A17).

References

Afify, A. A. (2007). Effects of variable viscosity on non-darcy MHD free convection along a non-isothermal vertical surface in a thermally stratified porous medium. *Applied Mathematical Modelling, 31*(8), 1621–1634.

Ahammad, N. A., & Krishna, M. V. (2021). Numerical investigation of chemical reaction, Soret and Dufour impacts on MHD free convective gyrating flow through a vertical porous channel. *Case Studies in Thermal Engineering, 28*, 101571.

Akram, M., Jamshed, W., Shankar, B. G., et al. (2022). Irregular heat source impact on carreaunanofluid flowing via exponential expanding cylinder: a thermal case study. *Case Stud Therm Eng., 36* (102171), 1–14.

Asogwa, K. K., Shankar, B. G., & Dharmendar, Y. R. (2022). Non-Newtonian electromagnetic fluid flow through a slanted parabolic started riga surface with ramped energy. *Heat Transfer, 51*(6), 4833–6026.

Asogwa, K. K., Shankar Goud, B., Ali Shah N., & Yook, S.-J. (2022). Rheology of electron magnetohydrodynamic tangent hyperbolic nanofluid over a stretching riga surface featuring dufour effect and activation energy. *Scientific Reports,12*. https://doi.org/10.1038/s41598-022-18998-914602.

Das, U. N., Deka, R. K., & Soundalgekar, V. M. (1999). Transient free convection flow past an infinite vertical plate with periodic temperature variation. *Journal of Heat Transfer, 121*(4), 1091–1094.

Hunt, J. C. R. (2016). *Some aspects of Magnetohydrodynamics (Thesis)*. University of Cambridge. doi:https://doi.org/10.17863/CAM.14141.

Hymavathi, T., & Shankar, B. (2009). A quasilinearization approach to heat transfer in MHD visco-elastic fluid flow. *Journal of Applied Mathematics and Computtation., 215*(6), 2045–2054.

Ibrahim, W., & Shankar, B. (2011). Unsteady boundary layer flow and heat transfer due to a stretching sheet by quasilinearization Technique. *World Journal of Mechanics, 1*(6), 288–293.

Kishore, P. M., Rajesh, V., & Vijayakumar Verma, S. (2010). Thermal stratification and viscous dissipation effects on MHD unsteady radiative natural convection flow past an infinite vertical accelerated plate embedded in a porous medium with variable surface temperature. *International Journal of Applied Mathematics and Mechanics, 6*(16), 1–18.

Krishna, M. V. (2021). Thermodiffusion, chemical reaction, and hall and ion slip impacts on MHD rotating flow past an infinite vertical porous plate. *Heat Transfer., 50*(8), 8426–8452.

Krishna, M. V. (2022a). Chemical reaction, heat absorption and Newtonian heating on MHD free convective Casson hybrid nanofluids past an infinite oscillating vertical porous plate. *International Communications in Heat and Mass Transfer, 138*, 106327.

Krishna, M. V. (2022b). Numerical investigation on steady natural convective flow past a perpendicular wavy surface with heat absorption/generation. *International Communications in Heat and Mass Transfer, 139*, 106517.

Reddy Machireddy, G., Polarapu, P., & Bandari, S. (2015). Effects of viscous dissipation and heat source on unsteady MHD flow over a stretching sheet. *Ain Shams Engineering Journal, 6*(4), 1195–1201.

Moreau, R. (1990). Magnetohydrodynamics. *Kluwer Academic Publishers, 121*, 1–31.

Ogulu, A., & Makinde, O. D. (2008). Unsteady hydromagnetic free convection flow of a dissipative and radiating fluid past a vertical plate with constant heat flux. *Chemical Engineering Communications, 196*(4), 454–462.

Pramod, P. K., Shankar, B. G., & Malga, B. S. (2020). Finite Element Study of Soret Number Effects on MHD Flow of Jeffrey Fluid through a Vertical Permeable moving Plate. *Part. Diff. Eqn Appl Math., 1*, 100005.

Raju, R. S. (2016). The combined Influence of thermal diffusion and diffusion thermo on unsteady hydromagnetic free convective fluid flow past an ınfinite vertical porous plate in presence of chemical reaction. *Journal of The Institution of Engineers (India): Series C, 97*(4), 505–515.

Ramya, D., Srinivasa Raju, R., Anand Rao, J. (2016). Influence of Chemical Reaction on MHD Boundary Layer Flow of Nanofluids over a Nonlinear Stretching Sheet with Thermal Radiation. *Journal of Nanofluids*, 5(6), 880–888.

Shankar, B. G. (2020). Heat Generation/absorption influence on steady stretched permeable surface on MHD flow of a micropolar fluid through a porous medium in the presence of variable suction/injection. *International Journal of Thermofluids, 100044*, 7–8.

Shankar, B. G., & Nandeppanavar, N. M. (2021). Ohmic heating and chemical reaction effect on MHD flow of micropolar fluid past a stretching surface. *Partial Differential Equations in Applied Mathematics, 4*, 00104.

Shankar, B. G., Kumar, P. M., & Malga, B. S. (2020). Effect of heat source on an unsteady MHD free convection flow of Casson fluid past a vertical oscillating plate in a porous medium using finite element analysis. *Partial Differential Equations in Applied Mathematics, 2*, 100015.

Shankar, B. G., & Dharmendar, R. Y. (2021). Finite element Soret, Dufour effects on an unsteady MHD Heat and Mass Transfer Flow past an Accelerated Inclined Vertical Plate. *Heat Transfer, 50*(8), 8553–8578.

Shankar, B. G., Dharmendar R. Y., & Mishra, S. R. (2022). Joule heating and thermal radiation impact on MHD boundary layer nanofluid flow along an exponentially stretching surface with the thermal stratified medium. *Proceedings of the Institution of Mechanical Engineers, Part N: Journal of Nanomaterials, Nanoengineering and Nanosystems*. https://doi.org/10.1177/23977914221100961.

Shankar Goud, B. (2020). Thermal radiation influences on MHD stagnation point stream over a stretching sheet with slip boundary conditions. *International Journal of Thermofluid Science and Technology, 7*(2), 1–11.

Shankar Goud, B., Srilatha Pudhari, Ramesh Babu, K., & Indira, L. (2020). Finite element approach on MHD flow through porous media past an accelerated vertical plate in a thermally stratified Fluid. *Journal of Critical Reviews, 7*(16), 69–74.

Shankar Goud, B., Pudhari Srilatha, Mahendar, D., Srinivasulu, T., & Dharmendar Reddy, Y. (2023). Thermal radiation effect on thermostatically stratified MHD fluid flow through an accelerated vertical porous plate with viscous dissipation impact. *Partial Differential Equations in Applied Mathematics, 7*, 100488.

Srinivasa Raju, R., Jithender Reddy, G., Anand Rao, J., & Rashidi, M. M. (2016a). Thermal diffusion and diffusion thermo effects on an unsteady heat and mass transfer MHD natural convection Couette flow using FEM. *Journal of Computational Design and Engineering, 3*(3), 349–362.

Srinivasa Raju, R., Aruna, G., Swamy Naidu, N., Varma, S.V.K., & Rashidi, M.M.: Chemically reacting fluid flow ınduced by an exponentially accelerated ınfinite vertical plate in a magnetic field and variable temperature via LTT and FEM. *Theoretical and Applied Mechanics, 43* (1), 49–83 (2016).

Thamizh Suganya, S., Balaganesan, P., & Rajendran, L. (2020). Marwan Abukhaled: Analytical discussion and sensitivity analysis of parameters of magnetohydrodynamic free convective flow in an inclined plate. *European Journal of Pure and Applied Mathematics, 13*(3), 631–644.

Feasibility and Insights into the Optimization and Characterization of Friction Welded Aluminum–Steel Dissimilar Joints

S. Hari Krishna Raj, G. Boopathy, S. Vijayananth, G. Balaji, and N. Ramanan

Abstract

This research article focuses on addressing the challenges associated with joining dissimilar metals through the application of solid-state welding techniques, specifically Friction Welding (FW). The study aims to develop optimal welding conditions, tools, and parameters for achieving a successful Aluminum–Steel (Al–Fe) butt joint. The resulting weld is extensively characterized through mechanical tests, microstructure analysis, and micro hardness measurements. Additionally, finite element analysis is conducted to simulate the behaviour of the prototype engine valve. The findings provide valuable insights into the feasibility and performance of friction welding for dissimilar metal joints, contributing to the further development and understanding of this welding technique.

Keywords

Characterization of dissimilar metals • Friction welding • Finite Element Analysis

S. H. K. Raj
Department of Mechanical Engineering, Vel Tech High Tech Dr. Rangarajan Dr. Sakunthala Engineering College, Chennai, India

G. Boopathy (✉)
Department of Aeronautical Engineering, Vel Tech Rangarajan Dr.Sagunthala R&D Institute of Science and Technology, Chennai, India
e-mail: boopathyg81@gmail.com

S. Vijayananth
Department of Mechanical Engineering, Vels Institute of Science, Technology & Advanced Studies, Chennai, India

G. Balaji
Aeronautical Engineering, Hindustan Institute of Technology and Science, Chennai, India

N. Ramanan
Mechanical Engineering, Sri Jayaram Institute of Engineering and Technology, Chennai, India

1 Introduction

Welding is a wonderfully adaptable and important procedure in the world of metalworking, serving as a skilled way to combine two distinct metal components using regulated heat. Its importance is felt across a wide range of sectors, where it not only ensures the assembly of sophisticated equipment and instruments, but also contributes to the construction of a plethora of critical structures and components. Because of its economic viability and widespread acceptance, the procedure has become a cornerstone of modern fabrication techniques. It is an art that involves a variety of processes, each precisely adapted to specific needs. To accomplish the necessary level of fusion, these strategies may include the use of gas torches releasing extreme heat or the strategic placement of external electrical supply. Welding emerges as a crucial method supporting various industries, whether it's the sophisticated accuracy necessary in automobile manufacturing, the aerospace domain with its demanding standards for aircraft and machinery frames, or even the sturdy building of railway waggons and boilers.

Furthermore, the impact of welding goes beyond original production and into critical maintenance and repair operations. Industries such as furniture manufacturing, structural work, and general machine maintenance rely largely on welding expertise to ensure the lifetime and reliability of their goods and equipment. Welding also plays an important role in secondary manufacturing processes, which are intricately linked to the development of refineries, oil tankers, and pipeline networks that propel the energy sector ahead.

Welding is based on the fundamental concept of physically connecting metal surfaces. This process entails the deliberate fusion of metals through controlled melting under external pressure, resulting in the development of a cohesive binding interface. Welding is a solid-phase joining method that has travelled through the ages, with a historical ancestry

M. Sumesh et al. (eds.), *2nd International Conference on Smart Sustainable Materials and Technologies (ICSSMT 2023)*,
Advances in Science, Technology & Innovation, https://doi.org/10.1007/978-3-031-49826-8_15

dating back around 3500 years to the rudimentary hammer welding procedures used by early blacksmiths.

Over ages, the art of welding has grown into a domain of varied approaches, each customized to unique purposes and limits. Brazing is a major contender among these techniques, relying on the idea of capillary action to suck molten filler material into the interstices of closely positioned metal surfaces, so cementing them into a cohesive whole. Meanwhile, friction welding is an innovation that capitalizes on the formation of heat by controlled mechanical friction, allowing metals to come together in a solid-phase state. Furthermore, reaching temperatures above 450 °C opens up the world of metal melting welding procedures, typified by operations involving controlled liquefaction followed by resolidification.

In place of traditional welding, a new technology known as soldering emerges, offering a parallel approach to material coalescence. Soldering makes use of a specialized alloy known as solder, which has a significantly lower melting point, frequently falling below the 450 °C mark. Unlike welding, which involves the metals themselves becoming molten, soldering works on the idea of aiding bonding through the liquefaction of the solder material. As the solder is carefully applied to the juncture during the soldering process, an intricate link is formed without the joined metals completely liquefying. This nuanced approach is especially important in situations where melting the base metals is either unwanted or impractical. Thus, soldering shows to be a flexible process, with applications spanning from electronics assembly to jewellery manufacturing.

The classification of weld techniques is inextricably linked to the intricacies of the joining process, which include aspects such as pressure application and the integration of filler materials. These parameters, taken together, determine how the connection is created and the resulting joint strength. Welding procedures range from those in which pressure is a critical component to those in which the addition of extra materials serves to reinforce and enhance the integrity of the connection.

Metal joining operations play a pivotal role across various sectors, demanding meticulous attention to material selection, techniques, safety protocols, and stringent quality standards (Li et al., 2016). In response to these challenges, the concept of solid-state welding emerges as a groundbreaking solution. It circumvents the need for complete material liquefaction by applying controlled pressure, inducing frictional heat, and meticulously maintaining temperatures beneath the melting threshold of the constituent materials. This innovative methodology serves as a direct response to the shortcomings inherent in fusion welding (Cai et al., 2019). When confronted with the task of amalgamating materials possessing disparate properties, the conventional method of fusion welding faces inherent limitations due to the hindrance imposed by their varying melting points. Consequently, this process may culminate in the formation of brittle intermetallic compounds (Mehta, 2019) The inherent adaptability of friction welding renders it an exceptionally promising solution across a myriad of industrial domains. Particularly in scenarios characterized by mass production necessitating metal joining, this technique shines as a beacon of innovation. Its unparalleled capability to forge steadfast connections between diverse materials, all while circumventing the limitations inherent in traditional fusion welding methods, underscores its transformative potential within the realm of modern industrial practices [4]. Friction welding (FW) exemplifies an exemplary paradigm of solid-state joining. This technique facilitates the fusion of materials by orchestrating rotational or translational motion between the workpieces through the application of precisely controlled compressive stresses. The ensuing kinetic energy generates heat, propelling plastic deformation and culminating in the establishment of a robust welding interface. Astonishingly, the efficacy of FW is achieved without the necessity for supplementary filler metals, flux, or shielding gases, thus amplifying its appeal and practicality (Maalekian, 2007).

The detailed study on joining of AA6061 (aluminum alloy) and AISI 4340 (low alloy steel) using continuous drive friction welding. Through metallographic examination, X-ray diffraction, electron probe microanalysis, tensile testing, and microhardness measurements, they found that dissimilar metal welds with a silver interlayer exhibited a maximum tensile strength of 240 MPa and 4.9% elongation. The presence of silver as an interlayer reduced the magnesium concentration at the weld interface by intermittently replacing it with silicon on the AA6061 side, limiting the interaction between iron and aluminum. The improved strength and ductility of these dissimilar metal welds were attributed to the formation of ductile phases such as Ag_3Fe_2, Ag_2Al, and Ag_3Al (Suresh & Meshram, 2015). Study on the continuous drive friction welding method used to solid-state join different materials, specifically AA6082 aluminium alloy and AISI 304 stainless steel. This process combines plastic deformation with heat produced by friction between two surfaces. Tensile testing, impact testing, Vickers micro hardness testing, fatigue testing, microstructural observation, and EDS measurements were among the tests and measurements used in the research of the welding process. The findings demonstrated that the austenitic stainless steel and aluminium parts' tensile strengths were greater than those of their respective base metals. As the friction time increased, the joint strength initially rose and subsequently fell after reaching a maximum value. The spectrum study of the alloying elements showed significant weight differences. The Steel-Al joints contained intermetallic compounds, according to EDX studies. In comparison to base metals, the hardness was increased close to the weld contact (Mohan & Gopi, 2017).

When endeavouring to unite steels with other materials through fusion welding processes, it becomes evident that unanticipated complexities can emerge. These challenges encompass a spectrum of potential issues, ranging from the propagation of unforeseen phases and vulnerability to grain boundary corrosion to the prospective emergence of delta and sigma ferrite phases at the weld interface. Particularly germane to scenarios where steel and aluminum constituents interplay within rotating systems and structural frameworks, the imperative to cultivate reliable, efficient, and economically viable joining methodologies is underscored (Chander et al., 2012). Against this backdrop of exigencies, the imposition of deliberate preventive measures assumes a paramount role. A notable tactical approach involves the judicious implementation of targeted heat treatments or the strategic adoption of accelerated welding speeds. These interventions wield the potential to markedly enhance the uniformity intrinsic to the welding interface. By adroitly choreographing the thermal kinetics and dynamics governing the welding process, a heightened degree of coherence and consistency can be achieved at the weld contact. This, in turn, substantiates the overarching structural soundness and operational efficacy of the assembled components (Padhy et al., 2015).

The experimental results indicated that the tensile strength of the welded material increased with an increase in upsetting force and heating time. The tensile strength also showed an initial increase with friction force up to a certain extent, but started decreasing when higher burn-off length, welding temperature, and flash were present. Optimal tensile strength was achieved at a friction force of $5kN/mm^2$ and a heating time of 12 s. The variation of these parameters also had a relative effect on the hardness of the weld fusion zone and heat-affected zone (HAZ). In particular, specimen number 5 exhibited a hardness value of 85 and 88 in the fusion zone and HAZ, respectively, while specimen number 3 had a hardness value of 90 and 92 in the fusion zone and HAZ, respectively. These findings demonstrate that, for maximum weld strength, it is desirable to have minimum hardness values in the fusion zone and HAZ (Sriram Ravi Ramados, 2015).

Researchers focused on welding Austenitic stainless steel 304 and Aluminium alloy 6082 T6. Tensile tests were conducted on the welded samples, and the welding parameters were optimized using a design of experiments approach. The study aimed to analyse the effects of these parameters on the tensile strength of the joints in Aluminium alloy-Stainless steel welds. Experiments were carried out using various combinations of friction pressure, forging pressure, friction time, and forging time. The tensile strength of the welded joints was then analysed. The results revealed that increasing the friction time and forging pressures led to an increase in the tensile strength. However, beyond a certain point, further increases in these parameters resulted in a decrease in strength. Conversely, decreasing the friction pressure and forging time led to a decrease in the tensile strength. However, beyond a certain threshold, further decreases in these parameters resulted in an increase in strength. Microstructure analysis played a vital role in understanding the observed variations in tensile strength. Differences in grain structure were identified, which correlated with the higher and lower tensile strengths observed in the welded joints (Deepak Kumar & Venkatakrishnan, 2014).

In order to comprehend the effects, it was necessary to compare and contrast the microstructure, tensile strength, micro hardness, and FESEM-EDS results of welds with and without a nickel interlayer in the study on properties of friction welded joints between austenitic stainless steel 304 and medium carbon steel AISI 1040. The findings lead to a number of important conclusions. It was discovered that the forging pressure directly affected the ultimate tensile strength of welds without an interlayer, with the greatest strength of 636 MPa being attained at the highest forging pressure of 1.884 tonnes. The greatest strength of 661 MPa was nonetheless attained at the highest burn-off length of 8 mm when a nickel interlayer was added. When welding was done without an interlayer, peak hardness at the interface was 454 HV, but microhardness data showed a drop, registering 391 HV. This decline can be linked to nickel's presence at the contact, which inhibits the precipitation of chromium carbide (James & Sudhish, 1040). Authors focused on the evaluation of Flash Butt Welded Aluminium Copper transition Joints. They found that flash butt welded joints between aluminum–copper (Al–Cu) strips exhibited sufficient strength. However, the joint strength deteriorated significantly when subjected to heat treatment at 673 K for 20 h. This indicates that these joints may perform well for electric applications. Additionally, controlled bend tests revealed that the joints exhibited adequate ductility (Kumar et al., 2021).

The corrosion rate of chalcolithic copper, as evaluated by Tafel extrapolation technique in 3.5% NaCl solution, was only slightly greater than that of modern copper, according to authors who concentrated on the material characterization of ancient Indian copper. The existence of second phase sulphide inclusions has been suggested as the cause of the greater rate of corrosion in archaeological Cu (Srivastava & Balasubramaniam, 2003). Effects of intermetallic compounds on the mechanical and electrical characteristics of friction welded Cu/Al bimetallic joints during annealing have been the subject of research. As a result, the consequences of these qualities were examined when annealing. As a result, AlCu and $AlCu_2$ were found where intermetallic layers were being formed at joints (Lee & Bang, 2005). Friction welding of steel to steel, steel to aluminium, and aluminium to aluminium. Welding process temperature rise and transient heat generation were modelled. Over the weld zone, tensile testing and microhardness measurements were made. SEM was used to evaluate HAZ. The yield, tensile, and breaking strengths as well as the width of the heat-affected zone on the Al side are considerably impacted by the interaction of weld parameters for Al-steel welds, according to the results (Boopathy et al., 2019; Sahin et al., 1996).

The tensile strength and fatigue strength of commercial/tough pitch copper friction-weld joints have been examined using deformation heat input. The relationships between upset burn-off length, joint strength, and deformation heat input at the upset stage were studied for friction-weld joints made of 1050 pure aluminium and C110 tough pitch copper. Joint quality was evaluated using tensile and fatigue strength. Joint efficiencies of sound joints may be attained when the deformation heat input during the upset stage or the burn-off length exceeded a predetermined value (Boopathy et al., 2017).

2 Friction Welding Process

In the friction welding process, one workpiece remains motionless while the other rotates within the chuck of a specialized friction welding machine. As the two workpieces are squeezed together and forced to interlace, friction generates heat. The energy generated by frictional rubbing causes a significant temperature increase within the rubbing cross section of the workpieces, which is crucial for accomplishing successful forging during welding. When the rubbing cross section reaches the required forging temperature, the process undergoes a key change. The rotation of the workpieces is abruptly stopped, and the axial pressure exerted between them is increased at the same time. This critical phase incorporates a forging action into the equation, which culminates in the welding of the workpieces.

This forging method encourages molecular mixing and diffusion, resulting in a strong and seamless welded junction between the two workpieces. The specialized equipment used for friction welding resembles a lathe, which is a machine used to shape materials, but it is intrinsically more durable and engineered to withstand the extreme demands of the friction welding process. The capabilities of this apparatus are defined by two essential characteristics. To begin, it must be capable of achieving a high spindle speed, often exceeding 12,000 rpm, in order to promote the quick frictional creation of heat. Second, the equipment must be extremely durable in order to bear the significant axial pressure, which can reach 50,000 N/cm^2.

3 Pull Load Test

Tension testing, often known as pull load testing, is a vital cornerstone in the field of materials science. This critical method includes subjecting a specimen to regulated tensile forces until it fails. Figure 1 meticulously depicts the specimens designated for the pull load test, providing a palpable view into the experimental setup. The information acquired from these tension tests is crucial, fulfilling three functions. Primarily, it aids in the careful selection of materials for specific purposes, influencing design choices in a variety of industries. Furthermore, the results contribute to a strict quality control regimen, ensuring that materials adhere to set standards and tolerances. Furthermore, the test results play an important part in unravelling the intricate behaviour of materials when subjected to a variety of different stresses, hence improving our understanding of material reaction under various mechanical conditions.

The ultimate pull load strength is a vital signal, directly ascertainable through the pull load test, among the multitude of mechanical qualities determined through tension testing.

Fig. 1 Dissimilar weld joints

Table 1 Testing machine specification

S. no.	Parameter	Value
1	Load range	5 Tonne Max
2	Pull load specimen dimension	L = 100 mm, D = 20 mm
3	Bend test specimen Dimension	L = 150 mm, W = 16 mm, T = 20 mm
4	Izod test specimen dimension	
5	Make	Associated Scientific Engg. Works, New Delhi
6	Software details	FIE make India

This property contains the maximum magnitude of tensile force that a material can withstand before failing, providing critical insights into material robustness and longevity. Furthermore, the tension test reveals a slew of additional material properties. The maximal elongation, a measure of how far a material can stretch before breaking, is important in applications that require durability and flexibility. The computed area reduction reveals the degree of localized deformation, offering light on material ductility and formability.

4 Testing Specification

Pull load experiments are conducted using a Universal Testing Machine; Table 1 lists the equipment and specimen parameters. The variations for stress are compiled in Table 2.

The Universal Testing Machine (UTM) has established as the standard apparatus for performing Pull Load Testing, playing an important role in material testing and research. These machines, known for their versatility and widespread usage, include two separate types of crossheads that work together to simplify the complex testing procedure. One of these crossheads is meticulously calibrated to suit the test specimen's length, while the other works in unison to apply regulated strain on the specimen. The precision and accuracy of the testing method are supported by the coordinated interaction of crossheads. Within the field of Universal Testing Machines, two prominent varieties stand out: hydraulic and electromagnetic machines. Both of these iterations use their own systems to generate the forces required for pull load tests. The hydraulic-powered iteration uses the hydraulic principle to create tension, demonstrating its ability to administer forces with remarkable fidelity. The electromagnetic-powered equivalent, on the other hand, harnesses electromagnetic force creation, providing an alternate pathway for applying tension with exact control and efficiency.

In the context of these devices, a critical factor governs their testing suitability: capability. To arrange the pull load testing of a wide range of specimens, these devices must excel in four key areas: force capacity, speed, precision, and accuracy. A UTM's force capacity incorporates its ability to produce sufficient force to cause specimen breakdown, which is critical for replicating real-world circumstances and revealing material behaviour under stress. The force capacity, like the capability of a powerful engine, accelerates the UTM towards accurate replication of a specimen's response to applied forces (Ochi et al., 2003).

The machine's ultimate goal is to take accurate measurements of gauge lengths and applied forces. Precise positioning of the test specimen within the apparatus is critical to avoid unwanted bending forces generated by misalignment, which is especially important for brittle materials. This is mitigated by using spherical seats or U-joints between the grips and the machine. When the beginning segment of the stress–strain curve deviates from linearity, this is an indication of specimen misplacement. Extensometers are usually used to measure strains, although strain gauges can also be used to measure Poisson's ratio or microscopic specimens. Digital time, force, and elongation measuring systems are used in modern testing equipment, with electronic sensors attached to data acquisition instruments, which are often computers, and software for processing and outputting results. Notably, analog machines,

Table 2 Pull load test results

Stress, kN/mm^2	Strain
106	6.2
102	6.3
101	6.5
101	6.6
104	6.7
103	6.9

adhering to ASTM, NIST, and ASM metal pull load testing standards, persist due to their capability to surpass these benchmarks, substantiating their ongoing relevance in the field.

5 Micro Hardness Test

Utilized for assessing weld hardness, a Micro Hardness Tester assumes a pivotal role in this endeavour. The determination of hardness entails two distinct approaches: longitudinal assessment, encompassing hardness along the weld's length, and transverse measurement, spanning hardness across the weld's width. In meticulous detail, hardness values were meticulously recorded from the inception to the culmination of the weld. Moreover, the weld's hardness was gauged through a linear traverse parallel to the welding direction. The evaluation of relative hardness, synonymous with hardness joint efficiency, hinged on a comparative analysis between the weld's hardness values and those characteristic of its parent metal. In strict adherence to ASTM guidelines, the test specimens were meticulously fashioned, mirroring the protocol illustrated in Fig. 2.

Hardness Test Result for Different Zones is shown in Table 3.

6 Morphological and Simulation Analysis

6.1 SEM Analysis

Scanning Electron Microscopy (SEM) has emerged as a critical tool in the field of solid inorganic material characterization, allowing for precise microanalysis and failure studies. SEM reveals a treasure trove of insights into the microcosmic complexity of materials, proving helpful in unravelling the mysteries of material behaviour. High magnification imaging, an avenue via which elaborate landscapes of minute features and objects are methodically revealed, is at the forefront of SEM's capabilities. SEM, which operates in the nanoscale domain, provides high-resolution pictures that serve as windows into the hidden regions of material structures. Furthermore, this approach isn't limited to just visual documentation; it also includes quantitative assessments that allow for the precise measurement of even the most insignificant qualities.

The coordination of a focused stream of high-energy electrons is the underlying premise of SEM. This electron beam, like a virtuoso's baton, choreographs an elaborate dance across the surface of solid objects, producing a symphony of interactions that produce a variety of signals. These signals are expertly manipulated to extract a lot of information. They provide information on topography, elemental composition, and even crystallographic orientation. Such diverse studies are a veritable toolkit for investigating the inner topography of materials. In practice, SEM is frequently used to scan across a specific section of a sample's surface. The data gathered from this spatial excursion is then skillfully put together to create a two-dimensional image. This image is a visual tapestry rich in detail, depicting differences in features that cover a wide range of aspects, from the specific fingerprint of chemical composition to the texture that characterizes the material's individuality.

To investigate the phases that form the welding interface, scanning electron microscopy and energy dispersive analysis of X-rays (EDAX) analyses were used. An energy dispersive analysis of X-rays (EDAX) study in conjunction with a 20 kV field effect scanning electron microscope was used to record the observations. The programme allows for beam piloting to scan along a surface or a line to produce X-ray mapping or element-specific concentration profiles. The interface zone in the friction welded joint between EN24 STEEL and AL6061 has a SEM microstructure.

The four layers' average grain sizes are as follows: BM > HAZ > TMAZ > NZ. BM (Base Metal, consisting of AL6061 and EN24 steel), TMAZ (Thermo mechanical

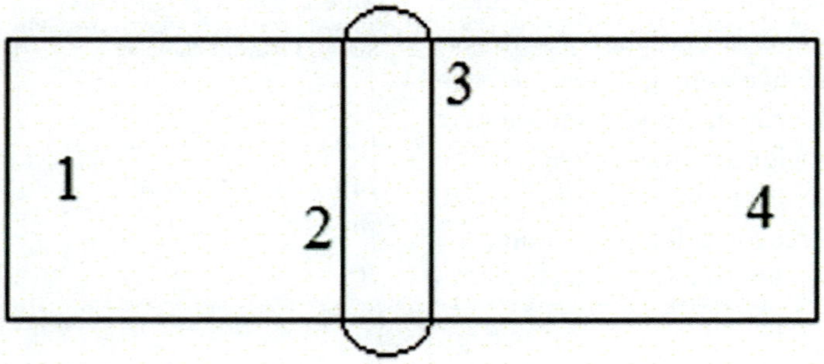

Fig. 2 Micro hardness test setup. 1. Zone of parent material Al.alloy 6061. 3. Zone of parent material EN24. 2. Zone of Al.alloy 6061 near the weld. 4. Zone of EN24 Steel near the weld

Table 3 Hardness test result for different zones

Sample no.	Zone-1	Zone-2	Zone-3	Zone-4
1	95.8	93.8	85.2	91.1
2	58.8	62.2	63.3	65.5
3	91.2	93.3	94.2	95.7
4	37.8	35.5	36.6	37.4
5	56.6	57.8	55.5	58.2
6	53.3	54.2	52.2	51.5

impacted zone), HAZ (Heat affected zone), and NZ (Nugget zone) are the four main FW zones that can be seen from Fig. 3. The base metal exhibits an agglomerate-like microstructure with lengthy grain sizes. Due to the homogenous nature of the entire region and the dispersion of such grains' crystal orientations, the tensile strength is affected by small grain size. The TMAZ area experiences plastic deformation, but because the deformation strain was insufficient, recrystallization did not occur there. The aluminium alloy underwent a heat cycle in HAZ, although minor plastic deformation happened. Thus, the tensile strength was reduced as grain size increased. The dislocation movement in the material will have an impact on the tensile strength as the grain size decreases. Due to the metal's plastic hardness to flex, minimizing grain dislocation movement (fine grain size) can increase the material's mechanical strength.

6.2 Simulation Analysis

The interaction (coupling) between two or more different types of phenomena (fields) is taken into account in a coupled-field analysis. (Boopathy et al., 2021) The fields involved in these assessments may be coupled directly or indirectly. ANSYS components that can be used in direct-coupled-field analyses are listed below and the captured image of stress variations depicted in Fig. 4 and Fig. 5 respectively. Thermal DOF is not present in all elements. An analysis that is directly connected computes DOF from numerous fields at once. The Direct Method, as it is known, is only required when the model's individual field responses depend on one another. Since the equilibrium condition must be met according to numerous criteria, directly coupled analyses are typically nonlinear. Matrix equations are larger and more expensive to analyse for multi-field models of comparable size because there are more DOF active per node.

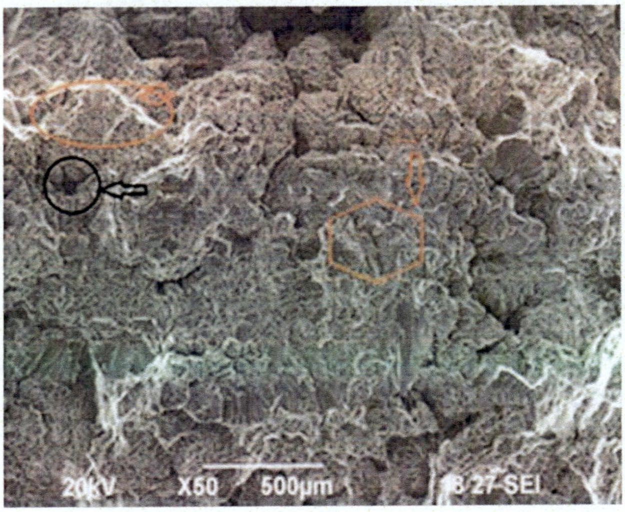

Fig. 3 Weld zone fracture surfaces

7 Conclusion

The difficulties of combining different metals, notably Aluminum and Steel, have been solved in this study article through the use of solid-state welding techniques like Friction Welding. The best welding setups, equipment, and settings have been created via considerable experimentation and analysis in order to produce good butt joins between aluminum and steel. Through mechanical tests, microstructure analysis, and measurements of the welds' hardness at the microscopic level, the finished joints have been fully characterized. The results of this investigation offer important new information about friction welding's performance and viability for joining dissimilar metals. The study emphasizes the significance of welding process optimization for producing durable junctions between steel and aluminum. Furthermore, the finite element study performed on the prototype engine valve has given researchers a better knowledge of how the weld will behave and perform under actual operating circumstances. The insights gathered from this research can direct future developments in welding technology and make it easier to successfully combine metals that are different from one another, creating new opportunities for a variety of industries and applications.

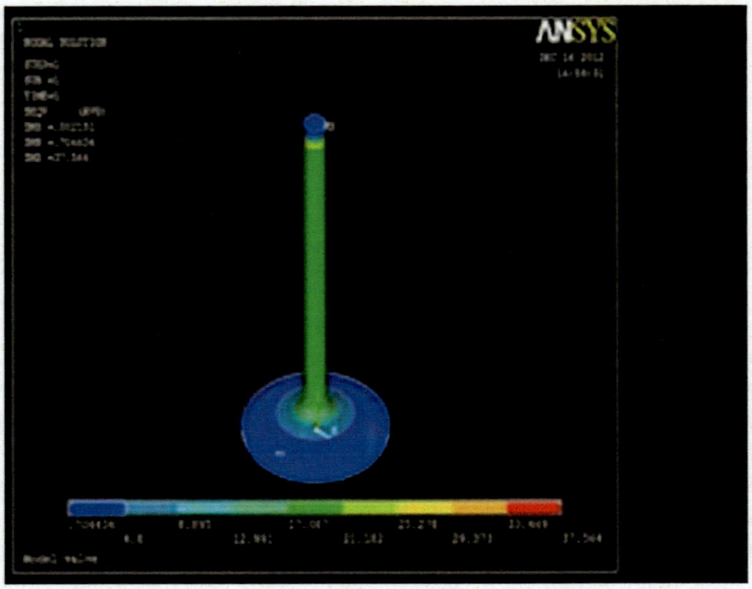

Fig. 4 Stress distribution

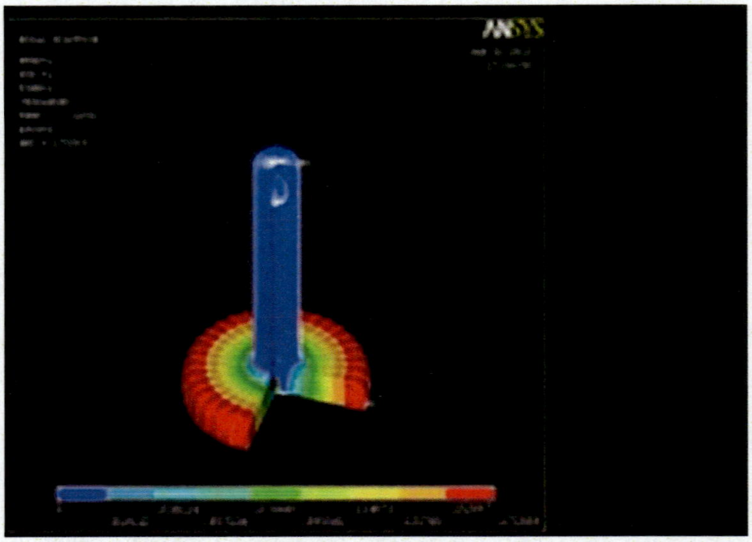

Fig. 5 Strain variations

References

Ananda Rao, G., & Ramanaiah, N. (2019). Dissimilar metals AISI 304 steel and AA 2219 aluminium alloy joining by friction welding method. *Mater Today Proc, 19*, 902–907.

Boopathy, G., Vijayakumar, K. R., & Chinnapandian, M. (2017). Fabrication and fatigue analysis of laminated composite plates. *International Journal of Mechanical Engineering and Technology, 8*(7), 388–396.

Boopathy, G., Vijayakumar, K. R., Chinnapandian, M., & Gurusami, K. (2019). Development and experimental characterization of fibre metal laminates to predict the fatigue life. *International Journal of Innovative Technology and Exploring Engineering, 8*(10), 2815–2819.

Boopathy, G., Ramanan, N., & Gopi, E. (2021). Design and analysis of a scooping engine valve. *International Journal of Ambient Energy, 42*(4), 383–388.

Cai, W., Daehn, G., Vivek, A., Li, J., Khan, H., Mishra, R. S., & Komarasamy, M. (2019). A state-of-the-art review on solid-state metal joining. *J Manuf Sci Eng Trans ASME, 141*, 1–35.

Chander, G. S., Reddy, G. M., & Rao, A. V. (2012). Influence of rotational speed on microstructure and mechanical properties of dissimilar metal AISI 304-AISI 4140 continuous drive friction welds. *Journal of Iron and Steel Research International, 19*, 64–73.

Deepak Kumar, M., & Venkatakrishnan, P. G. (2014). Optimization on friction welding of aluminium alloy 6082 T6 and austenitic stainless steel 304, 2(1), ISSN: 2321-9939.

James, J. A., & Sudhish, R. (2016). Study on Effect of Interlayer in Friction Welding forDissimilar Steels: SS 304 and AISI 1040. *Procedia Technology,* 1191–1198

Kumar, J. P., Raj, A., Arul, K., & Mohanavel, V. (2021). A literature review on friction stir welding of dissimilar materials. *Materials Today: Proceedings, 47*, 286–291.

Lee, W.-B., & Bang, K.-S. (2005). 'Effects of Intermetallic compound on the electrical and mechanical properties of friction welded Cu/Al bimetallic joints during annealing. *Journals of Alloys and Compounds, 390*, 212–219.

Li, W., Vairis, A., Preuss, M., & Ma, T. (2016). Linear and rotary friction welding review. *International Materials Reviews, 61*, 71–100.

Maalekian, M. (2007). Friction welding - critical assessment of literature. *Science and Technology of Welding and Joining, 12*, 738–759.

Mehta, K. P. (2019). A review on friction-based joining of dissimilar aluminum-steel joints. *Journal of Materials Research, 34*, 78–96.

Mohan, D. G., & Gopi, S. (2017). Study on the mechanical behaviour of friction stir welded aluminium alloys 6061 with 5052. In: *2017 8th Annual Industrial Automation and Electromechanical Engineering Conference (IEMECON)* (pp. 147–152). IEEE

Ochi, H., Ogawa, K., Yamamoto, Y., Kawai, G., & Sawai, T. (2003). Formation of intermetallic compounds in friction- welded joints of aluminium alloys to copper and their effect on joint efficiency. *Welding International, 21*(3), 381–388.

Padhy, G. K., Wu, C. S., & Gao, S. (2015). Auxiliary energy assisted friction stir welding – Status review. *Science and Technology of Welding and Joining, 20*, 631–649.

Sahin, A. Z., Yilbas, B. S., & Al-Garni, A. Z. (1996). Friction welding of Al-Al, Al-steel, and steel-steel samples. *Journal of Materials Engineering and Performance, 5*(1), 89–99.

Sriram Ravi Ramados, S. (2015). Study on mechanical metallurgical properties of friction weld dismillar material, 3(11).

Srivastava, & Balasubramaniam, R. (2003). Material characterization of ancient Indian copper. *Bulletin of Materials Science, 26*(6), 593–600

Suresh, D., & Meshram, G. (2015). Madhusudhan Reddy, Friction welding of AA6061 to AISI 4340 using silver interlayer. *Defence Technology, 11*, 292–298.

Friction Stir Welded Magnesium AZ31B Alloy Used to Evaluate Mechanical Properties at Various Rotational Speeds

Bhiksha Gugulothu, P. S. Satheesh Kumar, N. Srinivasa Rao, S. Vijayakumar, D. R. Rajkumar, T. V. Janardhana Rao, and M. Naga Swapna Sri

Abstract

In this research work, Tensile, hardness, % elongation, density and micro structure analysis on Friction Stir Welded (FSW) magnesium alloy (Mg AZ31B) specimen was examined as per ASTM E8/E8M11 and ASTM E3/E112 at various rotation speeds between 900 and 1440 rpm, with fixed 40 mm/min at 2° High Speed Steel tool tilt angle. The metal matrix plate was produced by stir casting technique. At 1120 rpm, the ultimate tensile strength of the fabricated specimen is higher than those at 900 and 1140 rpm. But higher hardness showed at 1440 rpm than at 900 and 1120 rpm. At 900 rpm, the density of the stir zone joint is lower than that of the other joints. From this comparative analysis, it is evident that the value of FSW specimen was dependent on involvement of heat and material flow rate, which are achieved by process parameters.

Keywords

Friction stir welding • Rotational speed • Tilt angle • Welding speed and Magnesium alloys

1 Introduction

FSW is a revolutionary manufacturing technique that presents many benefits than traditional welding methods. These benefits include great mechanical properties, low residual stresses, and defects. The rotation and translation of the tool together lead to movement of solid from forward-facing to the rear pin (Vijayakumar et al., 2022). Kumar et al. (2008) studied two distinct modes of the flow of material in friction stir welding on two different modes which are pin and shoulder flow. The geometric of pin-driven flow which combined with the vertical measure of material due to shoulder interaction, results in the creation of onion rings. Ugendera et al. (2014) concluded that rotational speed of 1120 rpm formed superior mechanical properties compared with other speeds. The five different tool pin profiles with 3 different tool shoulder sizes to produce the joints to make optimized parameters on FSW MgAZ31B alloy. (Padmanaban & Balasubramanian, 2009). Dhanesh and Mohan (2007) explained the tool profile and orientation heat on the material during welding at various speeds. Malarvizhi and Balasubramanian (2012) focused on tool shoulder diameter which is a key source of producing heat while welding dissimilar AA6061 and AZ31-Mg alloys. Mansoor et al. (2015) analyzed tool rotation speed/welding speed between Al and Mg alloys. The intermetallic phases formed in the stirred zone

B. Gugulothu
Department of Mechanical Engineering, Bule Hora University, Post Box No-144, Hagere Mariam, Ethiopia

P. S.S. Kumar
Department of Physics, NPR College of Engineering and Technology, Natham, Dindigul, Tamilnadu 624 401, India

N. S. Rao
Department of Mechanical Engineering, Shri Vishnu Engineering College for Women, Bhimavaram, India
e-mail: srinivasaraon@svecw.edu.in

S. Vijayakumar
Department of Mechanical Engineering, BVC Engineering College (Autonomous), Odalarevu, Andhra Pradesh, 533210, India

D. R. Rajkumar (✉)
Department of Mechanical Engineering, CARE College of Engineering and Technology, Tiruchirappalli, 620 009, India
e-mail: rajkumardevapiriam@gmail.com

T. V.J. Rao
Department of Electronics and Communications Engineering, BVC Engineering College (Autonomous), Odalarevu, Andhra Pradesh, 533210, India

M. Naga Swapna Sri
Department of Mechanical Engineering, P V P Siddhartha Institute of Technology, Vijayawada, India
e-mail: mswapnasri@pvpsiddhartha.ac.in

M. Sumesh et al. (eds.), *2nd International Conference on Smart Sustainable Materials and Technologies (ICSSMT 2023)*, Advances in Science, Technology & Innovation, https://doi.org/10.1007/978-3-031-49826-8_16

and evaluated their influence on the microhardness and overall mechanical properties of the weld. The welding process can cause liquation, resulting in the formation of intermetallics at the joint interface (McLean et al., 2003). Singh (2018) provide a general overview of the FSW basic principle, along with a description of various aspects related to the welding of magnesium alloys. Al-Ni mixed powder is evenly spread to the butt surface. The results showed that the welded surface was free from any apparent welding defects and cracks. The joint with self-reaction had a tensile strength of 149 MPa, which was 57% higher compared to a direct joint (Zheng et al., 2019). Dejia et al. (Liu et al., 2018) focused on the texture changes in weak zones using electron diffraction data and suggested techniques to change the texture and improve weak zone and high welding speed could be used to weld the material at high tool revolution rates for magnesium alloys. The thermo affected zone exhibited equiaxed grains, and grain size at the stir zone decreased with increase in welding speed due to lower heat input (Cao & Jahazi, 2009). Mishra and Mab (2005) presented the mechanisms involved in weld formation and microstructural refinement, as well as the effects of both parameters materials properties in FSW. Notably, the application of Friction Stir processing with different Tool rotational speeds resulted in improved hardness and tensile strength. Specifically, samples processed at 1160 rpm exhibited a significant increase of approximately 56% in Ultimate Tensile Strength (UTS) and a remarkable 118% enhancement in percentage elongation compared to the initial material (Tripathi et al., 2022). The analysis of the Friction Stir (FS) processed samples reveals the presence of consistent and even grain distribution due to the dissolution of the secondary phase (Mabuwa et al., 2022; Mehdi et al., 2022). In this work, we carried out the FSW welding joint for MgAZ31B Alloy at Different Rotational Speeds (900, 1120, and 1140 rpm), tensile, and hardness properties were determined and microscopic examinations were also conducted.

2 Materials and Methodology

MgAZ31B possesses excellent strength and ductility at room temperature and corrosion resistance and weldability properties as well. Figure 1 illustrates a plate made of AZ31B magnesium alloy by strir casting technology which has fabrication flexibility and removes casting flaws (Dao et al., 2012).For this literature study, the authors have selected MgAZ31B as our base metal. A general chemical composition is presented in Table 1. The plate was held in position by mechanical clamping as shown in Fig. 2. The welding was performed at rolled plates made of MgAZ31B alloy, with a thickness of 6 mm which is depicted in Fig. 3. A square butt joint measuring 150 mm × 60 mm × 6 mm

Fig. 1 Magnesium AZ31B alloy plate

was performed using the FSW welding technique for the investigation of mechanical properties.

In this experiment, we performed friction stir welding on MgAZ31 alloy using a range of tool rotational speeds. Specifically, we used rotational speeds of 900, 1120, and 1440 rpm, while keeping the welding rotational speed constant at 40 mm/min. The axial load applied during the process was 5KN, and 2° inclination of the tool was set with a tapered surface, which is depicted in Fig. 4.

3 Materials Testing

3.1 Tensile Test

To prepare the specimens for tensile testing, we cut the welded joints using milling cutters and machined them to the required dimensions (Anusha et al., 2023; Sharma et al., 2022). The tensile testing specimens were then fabricated following the ASTM E8/E8M11 standards, as shown in Fig. 5.

3.2 Hardness Test

We determined the hardness of the welded joint by measuring it with a Vickers hardness machine using a load of 100 g at 20 s time. The micro hardness evaluated depends upon area cross-section of the welded specimen and the stir zone.

3.3 Microstructure Analysis

Micro structure analysis of the weld zone in the square joint by FSW specimens was conducted as per ASTM E3/E112

Table 1 Composition of MgAZ31B

Metal	Al	Zn	Si	Cu	Fe	Mn	Mg
Composition	3.5	1	0.1	0.05	0.005	1	Balanced

Fig. 2 Mechanical clamping for the FSW process

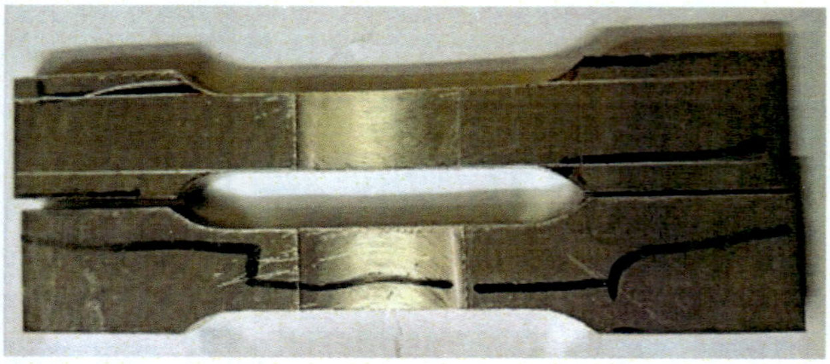

Fig. 5 FSW Tensile test specimen

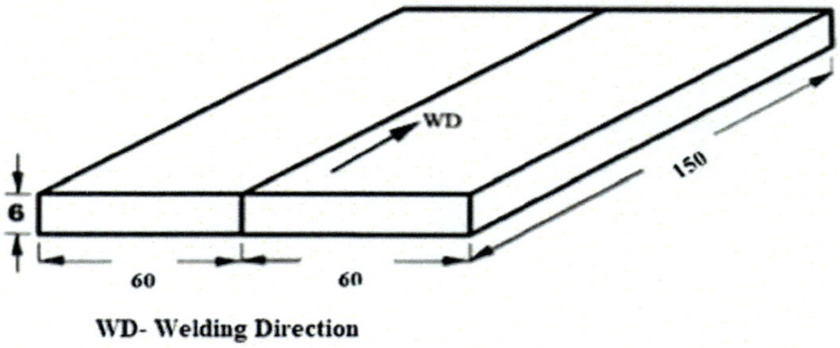

Fig. 3 Configuration of Square Butt Joint

Fig. 4 HSS Taper tool for FSW process

standards. After the specimen has been made to the required shape using grinding machine, the specimen is taken to 1/0, 2/0 grade emery papers. Immediately polishing operation is employed on the specimen which removes scores and yields fine finish surfaces. Finally, it was washed with 10% aqueous acetic acid, 2% oxalic acid in water on 4% aqueous picric chemicals (Sinha, 2023).

4 Result and Discussion

4.1 Tensile, Hardness Test

We evaluated several tensile properties of the fabricated welding, including tensile strength, hardness, density, % of elongation, and heat input. We prepared the tensile specimens according to the ASTM E8/E8M11 standards and conducted the tensile testing using a digital tensometer with a load capacity of 60 KN. Figure 6 shows tensile specimens after the test. The fabricated welded joint done at stir zone of welded joint subjected to micro hardness of metal is tested

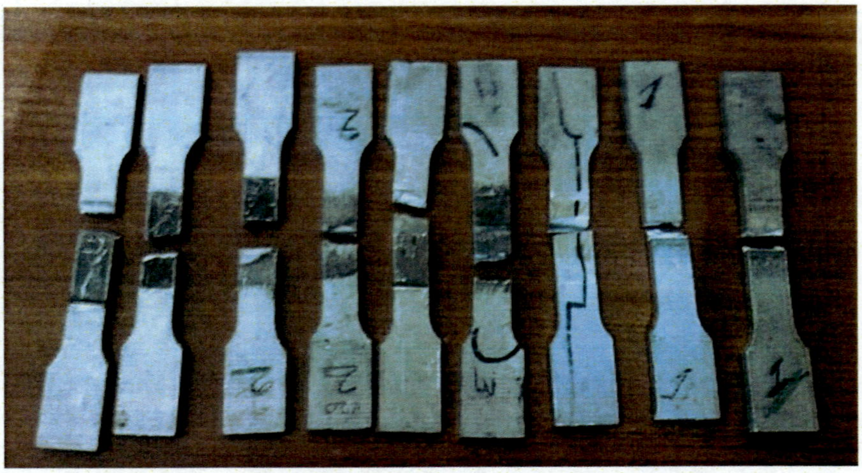

Fig. 6 Tensile specimens after tensile test

Table 2 Observations for mechanical properties

Speed (rpm)	Ultimate tensile stress (Mpa)	Hardness (Hv)	Elongation (%)	Density (g/cm^3)
900	106.85	87.57	8.10	1.87
1120	135.55	92.21	11.10	1.81
1440	134.90	96.53	11.43	1.82
Base metal	215.00	70.00	6.00	1.77

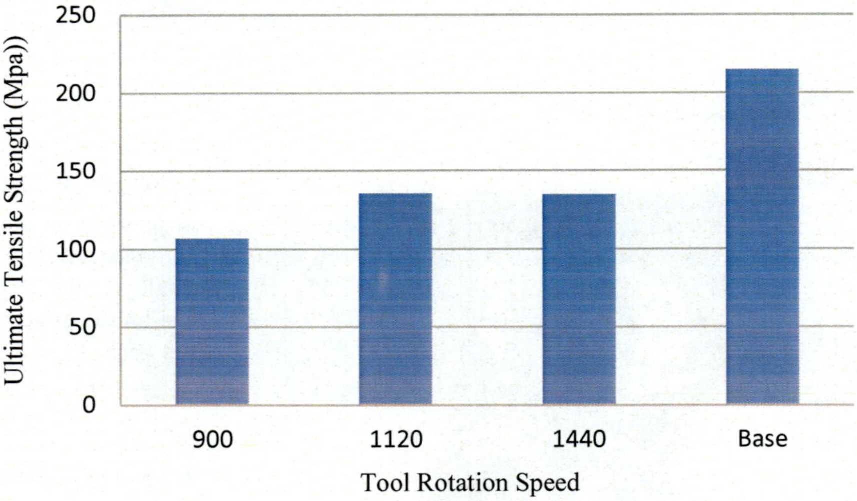

Fig. 7 Tool rotational speed versus ultimate tensile strength

on the Vickers hardness tester and result of tensile test and hardness test are in Table 2.

As a result, from Fig. 7, it is evident that the welded specimen with 1120 rpm speed is preferred for achieving high ultimate strength of 135.55 Mpa, hardness value also achieved at 1440 rpm which possesses 96.53Hv illustrated in Fig. 8, and the maximum elongation (11.43%) is at 1440 rpm shown in Fig. 9.

The density of the stir zone in the test specimen is also measured. The density of the base metal is 1.77 g/cm^3. But the specimen with a rotational speed of 900 rpm performs a higher density of 1.87 g/cm^3 which is shown in Fig. 10.

4.2 Microstructure

The microstructure analysis is conducted using optical microscopy to capture images of FSW specimen and the variation in grain size within the zone. The heat input exhibited an influence on the grain size, transmuting the

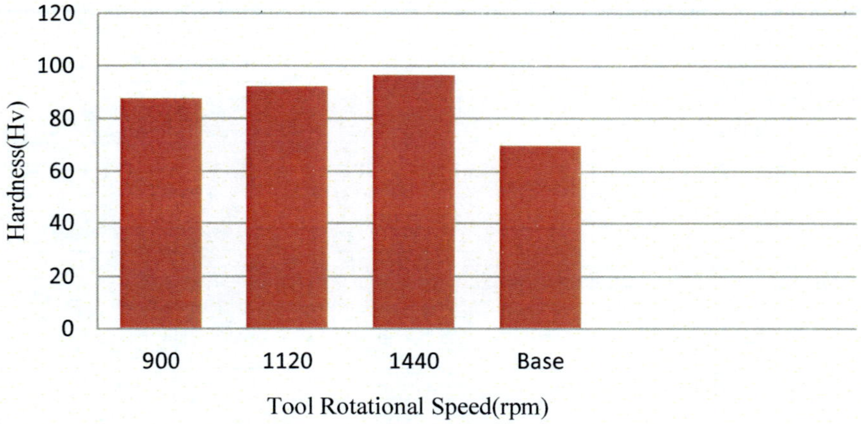

Fig. 8 Tool rotational speed versus hardness

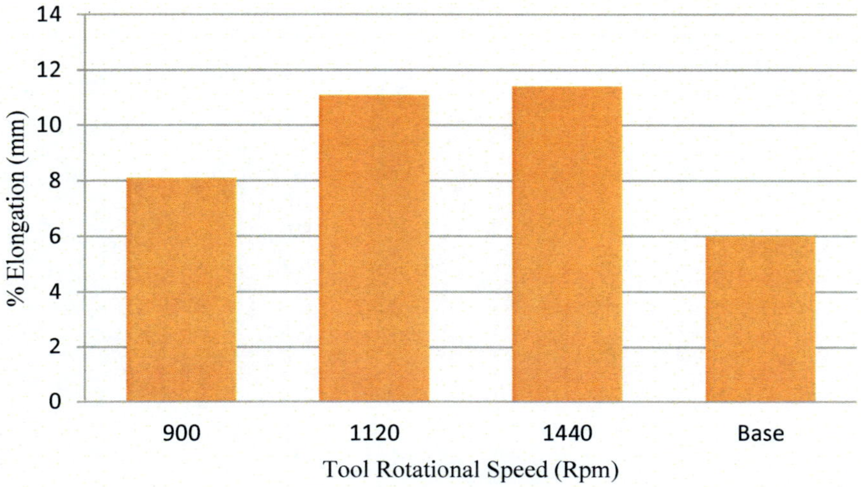

Fig. 9 Tool rotational speed versus % elongation

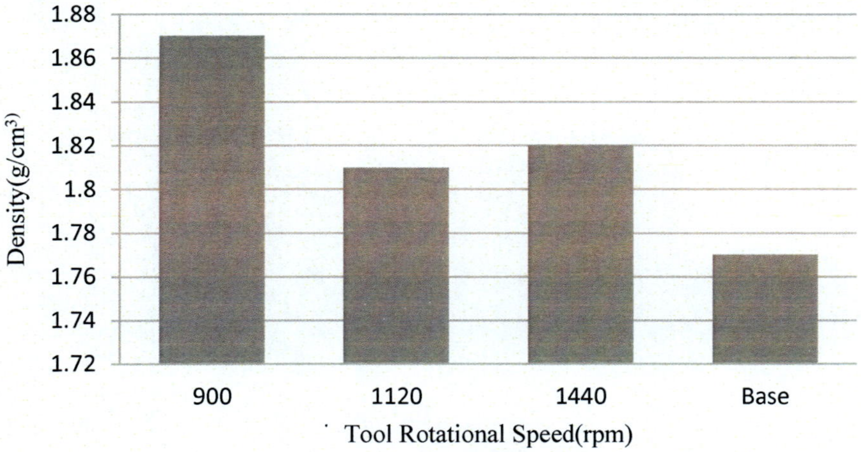

Fig. 10 Tool rotational speed versus density

granular grains in the base metal into fine grains, thereby contributing to the enhanced tensile strength. The change caused by the speed of tool rotation on the microstructure is illustrated in Figs. 11, 12, 13, 14. Higher rotational speed may cause abnormal stirring, resulting in a tunnel defect at the weld nugget due to increased strain rate and turbulence. The stress area expands and the point of greatest stress moves from the advancing side to the retreating side, which can affect the fracture location of the joints.

Fig. 11 Coarse grain dendrites in Base metal

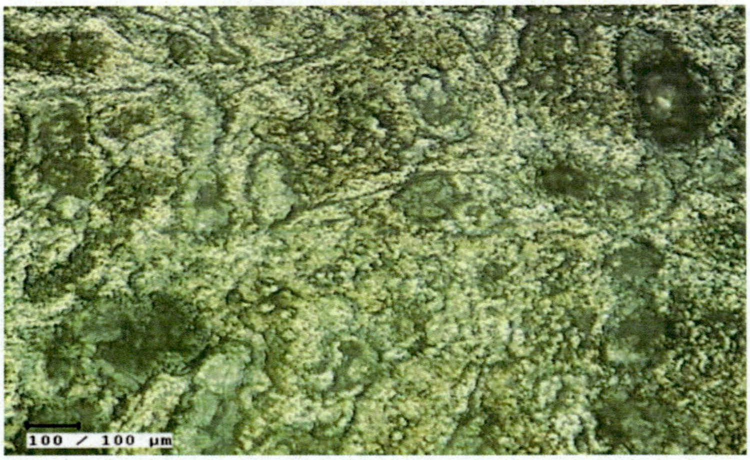

Fig. 12 HSS 900 rpm due to heat input the size of coarse grains are changed

Fig. 13 The coarse grains changed to fine grains structure at 1120 rpm

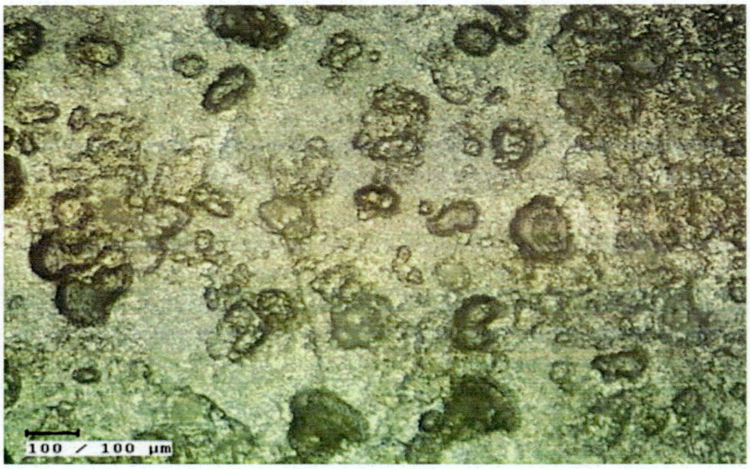

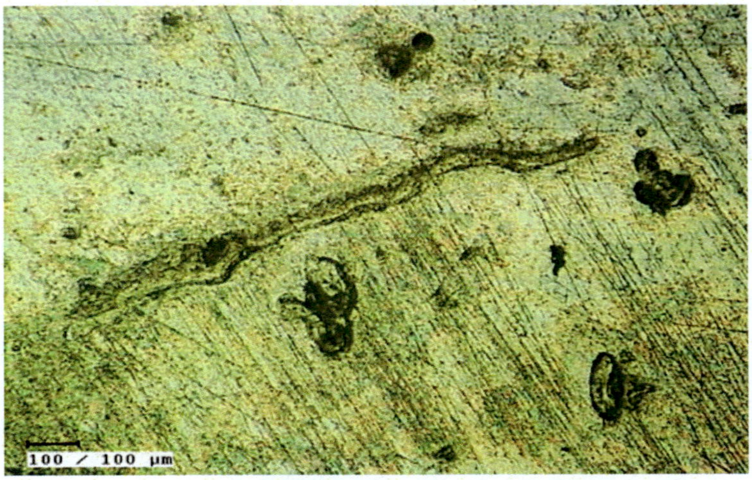

Fig. 14 The effect of tool which changes the size of the fine grains at 1440 rpm

5 Conclusion

Important factors affecting the mixing range of AZ31B alloy are tool material and rotational speeds, the tensile hardness, and micro structure analysis of the fabricated welding joints at different rotational speeds, namely 900, 1120, and 1440 rpm were compared in terms of tensile strength, % elongation, and hardness, with the same tool and welding speed of 40 mm/min.

The following conclusions can be drawn below,

- Lower rotational speed leads to higher hardness than the base material.
- Tool material and rotation speed achieve high strength properties in the stir zone. At 1120 rpm, the ultimate tensile strength of the fabricated specimen is higher than those at 900 and 1140 rpm.
- The fabricated joint at a rotational speed of 1440 rpm shows higher hardness compared to other joints at 900 and 1120 rpm.
- At 900 rpm, the density of the stir zone joint is lower than that of the other joints. At 1120 rpm with a welding speed of 40 mm/min, the specimen exhibits superior ductile property and moderate hardness and % elongation, making it the most feasible option for FSW of the metal.
- It can be used in the production of airplane fuselages, laptop and cell phone cases, speaker cones, and concrete tools and subjected to super forming techniques at high temperatures to create intricate components for the automotive industry.

References

Anusha, P., Sri, M. N. S., Vijayakumar, S., Rao, T. V. J., Paramasivam, P., Jeyakrishnan, S., & Saxena, K. K. (2023). Design and optimization the wear characteristics for Al7178/TiO_2/B4C/FA central hybrid composite. *IJIDeM*. https://doi.org/10.1007/s12008-023-01341-5

Cao, X., & Jahazi, M. (2009). Effect of welding speed on the quality of friction stir welded butt joint of Mg alloy. *Materials and Design, 30*, 2033–2042.

Dao, V., Zhao, S., Lin, W., & Zhang, C. (2012). Effect of process parameters on microstructure and mechanical properties in AlSi9Mg connecting-rod fabricated by semi-solid squeeze casting. *Materials Science and Engineering A, 558*, 95–102.

Dhanesh, G., & Mohan, S. G. (2007). Influence of induction heat of friction stir welding on corrosive behavior mechanical and microstructure of different metals. *Indian Journal of Engineering and Materials Science, 25*, 203–208.

Kumar, K., Kailas, V., & Srivatsan, T. S. (2008). Influence of tool geometry in friction stir welding. *Materials and Manufacturing Processes*, *23*(2), 188–194. https://doi.org/10.1080/10426910701774734

Liu, D., Tang, Y., Shen, M., Hu, Y., & Zhao, L. (2018). Analysis of weak zones in friction stir welded magnesium alloys from the viewpoint of local texture, a short review. *Metals*, *8*, 970. https://doi.org/10.3390/met8110970

Mabuwa, S., Msomi, V., Mehdi, H., & Saxena, K. K. (2022). Effect of material positioning on Si-rich TIG welded joints of AA6082 and AA8011 by friction stir processing. *Journal of Adhesion Science*, 1–19. https://doi.org/10.1080/01694243.2022.2142366

Malarvizhi, S., & Balasubramanian, V. (2012). Influences of tool shoulder diameter to plate thickness ratio (D/T) on stir zone formation and tensile properties of friction stir welded dissimilar joints of AA6061 aluminum–AZ31B magnesium alloys. *Materials & Design*, 453–460.

Mansoor, B., Dorbance, A., Ayouh, G., & Imad, A. (2015). Friction stir welding of AZ31B Mg alloy with Al6061-T6 influence of processing parametres of micro structure and mechanical properties. *JOM* 259–266.

McLean, A. A., Powell, G. L. F., Brownand, I. H., & Linton, V. M. (2003). Friction stir welding of magnesium alloy AZ31B to aluminium alloy 5083. *JOM* 462–464. https://doi.org/10.1179/136217103225009134

Mehdi, H., Mabuwa, S., Msomi, V., & Saxena, K. K. (2022). Influence of friction stir processing on the mechanical and microstructure characterization of single and double V-groove tungsten inert gas welded dissimilar aluminum joints. *Journal of Materials Engineering and Performance*. https://doi.org/10.1007/s11665-022-07659-7

Mishraa, R. S., & Mab, Z. Y. (2005). Friction stir welding and processing. *Materials Science and Engineering: R: Reports*.

Padmanaban, G., & Balasubramanian, V. (2009). Selection of FSW tool pin profile, shoulder diameter and material for joining AZ31B magnesium alloy—An experimental approach. *Materials & Design, 30*, 2647–2656.

Sharma, A. K., Goyal, D., Pabla, B. S., Saxena, K. K., Prakash, C., & Buddhi, D. (2022). Optimization of friction stir welding parameters for micro alloying of AA6082 alloy. *IJIDeM*. https://doi.org/10.1007/s12008-022-01105-7

Singh, K. (2018). A review on magnesium & its alloys. *Journal of Magnesium and Alloys*, 1–18.

Sinha, A. N. (2023). Specimen preparation for optical microscopy. National Metallurgical Laboratory, Jamshedpur 831 007.

Tripathi, H., Bharti, A., Saxena, K. K., & Kumar, N. (2022). Improvement in mechanical properties of structural AZ91 magnesium alloy processed by friction stir processing. *Advances in Materials and Science Engineering*, 81543–1556. https://doi.org/10.1080/2374068x.2021.1949540

Ugendera, S., Kumar, A., & Somi Reddy, A. (2014). Microstructure and mechanical properties of AZ31B magnesium alloy by friction stir welding. *Procedia Materials Science, 6*, 1600–1609.

Vijayakumar, S., Satheesh Kumar, P. S., Sampath Kumar, P., Manickam, S., Ramaiah, G. B., & Pydi, H. P. (2022). The effect of stir-squeeze casting process parameters on mechanical property and density of aluminum matrix composite. *Advances in Materials Science and Engineering*, 1–10. https://doi.org/10.1155/2022/3741718

Zheng, B., Xinbin, H., Lv, Q., Zhao, L., Cai, D., & Dong, S. (2019). Study of self-reaction assisted friction stir welding of AZ31B Mg/5052 Al alloys. *Materials Letters, 261*, 127138.

Evaluation of Mechanical and Thermomechanical Properties of Shape Memory Polymers Reinforced with Graphite, Graphene, and CNT

M. K. Ranganatha Swamy, Nagaraj Patil, U. S. Mallik, and V. Udayakumar

Abstract

Shape memory polymers (SMPs) are a subclass of responsive polymers that respond to environmental stimuli including heat, electricity, magnetic fields, light, and even wetness by changing from a fixed temporary shape to a memorised permanent shape. SMPs have recently attracted a lot of attention due to their numerous applications in the domains of space, automobiles, and biomaterials science. Following the synthesis and evaluation of a novel polymer with shape memory based on polyethylene glycol, the percentage of the polymer with outstanding shape memory was reached. Different mechanical characteristics of the SMPs were assessed using tests such as the tensile test, dynamic mechanical analysis, cyclic thermomechanical test, and hardness test. By adding graphite fillers, blank shape memory polymers' tensile strength and Young's modulus were increased at high temperatures, and the shape memory capabilities of the hardness-reinforced shape memory polymers at room temperature were sufficiently decreased. By altering the ratio of hard to soft segments, the glass transition temperature and viscoelastic characteristics were essentially enhanced. SMPs' mechanical qualities should be assessed for a variety of applications.

Keywords

Shape Memory Polymers • Shape Memory Effect • Dynamic Mechanical Analysis • Cyclic Thermomechanical Test

1 Introduction

Shape memory polymers (SMPs) have received increased attention in recent years for a variety of applications, including actuators, biomedical equipment, foam, smart fabrics, and deployable space constructions (Small et al., 2005). Polyethylene glycol-based shape memory polymers have drawn increasing research interest across all types of polymers because they are simple to manipulate in terms of the crucial recovery temperature and crystal structure. Due to the incompatibility of each segment, SMP is composed of hard and soft segments, which individually constitute hard and soft domains (Sokolowski et al., 1999). Through interactions including hydrogen bonds and dipole–dipole interactions, hard segment connects to one another and restores its original shape after distortion (Swiain, 1986). When elongating or compressing, a soft section absorbs external stress and adopts the phase transition temperature (Takahashi et al., 1996). The degree to which the hard and soft segments are aligned determines how effectively the shape may be recovered. Programming must be done to improve shape recovery. The stiffness and strength properties of shape memory polymers are one of their main drawbacks; these polymers bend quickly and lose rigidity after a few cycles (Tey et al., 2001). Shape memory polymers' mechanical qualities will be improved by the addition of

M. K. R. Swamy (✉)
Department of Mechanical Engineering, Faculty of Engineering and Technology, Jain Deemed to Be University, Bangalore, Karnataka, India
e-mail: mk.ranganatha@jainuniversity.ac.in

N. Patil
Department of Mechanical Engineering, Faculty of Engineering and Technology, Jain Deemed to Be University, Bangalore, Karnataka, India
e-mail: nagaraj.patil@jainuniversity.ac.in

U. S. Mallik
Department of Mechanical Engineering, Siddaganga Institute of Technology, Affiliated to VTU Belgaum, Tumkur, Karnataka, India
e-mail: usm@sit.ac.in

V. Udayakumar
Department of Chemistry, Siddaganga Institute of Technology, Affiliated to VTU Belgaum, Tumkur, Karnataka, India
e-mail: udayakr@gmail.com

M. Sumesh et al. (eds.), *2nd International Conference on Smart Sustainable Materials and Technologies (ICSSMT 2023)*,
Advances in Science, Technology & Innovation, https://doi.org/10.1007/978-3-031-49826-8_17

fillers. Due to their excellent thermal, mechanical, and electrical properties, graphite, graphene, and CNT are frequently utilised as fillers for structural shape memory polymers (Tobushi et al., 2006). For larger-scale manufacture, a melt mixing method can be used. In this study, we synthesised SMP with the aid of graphite, graphene, and CNT as fillers, as well as the solvents polyethylene glycol (PEG) and dimethyl formamide (Tobushi et al., 2004). In this study, along with the pertinent data, a brief explanation of the strength parameter, hardness, and viscoelastic properties is provided (Vaia, 2005). Studies on the use of fillers have revealed that SMPs have more strength, stiffness, and unique features depending on the fillers added, which can provide additional benefits over plain SMPs (WeiZ et al., 1998).

2 Materials and Method

The supplies used for this project were from Sigma-Aldrich. Polyethylene glycol (Mw 6,000) is the polyol employed here. 4,4′-diphenylmethane diisocyanate (MDI), isophorone diisocyanate (IPDI). Polyethylene glycol (Mw200) 1,4-butanediol (BDO) and dibutyltin dilaurate. As fillers, graphite, graphene, and CNT were employed.

3 Experimental

3.1 Sample Preparation

A 500 ml round-bottom, three-necked, separable flask with a nitrogen intake, a thermometer, and a mechanical stirrer. SMP was created using a prepolymer technique under dry nitrogen solution polymerization. To create a prepolymer with a terminal C = O group, mostly PEG (Mw6,000) and filler were stirred with IPDI in the presence of a catalyst for two hours at 90 °C. Following this process, MDI and PEG (200) are added as a crosslinking agent and heated to 90 °C for an hour to produce a polymer with an NCO group. After that, the prepolymer's chain was stretched by adding BDO for one hour at 6090 °C. The polymer resin was poured into a heated glass mould that had been coated with a releasing agent to create the SMP films. The films were then baked in an oven at a temperature of 60 °C for 12 h, 80 °C for 24 h, and 100 °C for 8 h, respectively (Xie & Xiao, 2008). Table 1 lists the chemical make-up of the Soft segment and the Hard segment.

3.2 Experimental Work

The link between the storage modulus and the loss modulus of the SMPs with respect to temperature was demonstrated using dynamic mechanical analysis (DMA), which can then be evaluated with differential scanning calorimetry (Yakacki et al., 2008).

Tensile testing devices were used to establish the correlation between the stress, strain, and temperature of SMPs. According to ASTM standard D638, dumbbell-shaped specimens with various elastic filler volume fractions were created (Yang et al., 2006).

Cyclic Thermomechanical Test: The only test that uses both a constant temperature chamber and a tensile tester to evaluate recovery stress is the cyclic thermomechanical test (Shizuo & Nobuo, 1991). The Shore A Hardness Scale, which ranges in hardness from very soft and flexible through medium and somewhat flexible to hard with almost no flexibility, quantifies the hardness of flexible mould rubbers and polymers. The top end of the Shore A Scale can also be used to assess semi-rigid polymers (Small et al., 2005).

4 Results

4.1 Dynamic Mechanical Analysis (DMA)

A dynamic mechanical analysis (DMA) test was performed on an RSA III tester from TA Instruments to evaluate the shape memory polymer's glass transition temperature and viscoelastic properties. In the DMA tensile grips, a rectangular bar with measurements of 15 mm, 8 mm, 1 mm was inserted. The bar was subjected to a minor dynamic load at 1 Hz, and the temperature ramped from room temperature to 80 °C at a rate of 3 °C/min. The 15 m (Sokolowski et al.,

Table 1 Composition of raw materials along with filler

Sample ID	Hard segment wt%	Soft segment wt%	Filler wt%		
			Graphite	Graphene	CNT
SMP-1	40	60	0.1	0.1	0.1
SMP-2	42	58	0.2	0.2	0.2
SMP-3	48	52	0.3	0.3	0.3
SMP-4	56	44	0.4	0.4	0.4
SMP-5	58	42	0.5	0.5	0.5

1999) amplitude was chosen. The loss modulus was used to calculate the glass transition temperature Tg. The sample including graphite, graphene, and CNT will have a higher storage modulus than a sample containing only SMP, according to DMA data.

The region beneath the curvature is larger which signifies that it has a 4.5 Mpa mode modulus of elasticity, giving it more stiffness for the SMP to maintain strength for the most number of cycles.

Figures 1 and 2 showed the storage modulus, loss modulus, and tan values and are tabulated in Table 2. The midway temperature at which the material is undergoing the extreme modification in polymer mobility is shown by the peak in the loss modulus (E″), which resembles the chemical definition of the glass transition temperature (Tg). Since it was the first DMA feature to be defined, the loss tangent (tan) peak has historical significance and denotes a material's damping properties. The tan peak temperature served as the foundation for the DMA Tg reference data. The PEG-SMP-1 and PEG-SMP-2 specimens had a clear single Tg at 45 and 40 °C, respectively, as was discovered from the peak of the loss modulus.

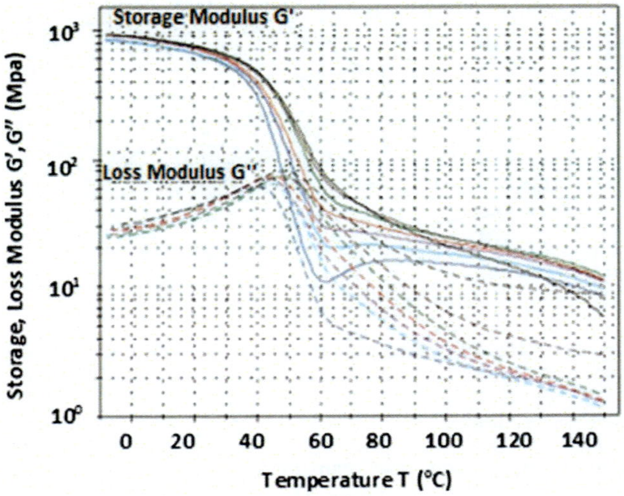

Fig. 1. Variation of the storage modulus with temperature

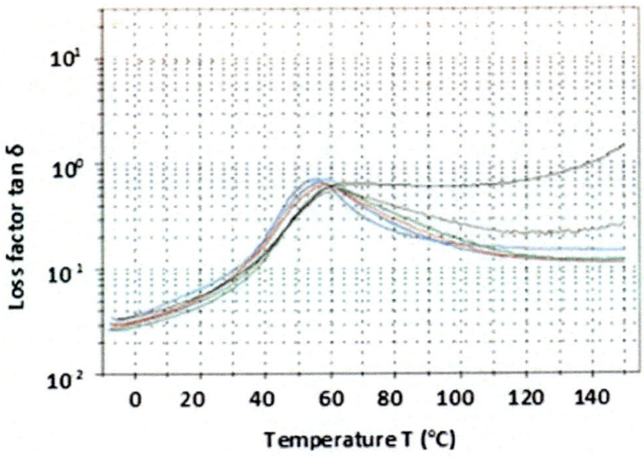

Fig. 2. Variation of loss shear moduli with temperature

4.2 Tensile Test

The PEG-SMP's tensile test graphs are displayed in Fig. 3. Table 3 provides a review of these materials' varied mechanical characteristics. Typically, as the MDI composition was increased, the toughness and % extension of the PEG-SMP increased, and as the IPDI composition was raised, the ultimate tensile strength was enhanced. This tendency was anticipated because pure PEG-SMP has the benefit of being both robust and strong. The yield strength and Young's modulus of the SMP-1 sample were both three times lower, while its toughness was more than ten times higher. It's noteworthy to notice that a modification in MDI composition led to an increase in tensile strength and toughness in the SMP-3 and SMP-4. The blend's higher tensile strength, toughness, and percentage extension were found to be in the 80% soft sector and 20% hard segment. Particularly suitable for load-bearing applications requiring great toughness, including biomedical applications, is the 80/20 soft/hard segment configuration (Swiain, 1986). The tensile strength up to 9.3 MPa, the elastic modulus up to 35.9 MPa, and the percentage of elongation at break up to 429.2 have all grown exponentially as a result of the addition of fillers to PEG matrix. The PEG-SMP's mechanical characteristics discovered in this study also offer a helpful framework for choosing materials.

4.3 Cyclic Thermomechanical Test

Using an MTS Insight 2 equipped with a thermal chamber, cyclic tensile testing was used to examine the shape memory capabilities. Four circles were evaluated in this investigation because the difference after the fourth circle was minimal. Graphite/PEG/BDO SMPUUs had a similar shape recovery and fixity to the Graphene/PEG/BDO copolymer system. (Takahashi et al., 1996) PEG-HS-SS-0.5%graphite-BDO-20% demonstrated good shape memory qualities after programming with both properties exceeding 90% and a slight variation (Takahashi et al., 1996). As seen in Fig. 4, the plain SMP's maximum stress (1.45 MPa) at 60 °C is almost half what it is at ambient temperature. Similar performance was regularly seen in all of the SMPs that had fillers. This is because the maximum stress declines when the shape memory polymer chains become flexible above the transition temperature.

Table 2 Showing the values of storage modulus, loss modulus, and Tan delta

Sample ID	Temp	E′(G′)	E″(G″)	
	Cel	Pa	Pa	tanD
SMP-1	36.93109	6.73E+08	1.39E+08	0.206434
SMP-2	36.92641	6.16E+08	1.31E+08	0.211851
SMP-3	36.91703	5.46E+08	1.2E+08	0.220142
SMP-4	36.90578	8.43E+08	1.64E+08	0.194459
SMP-5	36.9086	7.5E+08	1.5E+08	0.200447
SMP-5	36.89672	6.86E+08	1.41E+08	0.205135

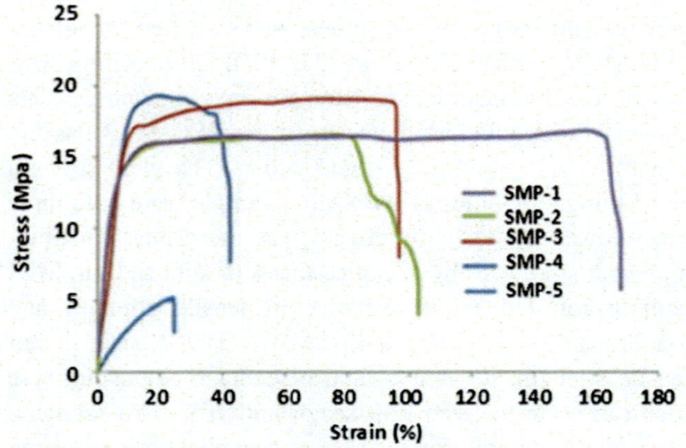

Fig. 3 Stress–strain curve of SMP-PEG for samples 1, 2, 3, 4, and 5

Table 3 Tensile test values of PEG-based shape memory polymers

Sample ID	Tensile strength (MPa)	Youngs modulus (MPa)	Elongation-at break (%)
SMP-1	7.8	31.5	406.7
SMP-2	7.4	32.9	299.5
SMP-3	8.3	33.7	412.9
SMP-4	7.9	35.7	422.8
SMP-5	9.3	35.9	429.2

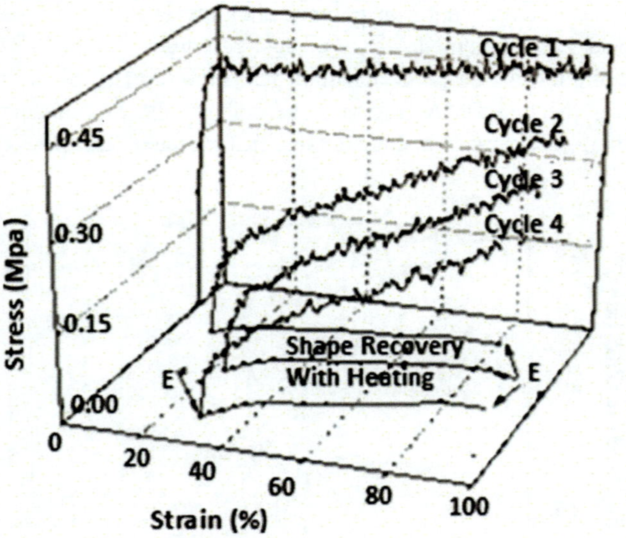

Fig. 4. Cyclic Thermomechanical test for SMP-PEG with various fillers

4.4 Hardness Test

The Shore A Hardness Scale was used to measure the hardness of the SMP samples. Flexible mould rubbers and polymers are measured using this scale. The hardness data indicate that sample IDs 1 and 2 have less hardness since there is more amorphous structure present during reaction time and sample. The presence of more crystalline phase in samples 3, 4, and 5 is causing them to exhibit greater hardness. Hardness of Shape Memory Polymers reinforced with graphite, graphene, and CNT is shown in Fig. 5. Hardness values with respect to durometer readings are shown in Table 4.

As a function of normal load, the plain PEG sample exhibits the most notable loss in hardness (Mallik & Sampath, 2008). The apparent elastic recovery of PEG, which can "artificially" increase the microhardness value at a low stress due to shrinkage of the indentations, may help to explain this. Since the indention is only moderately large, the artificially enhanced effect is not noticeable at greater loads.

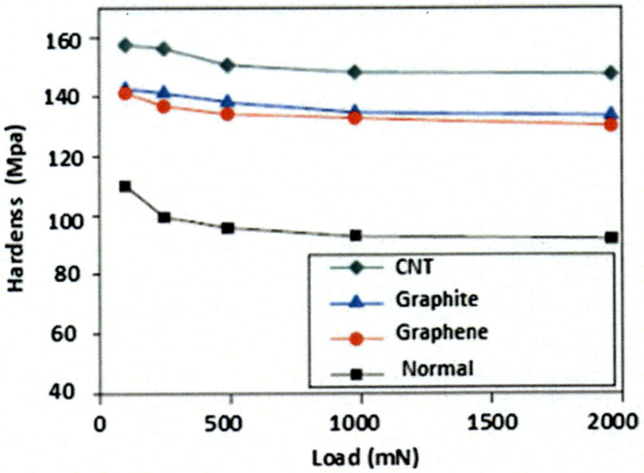

Fig. 5. Hardness of shape memory polymers reinforced with graphite, graphene, and CNT

Table 4 Hardness values with respect to durometer readings

Sl. no	Sample ID	Durometer	Applied mass in (kg)	Resulting force (N)	Hardness
1	SMP-1	Shore A	20 S	0.822	82
2	SMP-2	Shore A	20 S	0.822	84
3	SMP-3	Shore A	20 S	0.822	86
4	SMP-3	Shore A	20 S	0.822	88
5	SMP-3	Shore A	20 S	0.822	93

5 Conclusions

1. According to the aforementioned findings, the PEG-SMPs exhibit a better SME effect and have a higher recovery ratio of 95%, shape fixity ratio of 97%, and shape recovery duration of just 30 s.
2. The addition of fillers increased the strength hardness, which is suitable for use in specialised applications such as LEDs, transistors, and other microelectronic devices.
3. By adding fillers, the tensile strength was raised from 9 to 20 MPa.
4. Adding fillers raised the Vickers hardness rating from 20 to 93.
5. Dynamic mechanical investigation demonstrates that the material may be activated at 30 °C, and at 60 °C, the loss modulus is found to be 10 n/mm^2.

References

Mallik, U. S., & Sampath, V. (2008). Effect of composition and ageing on damping characteristics of Cu–Al–Mn shape memory alloys. *Materials Science and Engineering A, 478*, 48–55.

Shizuo K., & Nobuo, N. (1991). Cross-linked polymer having shape memorizing property. Method of its use and molded article having shape memory. US patent 5043396.

Small, W., IV., Wilson, T. S., Benett, W. J., Loge, J. M., & Maitland, D. J. (2005). Laser-activated shape memory polymer intravascular thrombectomy device. *Optics Express, 13*(20), 8204–8213.

Sokolowski W. M., Chmielewski A. B., Hayashi S., Yamada T. (1999). Cold hibernated elastic memory (CHEM) self-deployable structures. In *Proceeding of SPIE international symposium on smart structures and materials, Newport Beach, CA* (vol. 3669, pp. 179–185).

Swiain, M. V. (1986). Shape memory behavior in partially stabilized zirconia ceramics. *Nature, 322*(6076), 234–235.

Takahashi, T., Hayashi, N., & Hayashi, S. (1996). Structure and properties of shape memory polyurethane block copolymers. *Journal of Applied Polymer Science, 60*(7), 1061–1069.

Tey, S. J., Huang, W. M., & Sokolowski, W. M. (2001). Influence of long-term storage in cold hibernation on strain recovery and recovery stress of polyurethane shape memory polymer foam. *Smart Materials and Structures, 10*(2), 321–325.

Tobushi, H., Matsui, R., Hayashi, S., & Shimada, D. (2004). The influence of shape-holding conditions on shape recovery of polyurethane-shape memory polymer. *Smart Materials and Structures, 13*(4), 881–887.

Tobushi, H., Hayashi, S., Hoshio, K., & Miwa, N. (2006). Influence of strain-holding conditions on shape recovery and secondary-shape forming in polyurethane-shape memory polymer. *Smart Materials and Structures, 15*(4), 1033–1038.

Vaia, R. (2005). Nanocomposites: Remote-controlled actuators. *Nature Materials, 4*(4), 429–430.

WeiZ, G., Sandstorm, R., & Miyazaki, S. (1998). Shape-memory materials and hybrid composites for smart systems: Part I Shape memory materials. *Journal of Materials Science, 33*(15), 3743–3762.

Xie, T., & Xiao, X. (2008). Self-peeling reversible dry adhesive system. *Chemistry of Materials, 20*(9), 2866–2868.

Yakacki, C. M., Willis, S., Luders, C., & Gall, K. (2008). Deformation limits in shape memory polymers. *Advanced Engineering Materials, 10*(1–2), 112–119.

Yang, B., Huang, W. M., Li, C., & Li, L. (2006). Effects of moisture on the thermomechanical properties of a polyurethane shape memory polymer. *Polymer, 47*(4), 1348–1356.

Conductivity Tensor in an Infinite Potential Cylindrical Quantum Wire with Electron-Acoustic Phonon Interaction

Hoang Van Ngoc

Abstract

Low-dimensional semiconductors are currently widely used in various scientific applications. The properties of electrons in such systems undergo significant changes due to quantization of energy. A low-dimensional semiconductor system is the quantum wire, in which electrons can move without restrictions in one direction but are constrained in the other two directions. In this study, our study explores the conductivity tensor of an infinitely long cylindrical quantum wire in the presence of electrical field, specifically A laser beam and an electromagnetic wave field that is polarized linearly. Our analysis is limited to the case of interaction between electron and acoustic phonon. The equation of quantum kinetic for the carrier system is employed to calculate the expression of the tensor of conductivity in the presence of both fields. This study focuses on the quantum wire GaAs/GaAsAl and examines how the conductivity tensor varies with the external fields. Finally, figures are provided to demonstrate how the conductivity tensor varies with the external fields.

Keywords

Low-dimensional • Conductivity tensor • Infinite potential • Quantum wire

1 Introduction

The conductivity tensor is a crucial concept when studying low-dimensional semiconductors, as external fields carrying energy and momentum can significantly affect their electrical and magnetic properties (Hoang et al., 2017a, 2017b). Numerous studies have investigated the impact of external fields on materials, often leading to the emergence of new effects (Dao et al., 2017; Hoang et al., 2017a, 2017b). Among the low-dimensional semiconductor systems, the quantum wire (Bau & Trien, 2009; Blencowe & Shik, 1997; Dao et al., 2017; Hoang et al., 2017a, 2017b; Trien & Nhan, 2011), quantum well (Bau & Hoi, 2017; Bau et al., 2009, 2012; Dao et al., 2017; Grinberg & Luryi, 1998; Hung et al., 2012), superlattice (Bau & Hung, 2010; Bau et al., 2010; Kryuchkov et al., 1150; Shmelev et al., 1978), and quantum dot are the most commonly studied. Electrons in a quantum wire are constrained in two dimensions but can move freely in the other direction. The confinement of electrons leads to quantization of their energy and modification of their wave function based on the potential that constrains them (Hoang et al., 2017a, 2017b).

The electrical and magnetic characteristics of materials are considerably influenced by both laser beam and an electromagnetic wave field that is polarized linearly when considering the interaction between matter and external fields (Hoang et al., 2017a, 2017b). Specifically, the interaction between electrons and phonons is clear in the laser field (Dao et al., 2017; Hoang et al., 2017a, 2017b; Shmelev et al., 1982). In this study, we focus on the tensor describing the conductivity of a quantum wire with cylindrical geometry and infinite potential. under the influence of the laser beam and a wave field that is polarized linearly. Our analysis is limited to the scattering between electrons and acoustic phonons.

Conductivity varies according to the parameters of the external applied field, as well as the characteristic parameters of the quantum wire, and is determined by employing the equation of quantum kinetic for the carrier system in the presence of the two fields. The relationship between conductivity and the fields depicted by the figures and their significance is analyzed in detail.

H. Van Ngoc (✉)
Center for Forecasting Study, Institute of Southeast Vietnamese Studies, Thu Dau Mot University, Binh Duong, 820000, Vietnam
e-mail: ngochv@tdmu.edu.vn

M. Sumesh et al. (eds.), *2nd International Conference on Smart Sustainable Materials and Technologies (ICSSMT 2023)*, Advances in Science, Technology & Innovation, https://doi.org/10.1007/978-3-031-49826-8_18

2 Quantum Kinetic Equations in Bulk Semiconductors

Quantum momentum equation for electrons has the form:

$$i\frac{\partial f_{\vec{u}}}{\partial t} = \langle [a_{\vec{u}}^{+} a_{\vec{u}}, H] \rangle_t \quad (1)$$

With:

$$H = \sum_{\vec{u}'} \varepsilon_{\vec{u}'}\left(\vec{u}' - \frac{e}{c}\vec{A}(t)\right) a_{\vec{u}'}^{+} a_{\vec{u}'} + \sum_{\vec{q}} \omega_{\vec{q}} b_{\vec{q}}^{+} b_{\vec{q}} + \sum_{\vec{q},\vec{u}'} c_{\vec{q}} a_{\vec{u}'+\vec{q}}^{+} a_{\vec{u}'} (\vec{b}_{\vec{q}} + \vec{b}_{-\vec{q}}) \quad (2)$$

For simplicity, in the above Hamiltonian, we consider the electron placed in a laser field corresponding to the vector potential: $\frac{-1}{c}\frac{d\vec{A}(t)}{dt} = F_0 \sin \Omega t$

Replace the Hamiltonian in (2) into (1) and then calculate the exchange terms on the right-hand side (RHS) of (1):

$$i\frac{\partial n_{\vec{u}}}{\partial t} = -\sum_{\vec{q}} K_{\vec{q}} \left[a_{\vec{u}+\vec{q}}^{+} a_{\vec{u}} b_{\vec{q}} + a_{\vec{u}}^{+} a_{\vec{u}+\vec{q}} b_{-\vec{q}} - a_{\vec{u}}^{+} a_{\vec{u}-\vec{q}} b_{\vec{q}} - a_{\vec{u}-\vec{q}}^{+} a_{\vec{u}} b_{-\vec{q}} \right] \quad (3)$$

$$i\frac{\partial n_{\vec{u}}}{\partial t} = -\sum_{\vec{q}} K_{\vec{q}} \left[X_{\vec{u}+\vec{q},\vec{u},\vec{q}}(t) + X^{*}_{\vec{u},\vec{u}+\vec{q},-\vec{q}}(t) - X_{\vec{u},\vec{u}-\vec{q},\vec{q}}(t) - X^{*}_{\vec{u}-\vec{q},\vec{u},-\vec{q}}(t) \right] \quad (4)$$

To find X(t), we solve the equation:

$$i\frac{\partial X(t)}{\partial t} = \left\langle \left[a_{\vec{p}}^{+} a_{\vec{p}'} b_{\vec{k}}, H \right] \right\rangle_t \quad (5)$$

Next, we calculate each exchange term on the right-hand side of (5), and finally, we get:

$$X_{\vec{u},\vec{u}',\vec{k}}(t) = i \sum_{\vec{q}} K_{\vec{q}} \int_{-\infty}^{t} \left[\left\langle a_{\vec{u}}^{+} a_{\vec{u}'} (b_{\vec{q}} + b_{-\vec{q}}^{+}) b_{\vec{k}} \right\rangle_t - \left\langle a_{\vec{u}}^{+} a_{\vec{u}-\vec{q}} (b_{\vec{q}} + b_{-\vec{q}}^{+}) b_{\vec{k}} \right\rangle_t \right] \times \exp\left[i \left(\varepsilon_{\vec{u}} - \varepsilon_{\vec{p}'} - \omega_{\vec{q}} + \frac{e}{mc} (\vec{u} - \vec{p}') A(t)(t - t_1) \right) \right] dt_1 \quad (6)$$

Replace (10) into (4) and we get:

$$\frac{\partial n_{\vec{u}}}{\partial t} = 2\pi \sum_{\vec{q}} |K_{\vec{q}}|^2 N_{\vec{q}} \sum_{L=-\infty}^{+\infty} J_L^2(\vec{a}\vec{q}) \left[n_{\vec{u}\pm\vec{q}} - n_{\vec{p}} \right] \delta\left(\varepsilon_{\vec{u}\pm\vec{q}} - \varepsilon_{\vec{u}} - L\Omega \right) \quad (7)$$

The distribution function n(r,u) of the charged particle in such a system is:

$$\frac{\partial n(\vec{u},t)}{\partial t} + \left(e\vec{e}_0 + e\vec{e}(t) + \Omega_C \left[\vec{u}, \vec{h}(t) \right], \frac{\partial n(\vec{u},t)}{\partial \vec{u}} \right)$$
$$= 2\pi \sum_{\vec{q}} |K(\vec{q})|^2 N_{\vec{q}} \sum_{L=-\infty}^{+\infty} J_L^2\left(\frac{\wedge}{\Omega}\right) [n(\vec{u}+\vec{q},t) - n(\vec{u},t)]$$
$$\delta(\varepsilon_{\vec{p}+\vec{q}} - \varepsilon_{\vec{p}} - \hbar\omega_{\vec{q}} - L\hbar\Omega)[n(\vec{u}-\vec{q},t) - n(\vec{u},t)]$$
$$\times\ \delta(\varepsilon_{\vec{u}-\vec{q}} - \varepsilon_{\vec{u}} + \hbar\omega_{\vec{q}} - L\hbar\Omega) \quad (8)$$

where:

$\Omega_C = \frac{eH}{mc}$ is the Cyclotron frequency;

p represents the principal momentum of the charged particle, $J_L(x)$ is the Bessel function with real argument x, m denotes the effective mass, $n_0(u)$ stands for the phonon distribution function, and N(q) represents the electron–phonon interaction constant.

Multiplying both sides of Eq. (8) by $\frac{e}{m}\vec{u}\delta(\varepsilon - \varepsilon_{\vec{u}})$ and summing over p, we obtain the terms:

Examining the terms on the right-hand side (RHS) of (12):

In the linear approximation of the laser intensity, we only sum over L with three values L = 0 and $L = \pm 1$. Using the time recovery approximation, we have:

$$\frac{e}{m}\sum_{\vec{p}} (RHS)\delta(\varepsilon - \varepsilon_{\vec{u}}) = \frac{e}{m}\sum_{\vec{u}} (RHS|_{L=0})\delta(\varepsilon - \varepsilon_{\vec{u}}) + \frac{e}{m}\sum_{\vec{u}} (RHS|_{L=\pm 1})\delta(\varepsilon - \varepsilon_{\vec{u}}) \quad (9)$$

After some transformations, we have:

$$\frac{e}{m}\sum_{\vec{u}} (RHS|_{L=\pm 1})\delta(\varepsilon - \varepsilon_{\vec{u}}) = \vec{S}_0(\varepsilon_{\vec{u}}) + \vec{S}(\varepsilon_{\vec{u}})e^{-i\omega t} + \vec{S}^*(\varepsilon_{\vec{u}})e^{i\omega t} \quad (10)$$

With:

$$\vec{S}_0(\varepsilon_{\vec{u}}) = \frac{2\pi e}{m\hbar}\sum_{\vec{u},\vec{q}} |K_{\vec{q}}|^2 N_{\vec{q}} \vec{q} n_{10}(\vec{u})$$
$$\left\{ \left[\frac{\wedge^2}{4\Omega^2}\delta(\varepsilon_{\vec{u}+\vec{q}} - \varepsilon_{\vec{u}} - \hbar\omega_{\vec{u}} + \hbar\Omega) + \frac{\wedge^2}{4\Omega^2}\delta(\varepsilon_{\vec{u}+\vec{q}} - \varepsilon_{\vec{u}} - \hbar\omega_{\vec{q}} - \hbar\Omega) \right] - \left[\frac{\wedge^2}{4\Omega^2}\delta(\varepsilon_{\vec{u}+\vec{q}} - \varepsilon_{\vec{u}} + \hbar\omega_{\vec{q}} + \hbar\Omega) + \frac{\wedge^2}{4\Omega^2}\delta(\varepsilon_{\vec{u}+\vec{q}} - \varepsilon_{\vec{u}} + \hbar\omega_{\vec{q}} - \hbar\Omega) \right] \right\}$$
$$\delta(\varepsilon - \varepsilon_{\vec{u}}) = (\vec{S}_{01} + \vec{S}_{02} - \vec{S}_{03} - \vec{S}_{04})\delta(\varepsilon - \varepsilon_{\vec{u}}) \quad (11)$$

where:

$$S_{01} = \frac{\pi e^5 k_b T F^2}{4\hbar m^3 \varepsilon_0 \Omega^4}\left(\frac{1}{\chi_\infty} - \frac{1}{\chi_0}\right)\sum_{\vec{p}} n_{10}(\vec{u})\left[\sqrt{\vec{u}^2 + \frac{2m}{\hbar}\omega_{\vec{q}} - \frac{2m}{\hbar}\Omega} - u\right]^2 \tag{12}$$

$$S_{02} = \frac{\pi e^5 k_b T F^2}{4\hbar m^3 \varepsilon_0 \Omega^4}\left(\frac{1}{\chi_\infty} - \frac{1}{\chi_0}\right)\sum_{\vec{p}} n_{10}(\vec{u})\left[\sqrt{\vec{u}^2 + \frac{2m}{\hbar}\omega_{\vec{q}} + \frac{2m}{\hbar}\Omega} - u\right]^2 \tag{13}$$

$$S_{03} = \frac{\pi e^5 k_b T F^2}{4\hbar m^3 \varepsilon_0 \Omega^4}\left(\frac{1}{\chi_\infty} - \frac{1}{\chi_0}\right)\sum_{\vec{p}} n_{10}(\vec{u})\left[\sqrt{\vec{u}^2 - \frac{2m}{\hbar}\omega_{\vec{q}} - \frac{2m}{\hbar}\Omega} - u\right]^2 \tag{14}$$

$$S_{04} = \frac{\pi e^5 k_b T F^2}{4\hbar m^3 \varepsilon_0 \Omega^4}\left(\frac{1}{\chi_\infty} - \frac{1}{\chi_0}\right)\sum_{\vec{p}} n_{10}(\vec{u})\left[\sqrt{\vec{u}^2 - \frac{2m}{\hbar}\omega_{\vec{q}} + \frac{2m}{\hbar}\Omega} - u\right]^2 \tag{15}$$

With $\vec{S}(\varepsilon_{\vec{u}})$, we replace S_{01}, S_{02}, S_{03}, S_{04} with S_1, S_2, S_3, S_4

$$S_1 = \frac{\pi e^5 k_b T F^2}{4\hbar m^3 \varepsilon_0 \Omega^4}\left(\frac{1}{\chi_\infty} - \frac{1}{\chi_0}\right)\sum_{\vec{p}} n_1(\vec{u})\left[\sqrt{\vec{u}^2 + \frac{2m}{\hbar}\omega_{\vec{q}} - \frac{2m}{\hbar}\Omega} - u\right]^2 \tag{16}$$

$$S_2 = \frac{\pi e^5 k_b T F^2}{4\hbar m^3 \varepsilon_0 \Omega^4}\left(\frac{1}{\chi_\infty} - \frac{1}{\chi_0}\right)\sum_{\vec{p}} n_1(\vec{u})\left[\sqrt{\vec{u}^2 + \frac{2m}{\hbar}\omega_{\vec{q}} + \frac{2m}{\hbar}\Omega} - u\right]^2 \tag{17}$$

$$S_3 = \frac{\pi e^5 k_b T F^2}{4\hbar m^3 \varepsilon_0 \Omega^4}\left(\frac{1}{\chi_\infty} - \frac{1}{\chi_0}\right)\sum_{\vec{p}} n_1(\vec{u})\left[\sqrt{\vec{u}^2 - \frac{2m}{\hbar}\omega_{\vec{q}} - \frac{2m}{\hbar}\Omega} - u\right]^2 \tag{18}$$

$$S_4 = \frac{\pi e^5 k_b T F^2}{4\hbar m^3 \varepsilon_0 \Omega^4}\left(\frac{1}{\chi_\infty} - \frac{1}{\chi_0}\right)\sum_{\vec{p}} n_1(\vec{u})\left[\sqrt{\vec{u}^2 - \frac{2m}{\hbar}\omega_{\vec{q}} + \frac{2m}{\hbar}\Omega} - u\right]^2 \tag{19}$$

$$S_1^* = \frac{\pi e^5 k_b T F^2}{4\hbar m^3 \varepsilon_0 \Omega^4}\left(\frac{1}{\chi_\infty} - \frac{1}{\chi_0}\right)\sum_{\vec{p}} n_1^*(\vec{u})\left[\sqrt{\vec{u}^2 + \frac{2m}{\hbar}\omega_{\vec{q}} - \frac{2m}{\hbar}\Omega} - u\right]^2 \tag{20}$$

$$S_2^* = \frac{\pi e^5 k_b T F^2}{4\hbar m^3 \varepsilon_0 \Omega^4}\left(\frac{1}{\chi_\infty} - \frac{1}{\chi_0}\right)\sum_{\vec{p}} n_1^*(\vec{u})\left[\sqrt{\vec{u}^2 + \frac{2m}{\hbar}\omega_{\vec{q}} + \frac{2m}{\hbar}\Omega} - u\right]^2 \tag{21}$$

$$S_3^* = \frac{\pi e^5 k_b T F^2}{4\hbar m^3 \varepsilon_0 \Omega^4}\left(\frac{1}{\chi_\infty} - \frac{1}{\chi_0}\right)\sum_{\vec{p}} n_1^*(\vec{u})\left[\sqrt{\vec{u}^2 - \frac{2m}{\hbar}\omega_{\vec{q}} - \frac{2m}{\hbar}\Omega} - u\right]^2 \tag{22}$$

$$S_4^* = \frac{\pi e^5 k_b T F^2}{4\hbar m^3 \varepsilon_0 \Omega^4}\left(\frac{1}{\chi_\infty} - \frac{1}{\chi_0}\right)\sum_{\vec{p}} n_1^*(\vec{u})\left[\sqrt{p^2 - \frac{2m}{\hbar}\omega_{\vec{q}} + \frac{2m}{\hbar}\Omega} - u\right]^2 \tag{23}$$

In the linear approximation of the magnetic field intensity, we solve the system of equations above and obtain:

$$\vec{R}_0(\varepsilon_{\vec{u}}) = \tau(\varepsilon_{\vec{u}})\left[\vec{Q}_0(\varepsilon_{\vec{u}}) + \vec{S}_0(\varepsilon_{\vec{u}})\right] + \frac{2\omega_c\tau^2(\varepsilon_{\vec{u}})}{1+\omega_c^2\tau^2(\varepsilon_{\vec{u}})}\left[\vec{Q}(\varepsilon_{\vec{u}}), \vec{h}\right] + 2\omega_c\tau^2(\varepsilon_{\vec{u}})\mathrm{Re}\left\{\frac{\left[\vec{S}(\varepsilon_{\vec{u}}), \vec{h}\right]}{1 - i\omega_c\tau(\varepsilon_{\vec{u}})}\right\} \tag{24}$$

From the obtained results we can calculate the current density or conductivity of the bulk semiconductor structure.

3 Detailed Calculations

Since we are solely focusing on the scattering between electrons and acoustic phonons, the constant component of the distribution function of acoustic phonons is as follows:

$$N_q = \frac{k_B T}{v_s q_z} \tag{25}$$

and the constant for the interaction electrons-acoustic phonons is:

$$C_q^2 = \frac{\xi^2 q_z}{2\rho v_s V} \tag{26}$$

The system under consideration is a cylindrical conductor with an infinite potential and a radius of R:

$$\begin{cases} U_c(\vec{r}) = 0 \Leftrightarrow r < R \\ U_c(\vec{r}) = \infty \Leftrightarrow r > R \end{cases} \tag{27}$$

and

$$\begin{cases} \psi_{m,n,\vec{q}}(\vec{r}, \varphi, z) = 0 \Leftrightarrow r > R \\ \psi_{m,n,\vec{q}}(\vec{r}, \varphi, z) = \frac{1}{\sqrt{V_0}} e^{im\varphi} e^{i\vec{q}z} \psi_{m,n}(\vec{r}) \Leftrightarrow r < R \end{cases} \tag{28}$$

$$E_{m,n}(\vec{q}) = \frac{q_z^2}{2m} + \frac{B_{m,n}^2}{2mR^2} \tag{29}$$

here $\psi_{m,n}(\vec{r}) = \frac{1}{J_{m+1}(B_{m,n})} J_m\left(B_{m,n}\frac{r}{R}\right)$; the value of $B_{n,l}$ can be obtained as the solution to the Bessel function of order n that satisfies the equation $J_m(B_{m,n}) = 0$; $V_0 = \pi R^2 L_z$.

The laser beam: $\vec{f}_{laser}(t) = \vec{f}\sin(\Omega t)$, and electromagnetic field: $\begin{cases} \vec{e}_{wave}(t) = \vec{e}(e^{-i\omega t} + e^{i\omega t}) \\ \vec{h}_{wave} = [\vec{n}, \vec{e}(t)] \end{cases}$. The system's quantum kinetic equation is (Nguyen Quoc & Nguyen Thi, 1995; Shmelev et al., 1982):

$$\frac{\partial n_{m,n,\vec{q}_z}(t)}{\partial t} + \left(e\vec{e}(t) + \Omega_C\left[\vec{q}_z, \vec{h}(t)\right], \frac{\partial n_{m,n,\vec{q}_z}(t)}{\partial \vec{q}_z}\right) = 2\pi \sum_{m,n,m\prime,n\prime,\vec{u}} M_{m,n,m\prime,n\prime}(\vec{u}) \sum_{L=-\infty}^{+\infty} J_L^2(\vec{a},\vec{q}) \left[n_{m\prime,n\prime,\vec{q}_z+\vec{u}}(t) - n_{m,n,\vec{q}_z}(t)\right]\delta\left(\varepsilon_{m\prime,n\prime,\vec{q}_z+\vec{u}} - \varepsilon_{m,n,\vec{q}_z} - L\Omega\right) \tag{30}$$

here $I_{m,n,m\prime,n\prime}(\vec{q}_{h,j}) = \frac{2}{R^2}\int_0^R J_{|m-m\prime|}(q_{h,j}R)\psi_{m,n}(r)\psi^*_{m\prime,n\prime}(r)rdr$ is form factor;

$$M_{m,n,m\prime,n\prime}(\vec{p}) = \left|K_{\vec{p}}\right|^2 \left|I_{m,n,m\prime,n\prime}\right|^2 N_{\vec{p}} \tag{31}$$

$$\begin{aligned} n_{m,n}(\vec{q}_z,t) &= n_0(\varepsilon_{m,n,\vec{q}_z}) + n_1(\vec{q}_z,t); \\ n_0(\varepsilon_{m,n,\vec{q}_z}) &= \theta(\varepsilon_F - \varepsilon_{m,n,\vec{q}_z}) \\ n_1(\vec{q}_z,t) &= -\vec{q}_z\vec{X}(t)n_0\prime(\varepsilon_{m,n,\vec{q}_z}) \end{aligned} \tag{32}$$

$$\begin{aligned} n_0\prime(\varepsilon_{m,n,\vec{q}_z}) &= \frac{\partial n_0(\varepsilon_{m,n,\vec{q}_z})}{\partial \varepsilon_{m,n,\vec{q}_z}}; \\ n_1(\vec{q}_z,t) &= n_1(\vec{q}_z)e^{-i\omega t} + n_1^*(\vec{q}_z)e^{i\omega t}; \\ n_1(\vec{q}_z) &= -\vec{q}_z\chi n_0\prime(\varepsilon_{m,n,\vec{q}_z}) \end{aligned} \tag{33}$$

$$\vec{\chi}(t) = \vec{\chi}e^{-i\omega t} + \vec{\chi}^* e^{i\omega t}; \vec{\chi} = \frac{e}{m}\vec{e}\frac{t(\varepsilon)}{1 - i\omega t(\varepsilon)}; \vec{\chi}^* = \frac{e}{m}\vec{e}\frac{t(\varepsilon)}{1 + i\omega t(\varepsilon)} \tag{34}$$

$$\frac{\partial n_{m,n}(\vec{q}_z,t)}{\partial t} = \frac{\partial}{\partial t}\left[n_0(\varepsilon_{m,n,\vec{q}_z}) + n_1(\vec{q}_z)e^{-i\omega t} + n_1^*(\vec{q}_z)e^{i\omega t}\right] = -i\omega n_1(\vec{q}_z)e^{-i\omega t} + i\omega n_1^*(\vec{q}_z)e^{i\omega t} \tag{35}$$

with the simple case: $L = 0 \; and \; L = \pm 1$: $J_0^2 \approx 1$; $J_\pm^2 \approx \frac{(\vec{a},\vec{q})^2}{4}$; ; we have (Shmelev et al., 1982):

$$\vec{j}(t=0) = \int (\vec{R}_{pcd}(\varepsilon) + \vec{R}^*_{pcd}(\varepsilon))d\varepsilon = \vec{j}_1 + \vec{j}_1^* = \frac{4e^2 n_0}{m}\frac{t(\varepsilon_F)}{1+\omega^2 t^2(\varepsilon_F)}\left\{\left[\varepsilon_F - \frac{B^2_{m,n}}{2mR^2}\right]I + \lambda\frac{t(\Omega)[1 - \omega^2\tau(\Omega)t(\varepsilon_F)]}{1+\omega^2 t^2(\Omega)} + - A\frac{\tau(\varepsilon_F)[1 - \omega^2 t^2(\varepsilon_F)]}{1+\omega^2 t^2(\varepsilon_F)}\right\}\vec{e} \tag{36}$$

$$A = \frac{e^2F^2}{2m\Omega^3}\frac{\xi^2}{2\rho v_s^2 V}k_B T\left|I_{m,n,m\prime,n\prime}\right|^2\sqrt{2m\left(\varepsilon_F - \frac{B^2_{m,n}}{2mR^2}\right)} \tag{37}$$

$$\lambda = \frac{e^2F^2}{4m\Omega^3}\frac{\xi^2 k_B T}{\rho v_s^2 V}\left|I_{m,n,m\prime,n\prime}\right|^2\left\{\sqrt{2m\left(\varepsilon_F - \frac{B^2_{m,n}}{2mR^2}\right)}\left[\sqrt{2m\left(\Omega - \frac{B^2_{m,n}}{2mR^2}\right)} - 1\right]\right\} \tag{38}$$

with $\vec{j}(t=0) = \sigma\vec{e}(t=0) = \sigma 2\vec{e}$ (Nguyen Quoc & Nguyen Thi, 1995), so:

$$\sigma_{ik} = \frac{2e^2 n_0}{m}\frac{t(\varepsilon_F)}{1+\omega^2 t^2(\varepsilon_F)}\left\{\left[\varepsilon_F - \frac{B^2_{m,n}}{2mR^2}\right]\delta_{ij} + \frac{t(\Omega)[1 - \omega^2 t(\Omega)t(\varepsilon_F)]}{1+\omega^2 t^2(\Omega)}\lambda + - \frac{\tau(\varepsilon_F)[1 - \omega^2 t^2(\varepsilon_F)]}{1+\omega^2 t^2(\varepsilon_F)}A\right\} \tag{39}$$

This expression represents the tensor of conductivity for this quantum wire, in the scenario where electron-acoustic phonon scattering is considered (δ_{ik} is the Kronecker symbol).

4 Discuss the Results Obtained

The following parameters are utilized to graph the relationship between the tensor of conductivity and the external fields: (Bau & Trien, 2009; Blencowe & Skik, 1997; Dao et al., 2017; Hoang et al., 2017a, 2017b; Trien & Nhan, 2011): m = 0,0665m$_0$; $t(\varepsilon_F) \sim 10^{-12}\ s^{-1}$; $n_0 = 10^{23} m^{-3}$; $\rho = 5.3 \times 10^3\ kg/m^3$; R = 100 $\overset{0}{A}$; T = 300 K; $\omega_0 = 5 \times 10^{13} s^{-1}$.

Figure 1a, b show the relationship between the conductivity and laser field frequency for different temperature values. From Fig. 1a, it is observed that the conductivity is positive only when the laser field frequency exceeds $2*10^7\ s^{-1}$, and increases slightly when the frequency increases. However, when the laser field frequency surpasses $5*10^{12}\ s^{-1}$, the conductivity no longer increases. In fact, it is also evident that at the frequency $4*10^7\ s^{-1}$, the conductivity is no longer increased and further the increase in conductivity is now negligible. From the frequency $4*10^7\ s^{-1}$, it is also seen that all three graphs corresponding to three different temperatures are almost identical.

Figures 2a, b, and 3a display the relationship between σ and ω. As the frequency increases, the conductivity tensor decreases, with a rapid decrease when the frequency exceeds $10^{12}\ s^{-1}$. The plot form varies for each frequency range. At low frequencies, the graph has three separate plots for three temperature values. However, at high frequencies (greater than $10^9\ s^{-1}$), the three plots overlap for the three temperature values.

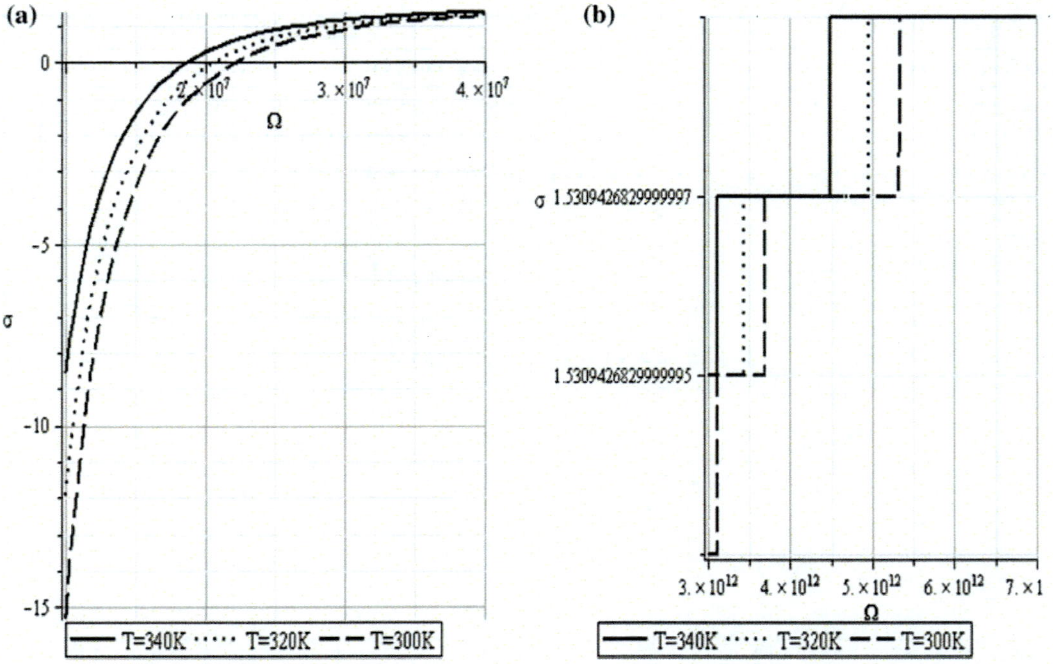

Fig. 1 **a** The relationship between conductivity and **Ω** with **Ω** < $4*10^7$ s^{-1}; **b** The relationship between conductivity and **Ω** with **Ω** < $4*10^7$ s^{-1}

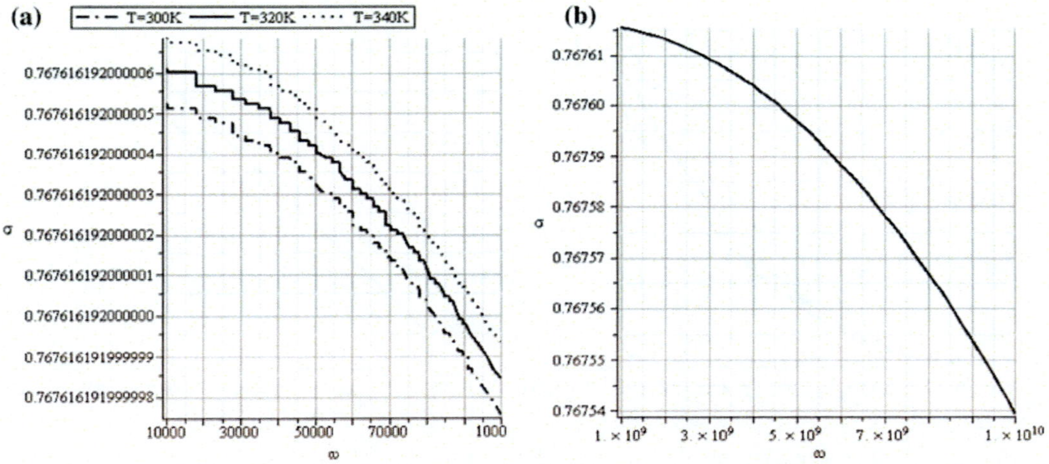

Fig. 2 **a** The relationship between conductivity and ω with ω < 10^5 s^{-1}; **b** The relationship between conductivity and ω with 10^9 s^{-1} < ω < 10^{10} s^{-1}

In the realm of low frequencies, the energy carried by electromagnetic waves remains insufficient to induce alterations in the conductive characteristics of the system. As the frequency escalates, the energy of the electromagnetic wave becomes substantial, fostering a noteworthy interplay with the crystal lattice. This substantial interaction subsequently leads to modifications in the electrical properties of the material's structure. However, as the frequency surpasses the threshold of 10^{13} s^{-1}, the system's conductivity becomes relatively independent of the electromagnetic wave frequency, resulting in a negligible correlation between conductivity and frequency fluctuations.

In addition, Figs. 3b and 4a depict the relationship between conductivity and F. The conductivity increases with the laser field amplitude. The plot has a serrated shape for small amplitudes, with a very slight increase in the conductivity as the amplitude increases. For the three different temperature values, there are three separate plots. Evidently, with the amplification of the laser field amplitude, the separation between the three degrees of freedom becomes more

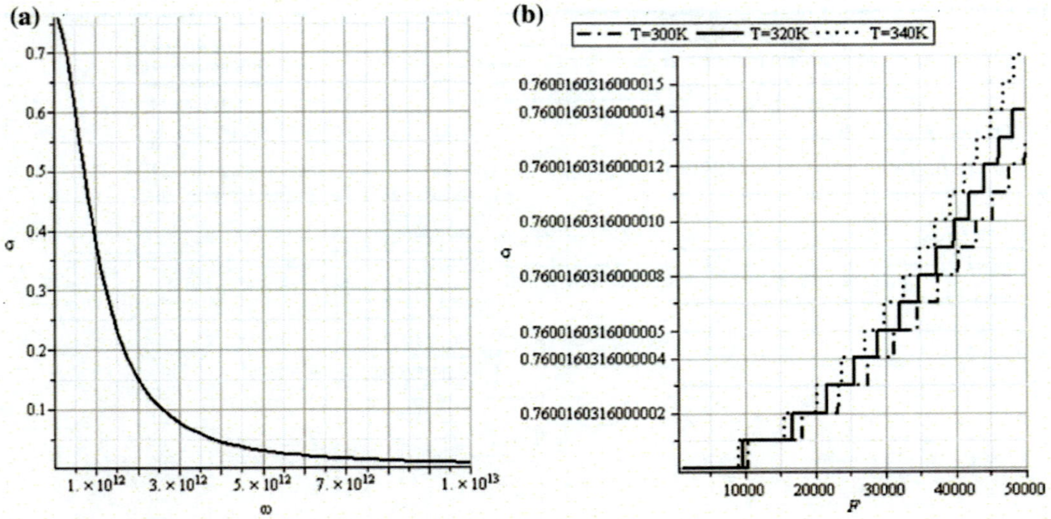

Fig. 3 **a** The relationship between conductivity and ω with $\omega > 10^{12}\ s^{-1}$; **b** The relationship between conductivity and F with $F < 10^5$

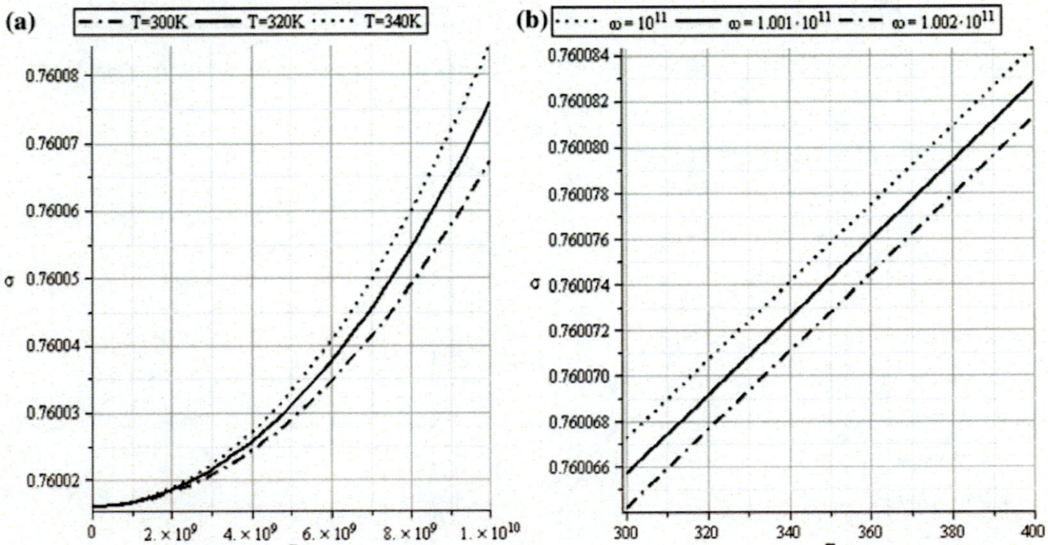

Fig. 4 **a** The relationship between conductivity and F with $F > 10^9$; **b** the relationship between σ and T

pronounced, thereby substantiating that the impact of temperature becomes increasingly prominent as the laser field amplitude is heightened.

Figure 4b illustrates the correlation between σ and the system temperature. The conductivity tensor exhibits a noticeable increase as the temperature rises, consistent with the theoretical expectations. As the temperature rises, the number of electrons increases, leading to more freedom and hence higher conductivity. Three separate plots are shown for three different values of the linearly polarized electromagnetic waves.

5 Conclusions

The research results in this work have shown the influence of externally applied fields on the conductivity of the system. The relationship between the electrical conductivity of the material and the frequency and amplitude of the external field has been investigated in different ranges of values. This dependence is nonlinear, and depending on the frequency range, the conductivity exhibits different behavior. The external fields have energy, and thus they interact with matter particles. This interaction affects the electrical conductivity of the system. This work contributes to the further development of the theory of low-dimensional quantum systems and provides insights for future applications.

References

Bau, N. Q., & Hoi, B. D. (2017). Influence of a strong electromagnetic wave (Laser radiation) on the Hall effect in quantum well with a parabolic potential. *Journal- Korean Physical Society, 60*, 59–64.

Bau, N. Q., & Hung, D. M. (2010). Calculating of the nonlinear absorption coefficient of a strong electromagnetic wave by confined electrons in doping superlattices. *PIER B, 15*, 175–185.

Bau, N. Q., Hung, D. M., & Ngoc, N. B. (2009). The nonlinear absorption coefficient of a strong electromagnetic wave caused by confined electrons in quantum wells. *Journal- Korean Physical Society, 54*, 765–773.

Bau, N. Q., Hung, D. M., & Hung, L. T. (2010). The influences of confined phonons on the nonlinear absorption coefficient of a strong electromagnetic wave by confined electrons in doping superlattices. *PIER Letters, 15*, 175–185.

Bau, N. Q., & Trien, H. D. (2009). The nonlinear absorption coefficient of strong electromagnetic waves caused by electrons confined in quantum wires. *Journal- Korean Physical Society*, *56*, 120–127.

Bau, N. Q., Hieu, N. V., & Nhan, N. V. (2012). Calculations of acoustoelectric current in a quantum well by using a quantum kinetic equation. *Journal of the Korean Physical society*, *6*, 2026–2031.

Blencowe, M., & Skik, A. (1997). Acoustoconductivity of quantum wires. *Superlattices and Microstructures*, *22*, 209–212.

Dao, T. H., Dao, T. H., Duong, T. T. T., & Nguyen, Q. B. (2017). The Ettingshausen coefficient in quantum wells under the influence of laser radiation in the case of electron-optical phonon interaction. *Photonics leters of Poland*, *8*, 79–81.

Grinberg, A., & Luryi, (1998). Theory of the photon—drag effect in a two-dimensional electron gas. *Physical Review B*, *38*, 87–96.

Hoang, V. N., Nguyen, V. N., Dinh, Q. V. (2017). The photon—drag effect in rectangular quantum wire with an infinite potential. *VNU Journal of Science*, *33*, 53–59.

Hoang, V. N., Nguyen, V. N., & Dinh, Q. V. (2017). The photon—drag effect in cylindrical quantum wire with an infinite potential for the case of electrons—acoustic phonon scattering. *VNU Journal of Science*, *33*, 80–86.

Hung, B. D., Nhan, N. V., Tung, L. V., & Bau, N. Q. (2012). Photostimulated quantum effects in quantum wells with a parabolic potential. *Proceeding National Conference Theoretical Physics, 37*, 168–173.

Kryuchkov, S. V., Kukhar, E. I., & Sivashova, E. S. (2008). Radioelectric effect in a superlattice under the action of an elliptically polarized electromagnetic wave. *Physics of the Solid State*, *50*, 1150.

Nguyen Quoc, A., Nguyen Thi, T. U. (1995). Electric conductivity tensor of semiconductors in a variable magnetic field. *VNU Journal of Science*, *01*, 19–23.

Shmelev, G. M., Chaikovskii, L. A., & Bau, N. Q. (1978). HF conduction in semiconductors superlattices. *Soviet Physics and Technics of Semiconductors, 12*, 1932.

Shmelev, G. M., Tsurkan, G. I., & Épshtein, É. M. (1982). Photostumilated radioelectrical transverse effect in semiconductors. *Physica Status Solidi (b)*, *109*, 53.

Trien, H. D., & Nhan, N. V. (2011). The nonlinear absorption of a strong electromagnetic waves caused by confined electrons in cylindrical quantum wire. *Progress in Electromagnetics Research Letters, 20*, 87–96.

Manufacturing of Eco-Friendly Brick Using Wasted Textile Apparel

S. Ispara Xavier, R. Divahar, P. S. Aravind Raj, P. J. Adarsh, Akshay Manoj, and Badusha Jalal

Abstract

Bricks have been a predominant building material throughout history, known for their exceptional strength and resilience. With the increasing emphasis on sustainable and locally sourced construction materials, this study explores the incorporation of waste apparel materials into conventional brick manufacturing processes. The waste apparel materials investigated in this research include polyester, polyester cotton (PC), and polyester viscose (PV). The manufacturing process involved subjecting a clay mixture to various treatments, with modifications based on the material type, manufacturing technique, and desired characteristics of the final product. Different weight measurements of waste apparel were added to the bricks to assess their impact on physical properties. This study focused on the physical characteristics of the resulting bricks, with a particular emphasis on compressive strength and water absorption. The experimental results revealed that the addition of 25 g of PC, PV, or polyester to a brick resulted in a remarkable 45 percent increase in compressive strength while simultaneously reducing water absorption. However, the investigation also demonstrated that exceeding a certain amount of textile waste in the brick composition can lead to a decline in compressive strength compared to the original results. Thus, a careful balance must be struck when incorporating waste apparel to achieve optimal improvements in brick performance. These findings open up possibilities for utilizing discarded clothing as a reinforcement in conventional brick production, highlighting its potential as a viable and environmentally friendly solution. By diverting waste apparel from landfills and utilizing it in construction materials, this approach offers a sustainable method for enhancing the properties of bricks while reducing environmental impacts. In conclusion, this study provides valuable insights into the incorporation of waste apparel into conventional brick manufacturing, presenting a novel avenue for enhancing brick performance and promoting a more eco-conscious approach to construction practices. Further research and development in this area can pave the way for more widespread adoption of sustainable and locally sourced building materials.

Keywords

Polyester cotton • Bricks • Polyester • Polyester viscose

1 Introduction

Over the last century, the construction industry has witnessed a surge in demand for building materials due to the rapid increase in global population and urbanization. As a result, there has been an acute shortage of traditional building supplies, particularly the most typical brick used in construction—burned clay bricks. To address this challenge, civil engineers and researchers have been exploring innovative ways to convert industrial wastes into workable and sustainable building and construction materials. One promising solution that has emerged is the recycling of construction waste. Construction activities generate massive amounts of waste materials, such as concrete, wood, metals, plastics, and more. Instead of sending these materials to landfills or incineration facilities, the idea of repurposing them for new construction projects presents an environmentally friendly and cost-effective approach. Recycling construction waste not only tackles the pollution problem associated with waste disposal but also helps to mitigate the strain on natural resources. By diverting waste materials

S. Ispara Xavier (✉) · R. Divahar · P. S. Aravind Raj · P. J. Adarsh · A. Manoj · B. Jalal
Department of Civil Engineering, Aarupadai Veedu Institute of Technology, Vinayaka Mission's Research Foundation, Paiyanoor, Chennai, India
e-mail: isparaxavier@gmail.com

M. Sumesh et al. (eds.), *2nd International Conference on Smart Sustainable Materials and Technologies (ICSSMT 2023)*, Advances in Science, Technology & Innovation, https://doi.org/10.1007/978-3-031-49826-8_19

from landfills, it reduces the demand for raw materials, conserving energy and reducing greenhouse gas emissions associated with extraction and manufacturing processes. One notable example of recycled construction material is reclaimed concrete. Crushed concrete can replace traditional aggregates in new concrete mixes, providing structural integrity while reducing the need for virgin materials. Additionally, incorporating recycled materials in concrete mixtures can lead to lower production costs, thus contributing to cost-effective building design. Another avenue for recycling is the use of industrial by-products or waste materials as raw ingredients for construction products. For instance, coal combustion fly ash, a by-product of coal-fired power plants, can be incorporated into cement production, enhancing its durability and reducing its carbon footprint. Similarly, slag, a by-product of iron and steel manufacturing, can be transformed into a valuable construction material that offers various engineering benefits. However, challenges remain in widespread adoption. Standardizing the quality and properties of recycled construction materials poses a significant obstacle. Engineers and regulatory bodies must work together to establish clear guidelines and performance standards to ensure the safety and reliability of such materials in construction applications.

The use of ecologically friendly, low-cost, lightweight, and sustainable construction materials has grown in favor in the construction sector. Some examples of waste that have been used include paper processing waste, smoking butts, slag, textile treatment and Effluent Treatment Plant (ETP), polystyrene foam, plastic soluble and insoluble dietary, straw, polystyrene fabric, domestic waste, dried effluent collected from an industrial wastewater treatment plant, rice husk ash, silica fumes slag, rubber, Kraft Pulp production waste, limestone dust, and wood sawdust. Civil engineering also makes use of textile waste; for example, textile material is used as a thermal insulating construction material. The contemporary period is worried about the environment's long-term viability, and the materials created should be environmentally benign. Non-biodegradable materials are employed in utility-based products. Garment waste is used as a filler material in places where there isn't a utility-based production material that has been engineered. The study aims to use mixed and 100 percent waste from synthetic and natural fabric in traditional brunt clay brick for technological advancements as a building material. An effort is being made to mix textile waste into the clay and reconstruct novel brick.

The research conducted by G. J. Kulkarni et al. in 2012 focused on exploring the possibility of using sludge from textile mills as fine aggregates in concrete with a specified M:20 grade. The study aimed to assess the impact of replacing traditional fine aggregates with sludge on the workability, density, and compressive strength of the concrete. The findings of the study indicated that as the replacement of fine aggregates with textile mill sludge exceeded 32 percent, certain negative effects on the concrete were observed. Specifically, the workability and density of the concrete were affected. Workability refers to the ease with which concrete can be mixed, placed, and finished, while density relates to its mass per unit volume. The decrease in workability and density might be attributed to the differences in properties between conventional fine aggregates (such as sand) and textile mill sludge. Sludge, being a waste material, may have different particle size distributions, shape, and surface characteristics compared to natural fine aggregates. These differences can result in difficulties in achieving the desired workability and density in the concrete mix. Furthermore, the study found that the compressive strength of the concrete was negatively impacted as the percentage of fine aggregate replacement by textile mill sludge increased beyond 32 percent. Compressive strength is a crucial parameter for assessing the structural integrity of concrete. The decline in compressive strength could be attributed to the lower quality of sludge as a fine aggregate replacement and its inability to provide the necessary strength characteristics in the concrete matrix. To address the decrease in compressive strength, the researchers introduced fly ash as an additional component in the concrete mix, along with 32 percent sludge, to partially substitute cement. Fly ash is a by-product of coal combustion and is known to possess pozzolanic properties, which can enhance concrete strength when used as a supplementary cementitious material. While the inclusion of fly ash helped mitigate the reduction in compressive strength to some extent, it could not completely restore the concrete's strength to the level of traditional concrete without sludge replacement. Therefore, the study suggests that there are limitations to the amount of textile mill sludge that can be used as a fine aggregate replacement in M20 grade concrete, as exceeding the 32 percent threshold adversely affects both workability and compressive strength.

Paki Turgut et al. focused on exploring the physical and mechanical characteristics of concrete mixtures incorporating high levels of Waste Silica Water (WSW) and Low-Plastic Waste (LPW). In comparing these mixtures to conventional concrete bricks readily available in the market, the researchers have found that all reported values for unit weight, compressive strength, flexural strength, ultrasonic pulse velocity (UPV), and water absorption fall well within the permissible range set by global standards. One of the standout findings is that the concrete mixtures containing elevated levels of WSW and LPW do not succumb to structural failure swiftly, even when subjected to loads surpassing the point of failure. This indicates a high degree of durability and stability under stress, which are essential qualities for construction materials. Moreover, these mixtures exhibit an exceptional capacity to absorb and dissipate energy when confronted with external forces, making them desirable for engineering applications requiring resilience and impact resistance. Another advantage is the significant reduction in unit weight achieved by incorporating WSW and LPW, which offers several benefits, including cost-effective transportation and ease of handling during construction. Furthermore, the introduction of these waste materials results in a smoother surface finish, not only enhancing the visual appeal but potentially improving the overall performance of the structure. In a study conducted by Nutan C. Patel et al. in 2013, various materials including fly ash, lime, sand, Kheda dust, and glass fiber were utilized in experiments to develop bricks. The focus of the investigation was on fiber fly ash bricks, and two types of testing were conducted: compressive strength tests and water absorption tests carried out after 7, 14, and 21 days. The results of the study revealed interesting findings. As the proportion of glass fiber increased in the fiber fly ash bricks, the compressive strength of the bricks showed a corresponding increase. This suggests that incorporating glass fibers in the mixture improves the bricks' ability to withstand compression forces, making them structurally stronger and more robust. Furthermore, the study found that the water absorption capacity of the fiber fly ash bricks decreased with the increasing fraction of glass fiber. This reduction in water absorption is highly beneficial as bricks with lower water absorption rates are less susceptible to moisture-related damage and have improved weather resistance. The utilization of various materials in the production of these bricks highlights the potential for enhancing the properties of conventional bricks through the incorporation of supplementary materials like glass fiber. By doing so, it is possible to develop bricks with improved mechanical strength and reduced water absorption, contributing to the construction of more durable and sustainable structures. It's important to note that the specific proportions and combinations of materials used in the study may vary depending on the desired characteristics and intended applications of the bricks. Therefore, further research and testing would be valuable to optimize the formulation of fiber fly ash bricks and explore their potential for widespread use in construction projects. Siram K. Krishna et al. conducted a thorough evaluation of two construction materials: Cellular Lightweight Concrete (CLC) Blocks and Clay Bricks. The overarching goal of this investigation is to identify and propose a potential replacement material for traditional red clay bricks in the construction sector. The evaluation process likely involves a comprehensive analysis of various essential properties and characteristics of both CLC Blocks and Clay Bricks. Factors such as mechanical properties (compressive strength, flexural strength, and tensile strength), thermal insulation capabilities, environmental impact (carbon footprint, energy consumption, and resource utilization), cost considerations, durability, longevity, and water absorption characteristics are likely to be carefully examined and compared. By methodically assessing these key aspects, the study aims to provide valuable insights and evidence-based recommendations for builders, construction professionals, and policymakers. If the research demonstrates that CLC Blocks offer superior performance and sustainability compared to conventional clay bricks, it could potentially pave the way for their wider adoption in the construction industry, leading to more eco-friendly and efficient building practices. However, it is crucial to consider the specific requirements and regional context of construction projects to determine the most suitable material for each application (Aravind Raj et al., 2020; Booth xxxx; Bouchhima et al., 2013; Briga, 2013; Divahar et al., 2021; Halil Murat Algin, 2008; IS, 1077:1992 1077, xxxx; IS3495:, 1992 1992, xxxx; Jeyaraj et al., 2012; Kim, 2013; Krishna Reddy & Reddy, 2013; Marjory xxxx; Murthy xxxx; Priti et al., 2012; Raut et al., 2011).

2 Materials Used

In this work, several measurements of polyester (Fig. 1), polyester cotton (Fig. 2), and polyester viscose (Fig. 3) were used, as well as clay, soil, and fine aggregate.

2.1 Polyester

Long-chain polymers that contain at least 85% by weight of an ester, a dihydric alcohol, and terephthalic acid are referred to as polyester. In other words, it describes how various esters are entangled within the fibers. Esters are created when an alcohol reacts with a carboxylic acid. The

Fig. 1 Polyester

Fig.2 Polyester cotton

Fig. 3 Polyester viscose

backbones of polyester polymers are generated by "esterification condensation of polyfunctional alcohols and acids." Polyester's physical attributes are as follows: In comparison to cotton, it has greater strength and luster. Polyester, on the other hand, absorbs less moisture than cotton and viscose. It is made entirely of 100% polyester.

2.2 Polyester Cotton

Polyester is a type of long-chain polymer that contains at least 85% by weight of an ester, a dihydric alcohol, and terephthalic acid. The term "polyester" refers to how various esters are entangled within the fibers of the material. Esters are formed when an alcohol reacts with a carboxylic acid. In the case of polyester, the backbones of the polymer are generated through "esterification condensation of polyfunctional alcohols and acids."

Polyester is known for its specific physical attributes when compared to other materials like cotton and viscose:

Strength: Polyester has greater strength compared to cotton and viscose. This property makes polyester fabrics durable and resistant to tearing or stretching.
Luster: Polyester often exhibits a natural shine or luster, giving the fabric a glossy appearance. This characteristic can enhance the aesthetic appeal of garments and other products made from polyester.
Moisture Absorption: Polyester has lower moisture absorption when compared to cotton and viscose materials. This makes polyester less prone to retaining moisture and also makes it a less breathable fabric, which is not suitable for hot and humid conditions.
100% Polyester: The polyster material does not contain any other fibers or materials even in a small portion.

Polyester is a material widely used in clothing, home textiles, industrial fabrics, and packaging.

2.3 Polyester Viscose

Polyester viscose fabric is mostly preferred for creating high-quality suits and garments. This fabric type combines the best characteristics of both polyester and viscose fabrics. The polyester component provides strength, durability, and wrinkle resistance, ensuring that the garment maintains its shape and appearance. On the other hand, the viscose component offers a soft and smooth texture, making the fabric comfortable to wear. Moreover, the blend of these two fabrics enhances the fabric breathability, which is crucial for ensuring comfort in various weather conditions. The mixed fabric is strong, stretchable, and long-lasting, making it a versatile choice for stitching elegant suits without compromising the style. Whether for formal occasions or everyday wear, polyester viscose blend fabric continues to be favored by designers and tailors for its winning combination of practicality and aesthetics.

3 Manufacturing Process

3.1 Specimen Details

Clay, soil, and water in a 2:1:1 ratio are the ingredients of Controlled Specimens. The mixed proportion of specimen

Table 1 Specimen details

Sample no	Constituent	Weight Gms
A1	POLYSTER	25
A2	POLYSTER	35
A3	POLYSTER	55
B1	POLYSTER COTTON	25
B2	POLYSTER COTTON	35
B3	POLYSTER COTTON	55
C1	POLYSTER VISCOSE	25
C2	POLYSTER VISCOSE	35
C3	POLYSTER VISCOSE	55
D1	P/V + P/C + POLYSTER	25
D2	P/V + P/C + POLYSTER	35
D3	P/V + P/C + POLYSTER	55

bricks used in this investigation was calculated using IS code 5454: 1976. Specimen Details are shown in Table 1.

3.2 Mixing

In the project's mixing process, both the control and specimen bricks are prepared manually. The initial wet brick obtained from the molding stage contains a moisture content ranging from 7 to 30%, depending on the specific formation procedure employed. Before proceeding to the firing process, a significant portion of the water is removed by draining the bricks in a dryer chamber. This drying step is crucial to reduce the moisture content and prepare the bricks for firing. The drying chamber operates at temperatures ranging from 100 to 400 degrees Fahrenheit (38 °C to 204 °C). The exact temperature used may vary based on the specific requirements of the clay and the type of bricks being produced. This drying process can take anywhere from 24 to 48 h, depending on the characteristics of the clay and the desired moisture level. Proper drying is essential to ensure uniform and controlled shrinkage during the firing process. If the bricks contain excessive moisture during firing, it could lead to cracking and deformations, compromising the quality and structural integrity of the final product. Therefore, the careful management of the drying process is critical to producing high-quality bricks with consistent properties. Overall, the manual mixing and controlled drying process described in the project are fundamental steps in brick manufacturing, contributing to the production of bricks that meet the required standards and specifications for use in construction applications.

3.3 Drying

To enhance thermal efficiency in brick manufacturing, kiln exhaust heat is often redirected to supply the necessary heat for dryer chambers. Proper temperature and humidity control are essential during the drying process to prevent brick cracking. This efficient use of heat not only reduces energy consumption but also ensures the production of high-quality bricks, optimizing the overall manufacturing process and promoting sustainable practices in the construction industry.

3.4 Burning

The burning process of bricks typically spans a duration of 10 to 40 h, varying based on kiln type and other influencing factors. Manufacturers employ various methods for kiln operation. The bricks are subjected to temperatures ranging from 800 to 1200 degrees Celsius during the firing process. This high-temperature treatment is crucial to achieve the desired structural strength and durability of the bricks, allowing them to withstand the demands of construction applications and ensure their suitability for diverse building projects.

3.5 Cooling

After reaching the peak temperature during the firing process, the cooling stage commences in brick manufacturing. The duration of the cooling process varies depending on the type of kiln used. In periodic kilns, the cooling times typically span between 5 to 24 h, allowing for a gradual reduction in temperature. On the other hand, tunnel kilns generally feature shorter cooling periods, usually not exceeding 10 h. The rate of cooling is a crucial aspect of the brick production process, as it directly influences the color and final properties of the bricks. A controlled and gradual cooling process helps to mitigate thermal stresses and prevent cracking, ensuring the bricks retain their structural integrity. Moreover, the rate of cooling can affect the color of the bricks. A slower cooling rate often results in darker-colored bricks, while a faster cooling rate may produce lighter-colored bricks. Brick manufacturers carefully tailor the cooling process to achieve the desired appearance and characteristics of the final product, meeting specific architectural and aesthetic requirements. Proper temperature management during the cooling stage is essential for producing high-quality bricks with uniform properties and consistent colors. By carefully controlling the cooling process, manufacturers can ensure that the bricks meet the required standards, making them suitable for diverse construction applications while achieving the desired visual appeal. Figures 4, 5 and 6 shows the process flow.

Fig. 5 Drying

Fig. 4 Mixing

Fig. 6 Burning

4 Result and Discussions

The study involved conducting tests and comparisons on controlled specimens along with various other types of bricks, namely Polyester bricks (PB), Polyester cotton bricks (PCB), Polyester viscose bricks (PVB), and a combination of Polyester, polyester cotton, and polyester viscose (P + PC + PV). All specimens, including the controlled ones, had uniform dimensions measuring 0.23 m in length, 0.1 m in width, and 0.07 m in height. The researchers likely conducted a series of tests to evaluate the physical and mechanical properties of these different brick types. Such analyses would provide valuable insights into the suitability and performance of these polyester-based bricks and their combinations for potential use in construction applications.

Fig. 7 Water absorption test

4.1 Water Absorption Test

Weigh each brick separately in a balance and record the weight as (W1) kg. In 24 h, completely submerge in water. Then take out all of the bricks and wipe down the area with a clean cloth. The wet brick (w2) is then weighed in kilograms. The weight differential (w2-w1) represents the amount of water detected by each brick. ((w2-w1)/w1)*100 is the proportion of water absorption.

Water Absorption: The data shows that each brick type exhibits a certain degree of water absorption, which is calculated as the difference between wet weight and dry weight. The percentage of water absorption varies between the different brick types, with values ranging from approximately 9% to 19%. Brick type C3 (17.01%) and D3 (19%) seem to have the highest water absorption, while brick type A3 (9.23%) has the lowest water absorption among the measured specimens. Water Absorption Test is shown in Fig. 7.

Consistency of Measurements: Each brick type has been measured three times, and the recorded values for dry and wet weights exhibit slight variations across these measurements. It is essential to consider such variations when assessing the properties of bricks to account for any potential inconsistencies during manufacturing or testing processes.

Impact of Brick Type: The data demonstrates that different brick types have distinct characteristics in terms of their water absorption properties. These variations might be attributed to differences in the composition, manufacturing techniques, or the raw materials used in producing each type of brick.

Controlled Specimen: The "Controlled Specimen" is likely a reference sample used for comparison purposes. Its water absorption percentage falls within the range of 15%, showcasing its standard behavior and serving as a benchmark for evaluating other brick types.

Water Absorption test values for various types of bricks is shown in Fig. 8 and the values are tabulated in Table 2.

4.2 Density Test

Each brick should be weighed separately. Then take measurements of the length, width, and depth of the brick samples. Finally, using the formula below, determine the density of bricks. Density for various types of bricks is shown in Fig. 9 and values are tabulated in Table 3.

$$\text{DENSITY} = \text{WEIGHT}/\text{VOLUME}$$

$$\text{VOLUME} = \text{LENGTH} \times \text{BREADTH} \times \text{DEPTH}$$

4.3 Compressive Test

Placed the specimen horizontally on the testing equipment with flat faces. Apply axial load at a constant pace until failure occurs. In addition, the maximum load at failure was reported. The load at failure is the highest load at which the specimen fails to continue increasing the indicative reading on the testing apparatus. Then use this formula to calculate the compressive strength of clay bricks: Compressive strength = compressive load/area. Compressive Test is

Table 2 Water absorption test for various types of bricks

Brick type	Dry weight	Wet weight	% of watre absorption
A1	3.450	3.780	10.36
	3.175	3.480	
	3.240	3.630	
A2	3.630	3.930	10.13
	3.125	3.560	
	3.165	3.375	
A3	3.180	3.520	9.23
	3.115	3.380	
	3.175	3.450	
B1	3.560	3.935	11.4
	3.045	3.480	
	3.085	3.384	
B2	3.165	3.485	9.8
	3.045	3.365	
	3.085	3.360	
B3	3.005	3.295	9.3
	3.135	3.385	
	2.960	3.268	
C1	3.100	3.335	12
	2.665	3.050	
	2.850	3.250	
C2	2.790	3.140	11.16
	2.875	3.185	
	2.815	3.105	
C3	2.790	3.250	17.01
	2.620	3.120	
	2.590	2.985	
D1	2.675	2.920	10.8
	2.895	3.255	
	2.975	3.236	
D2	2.790	3.250	17
	2.620	3.120	
	2.590	2.985	
D3	2.605	3.235	19
	2.695	3.128	
	2.615	3.058	
Controlled specimen	2.900	3.400	15
	2.980	3.345	
	3.145	3.645	

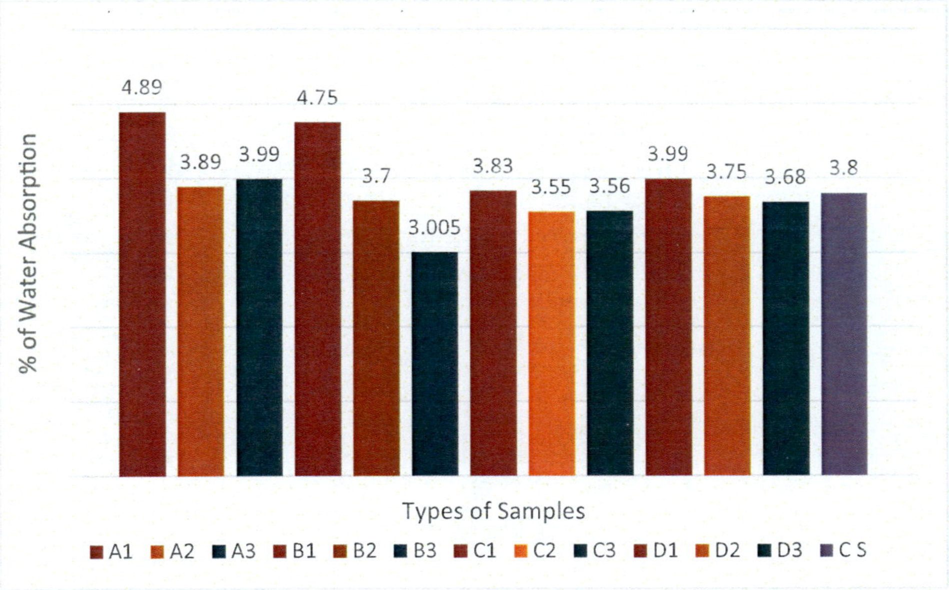

Fig. 8 Water absorption test values for various types of bricks

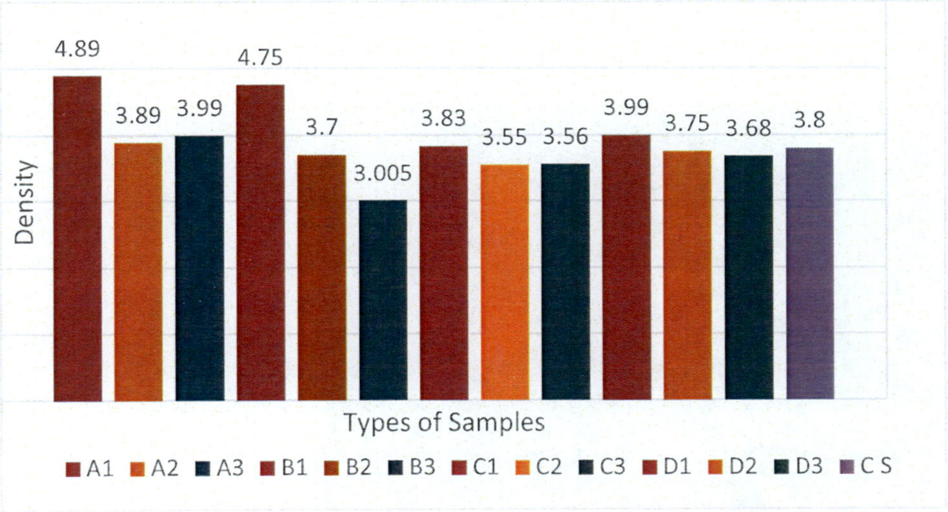

Fig. 9 Density for various types of bricks

shown in Fig. 10 and the values are tabulated in Table 4. Compressed strength analysis is shown in Fig. 11.

5 Conclusions

The study observed distinct burning behaviors of the different textiles used in the bricks. Cotton fibers burned and left a small quantity of fluffy gray ash, whereas polyester melted, becoming hard and non-crushable. Viscose demonstrated a similar burning tendency to cotton. The researchers investigated various physical properties of the bricks. The findings revealed that the addition of 20 g of polyester cotton (PC), polyester viscose (PV), or pure polyester to a brick increased its compressive strength by an impressive 45%, while also reducing water absorption. However, it was noted that increasing the quantity of textiles beyond this point led to diminishing strength compared to the original results.

Based on the study, it was suggested that used clothing, in the form of reinforced fabric, could be effectively utilized in conventional bricks. This approach has the potential to enhance the mechanical properties of the bricks while also promoting sustainable practices by recycling textile waste into construction materials. The research provides valuable insights into the potential benefits of incorporating textiles into brick manufacturing and presents a promising avenue for sustainable construction practices that reduce waste and improve the overall performance of building materials.

Table 3 Density test

Sample	Length	Breadth	Depth	Avg. Weight	Volume	Density	
	m	m	m	Kg	m^3	Kg/m^3	
A1	0.23	0.1	0.07	3.288	0.00161	2042.4	
A2	0.23	0.1	0.07	3.315	0.00161	2058.95	
A3	0.23	0.1	0.07	3.156	0.00161	1960.6	
B1	0.23	0.1	0.07	3.23	0.00161	2006.18	
B2	0.23	0.1	0.07	3.098	0.00161	1924.4	
B3	0.23	0.1	0.07	3.061	0.00161	1901.6	
C1	0.23	0.1	0.07	2.871	0.00161	1783.5	
C2	0.23	0.1	0.07	2.826	0.00161	1755.6	
C3	0.23	0.1	0.07	2.826	0.00161	1792.9	
D1	0.23	0.1	0.07	2.826	0.00161	1755.6	
D2	0.23	0.1	0.07	2.848	0.00161	1769.1	
D3	0.23	0.1	0.07	2.948	0.00161	1831.24	
ControlledsSpecimen	0.23	0.1	0.07	3.125	0.00161	1940.9	1915
	0.23	0.1	0.07	3.145	0.00161	1953.4	
	0.23	0.1	0.07	2.99	0.00161	1950.9	

Fig. 10 Compressive test

Table 4 Compressive strength

Sample	Length m	Breadth m	Depth m	Area mm^2	Compression load kN	Compressive strength N/mm^2	
A1	0.23	0.1	0.07	23,000	112.47	4.89	
A2	0.23	0.1	0.07	23,000	89	3.89	
A3	0.23	0.1	0.07	23,000	91.77	3.99	
B1	0.23	0.1	0.07	23,000	108.1	4.75	
B2	0.23	0.1	0.07	23,000	85.1	3.7	
B3	0.23	0.1	0.07	23,000	69.1	3.005	
C1	0.23	0.1	0.07	23,000	88.05	3.83	
C2	0.23	0.1	0.07	23,000	89.65	3.55	
C3	0.23	0.1	0.07	23,000	91.08	3.56	
D1	0.23	0.1	0.07	23,000	91.77	3.99	
D2	0.23	0.1	0.07	23,000	86.25	3.75	
D3	0.23	0.1	0.07	23,000	84.64	3.68	
Controlled specimen	0.23	0.1	0.07	23,000	89.5	3.84	3.80
	0.23	0.1	0.07	23,000	87.8	3.81	
	0.23	0.1	0.07	23,000	85.4	3.71	

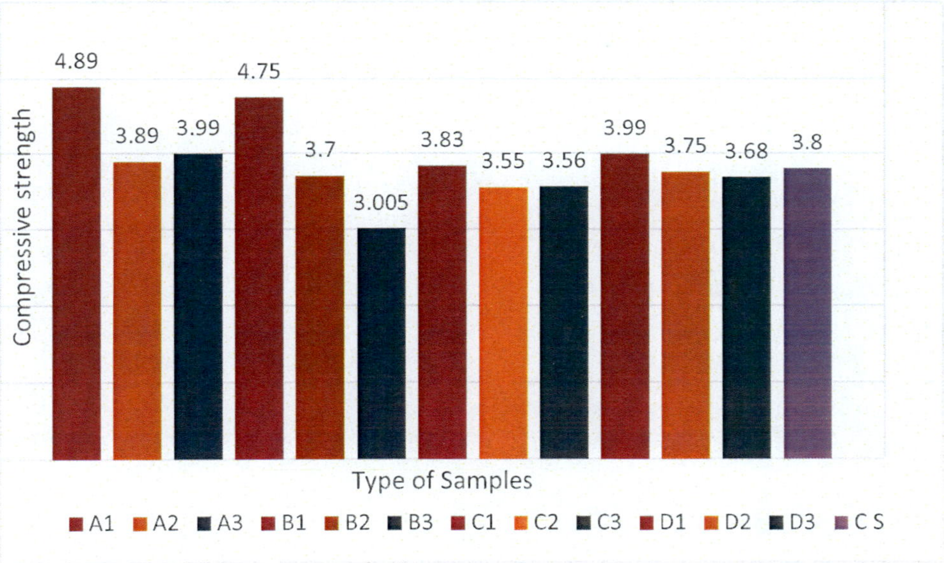

Fig. 11 Compressive strength for various types of bricks

References

Algin, H. M., & Turgut, P. (2008). Cotton and limestone powder wastes as brick material. *Construction and Building Materials, 22*, 1074–1080.

Aravind Raj, P. S., Divahar, R., Sangeetha, S. P., Naveen Kumar, K., Ganesh, D., & Sabitha, S. (2020). Sustainable development of structural joint made using high volume fly-ash concrete. *International Journal of Advanced Science and Technology, 29*(10S), 6850–6857.

Booth, J. E. *Textile mathematics* (vol. 1, pp. 154–156).

Bouchhima, L., Rouis, M. J., Choura, M. (2013). A study of pressure influences of phosphogypsum-based bricks. *International Journal of Civil Engineering & Technology (IJCIET), 4*(3), 143–154.

Briga, A. (2013). Textile waste as an alternative thermal insulation building material solution. *Construction and Building Materials, 38*, 155–160.

Divahar, R., Aravind Raj, P. S., Siva, M., Ispara Xavier, S. (2021). Durability performance of self-healing bacterial impregnated concrete with M-sand for sustainable environmental. *Indian Journal of Environmental Protection, 41*(10), 1120–1125.

IS 1077:1992, "Common burnt clay building brick-specification". Fifth revision.

IS 3495:1992, "Methods of tests of burnt clay bricks". Third Revision.

Jeyaraj, K. L., Muralidharan, C., Senthilvelan, T., & Deshmukh, S. G. (2012). A hybrid business strategy selection process for a textile company using SWOT and fuzzy ANP–a case study. *International Journal of Management (IJM), 3*(2), 124–143.

Kim, H. J., Mission, J. L., & Ko, J. H. (2013). Artificial neural network (Ann) prediction of one-dimensional consolidation in homogeneous clay under uniform applied load. *International Journal of Civil Engineering & Technology (IJCIET), 4*(4), 135–145.

Marjory. *Textbook on textile fibre* (pp. 11–13).

Murthy, H. V. S. *Textbook on textile fibre* (vol. 122).

Patel, P. A., Desai, A. K., Desai, J. A. (2012). Upgradation of non-ductile reinforced concrete beam column connections using fibre. *International Journal of Civil Engineering & Technology (IJCIET), 3*(2), 241–250.

Raut, S. P., Ralegaonkar, R. V., & Mandavgane, S. A. (2011). Development of sustainable construction material using industrial and agricultural solid waste: A review of waste-create bricks. *Construction and Building Materials, 25*, 4037–4042.

Krishna Reddy K. V., & Reddy K. P. (2013). Maturity period and curing as important quality control parameters for lime stabilized clay subgrades. *International Journal of Civil Engineering & Technology (IJCIET), 4*(2), 393–401.

Performance Analysis of Savonius VAWT with and Without Dimples and Fins by Using CFD

Shivaji Suryabhan Aher and Netra Pal Singh

Abstract

In an energy-demanding world with a growing population, effective energy methods to increase the extraction of work from various renewable energy sources are always encouraged. The primary goal of this study is to improve the performance of the Savonius vertical axis wind turbine, which is commonly used in areas with low wind speed. Dimples are added to the blade profile, resulting in a more aerodynamic design that reduces drag on wind turbine blades. In addition, to save space and accelerate pressure along the blade width, fins are added. This influences the wind turbine's ability to generate more power. The Computational Fluid Dynamics (CFD) analysis is used to understand the performance of the Savonius VAWT with and without Dimples and Fins when subjected to low incoming wind velocity. In this study, wind turbine blade is tested over air velocity 5, 8, 10, and 12 m/s with the use of CFD analysis. Result shows the combination of Dimples and Fins on blade surface of Savonius VAWT performs better at the air velocity of 10 m/s to precipitate more torque and output power for the generation of electricity.

Keywords

Renewable Energy • Savonius Vertical Axis Wind Turbine • Dimples • Fins • Computational Fluid Dynamics

S. S. Aher (✉) · N. P. Singh
Department of Mechanical Engineering, Oriental University, Indore, M.P, India
e-mail: shiva.mechboy@gmail.com

1 Introduction

In the twenty-first century, one of the greatest challenges we face is the energy resources that limit social progress and sustainable development. The lack of mineral resources and rise in environmental pollution pushes people to investigate and utilize advanced energy resources. In today's entire world renewable energy and sustainable invention is a unique subject to produce energy. Wind power is one of the foremost encouraging resources of inexhaustible energy. Several countries have the permit to explore and use it to its nice advantage of contamination free, plethoric convenience and conversion domestically, etc. It will therefore facilitate to scale back environmental contamination and also rely on fossil fuels.

Savonius wind turbine is the vertical axis wind turbine inaugurated by Sigurd J. Savonius in the 1920s which is simple in construction and operates with wind flow from all directions (Zemamou et al., 2017). Torque and power producing capacity of wind turbine could rise by 16.08% when radius, quantity, and location of Dimples are designed appropriately on the surface of wind turbine blade (Sedighi et al., 2020). Efficiency of wind turbine gets increased when size and arrangements of Dimples are properly located on the surface of the wind turbine blade (Arun et al., 2018). Implementation of Fin improves the production of Savonius wind turbine by 11%, and is qualified to deliver supreme performance in Savonius-type wind turbine (Utomo et al., 2018). Reducing drag and increasing lift utilizing Dimples on the blade surface helped in the increment of power of wind turbine and CFD results show efficiency increased by 15% of wind turbine (Rajaram Narayanana et al., 2018).

The wind turbine blade was analyzed with one fin, two fins and in the absence of fin at wind velocity of 5 and 7 m/sec. The outcome of simulation showed that with 2 fins wind turbine gives better performance than 1 and without fin (Ridwan & Setiyono, 2019). To trap the area in the blade in the direction of wind flow the fin is added. Blade with 1 Fin

M. Sumesh et al. (eds.), *2nd International Conference on Smart Sustainable Materials and Technologies (ICSSMT 2023)*, Advances in Science, Technology & Innovation, https://doi.org/10.1007/978-3-031-49826-8_20

generates electrical power more by 22.71% which is the highest compared with 2, 3, and 4 Fins at wind velocity of 5 m/s (Pamungkas et al., 2018).

Since its establishment, many researchers have studied Savonius vertical axis wind turbines. Some alterations improve wind turbine production by addition of Fin, end plates, aspect ratio, dispersed buckets, bucket spacing, obstacles, number of stages, blade angle, blade shape, guide blade and also many more (Chowdhury et al., 2016; Kalluvila & Sreejith, 2018; Salih Meri et al., 2019; Zheng et al., 2015). All of these alterations are made only to improve the angular speed and generations of high torque that the Savonius turbine can produce. This groundwork study will analyze combined effect of Dimples and Fins on Savonius wind turbine blade and compare the wind turbine with turbine blade without Dimples and Fins to generate more torque and output power.

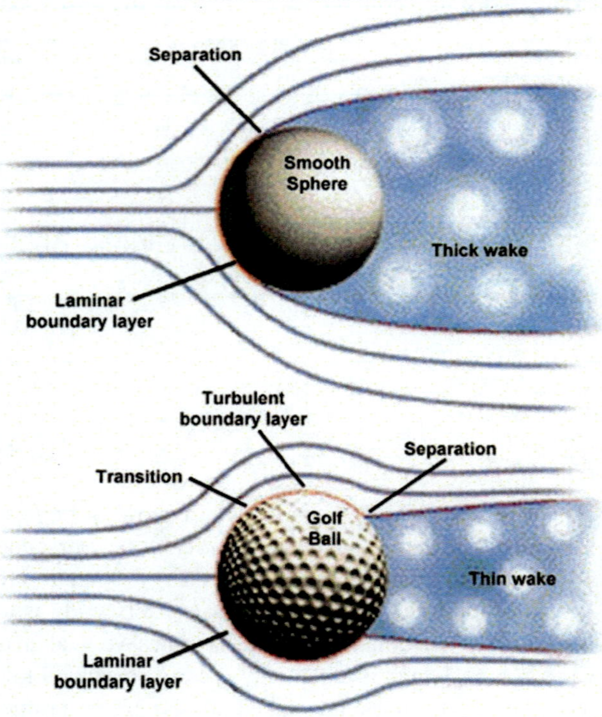

Fig. 1 Effect of Dimples on golf ball (Arun et al., 2018)

1.1 Dimples

Dimples are very powerful to lessen air drag as it is clean from the discount of drag and boom in carry-in golfing balls and covers lengthy flight interval (Alam et al., 2011; Prabowo et al., 2018; Ting, 2003). The bulk of wake vicinity at the back of the ball in motion significantly impacts drag pressure on it with the aid of the wind. If this wake vicinity is diminished through slowing down the segregation of air from surface of ball the overall pressure behind the ball will grow as shown in Fig. 1. Thus, the stress discrepancy in front of the ball and at the back of the ball will lower resulting in a reduction of drag force constituted via air. Subsequently, assimilation of Dimples at the wind turbine blade surface has assisted in growing the general performance of the wind turbine.

For this reason, the usage of CFD and ANSYS analysis an attempt changed into made to growth the torque and power of Savonius VAWT and subsequently the performance by means of changes within the floor topology of wind turbine blades.

1.2 Fins

Fins are a reformed segment of a turbine blade which saves space while improving friction across its width. Also, the power generation capacity of wind turbine is increased by adding fins on turbine blade (Aher & Singh, 2021a, 2021b). Turbine blade with Fins is shown in Fig. 2.

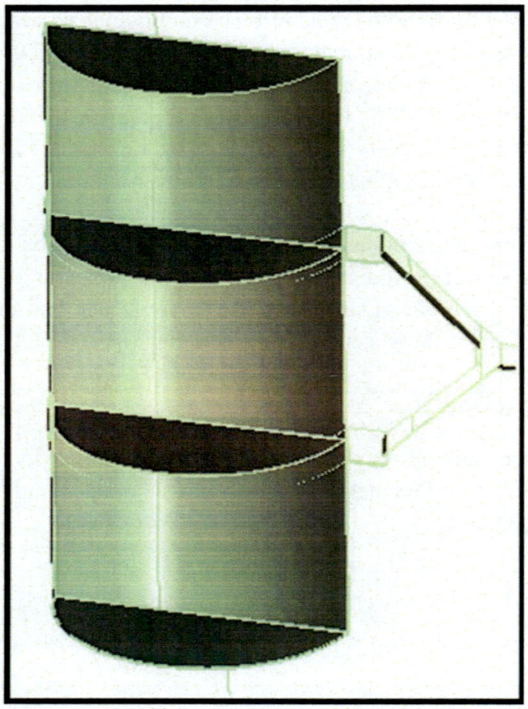

Fig. 2 Turbine blade with fins

2 Methods

Dimples on Blade Surface are shown in Fig. 3. This research examines the performance of S-VAWT with Dimples and Fins and without Dimples and Fins blades using specifications as follows (Tables 1 and 2).

The S-VAWT blades are incorporated with Dimples to raise the turbine efficiency and also the power output. The Dimples on the blade surface are used with the following specifications;

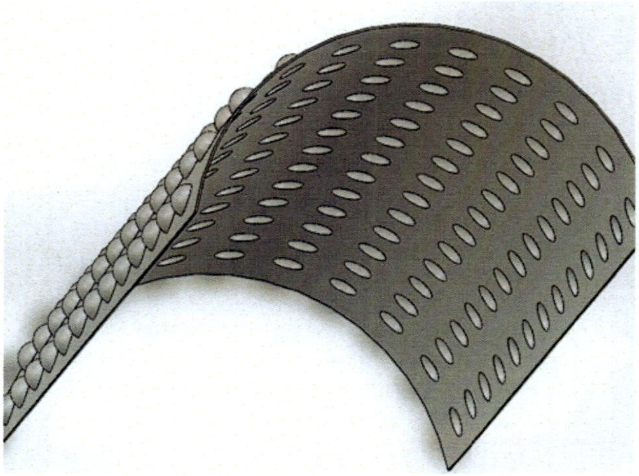

Fig. 3 Dimples on blade surface

Table 1 Specifications of the S-VAWT

Specifications	Value
Height of wind mill [H]	2.325 m
Diameter of wind mill [D]	2.866 m
Area of wind mill [A]	6.667 m^2
Width of blade [C]	0.477 m
Number of blades	3
Power co-efficient [Cp]	0.25
Tip speed ratio [TSR]	0.75

Table 2 Dimensions of Dimples on turbine blade surface

Specifications	Value
Diameter of Dimple	50 mm
Depth of Dimple	12.5 mm
Pitch	50 mm
Number of columns	11
Number of Dimples in column	13

3 CFD Analysis (Tool Used)

For creating CAD models, Ansys Design Modeler is adopted and to compute pressure distribution and velocity profile, Ansys CFX 19.2 is used and utilization of torque enumerates power output by performing CFD analysis as in Fig. 4.

3.1 Geometry Details

Geometry details are shown in Fig. 5. Air volume is divided into two parts for easy solving, such as stator and rotor. Rotor domain is defined as rotating part with 3000 RPM and stator is defined with stationary fluid domain.

3.2 CFD Analysis Without Dimples and Fins for Air Velocity at 5, 8, 10 and 12 m/s

The torque output on the blade obtained as a result of pressure distribution is calculated by utilizing a calculator embraced in the CFX article. After the simulation of the turbine without Dimple and Fins a torque of 74,919 N-m, 95,684 N-m, 116,293 N-m and 70,919 N-m is recorded at 5 m/s, 8 m/s, 10 m/s and 12 m/s wind velocity, respectively. The maximum torque 116,293 N-m is produced at 10 m/s wind velocity.

3.3 CFD Analysis with Dimples and Fins for Air Velocity of 5, 8, 10 and 12 m/s

Figures 6, 7, 8, 9, 10, 11, 12, 13, 14, 15, 16, 17, 18, 19, 20 show the CFD analysis. The torque output on the blade obtained as a result of pressure distribution is calculated by utilizing a calculator embraced in the CFX article. After the simulation of turbine without Dimple and Fins a torque of 84,919 N-m, 105,684 N-m, 126,293 N-m and 80,919 N-m is recorded at 5 m/s, 8 m/s, 10 m/s and 12 m/s wind velocity, respectively. The maximum torque 126,293 N-m is produced at 10 m/s wind velocity.

4 Results and Discussion

4.1 Result

The analysis of S-VAWT is preceded in two steps. The first step was done without Dimples and Fins which is compared with the results obtained for S-VAWT blades with Dimples and Fins. The results obtained for S-VAWT with and without Dimples and Fins are tabulated as follows (Table 3).

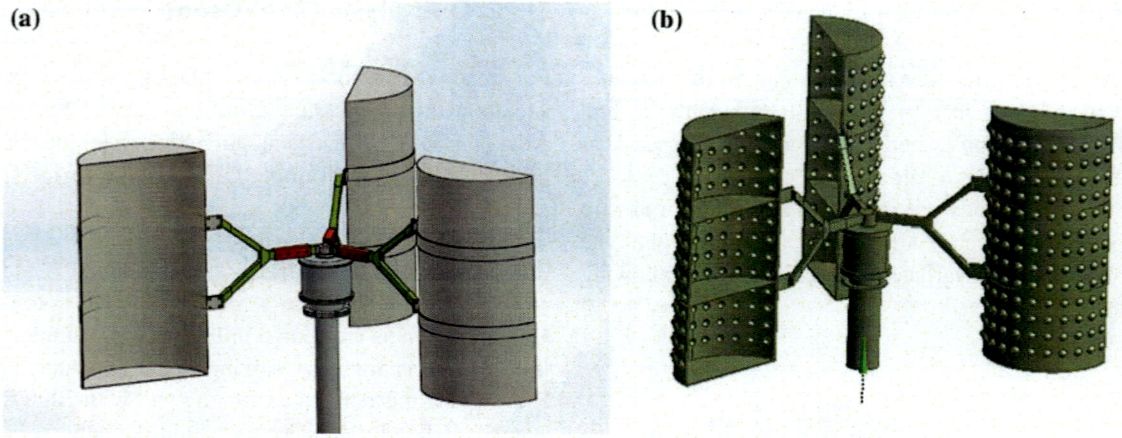

Fig. 4 CAD model of Savonius VAWT **a** without Dimples and Fins **b** with Dimples and fins

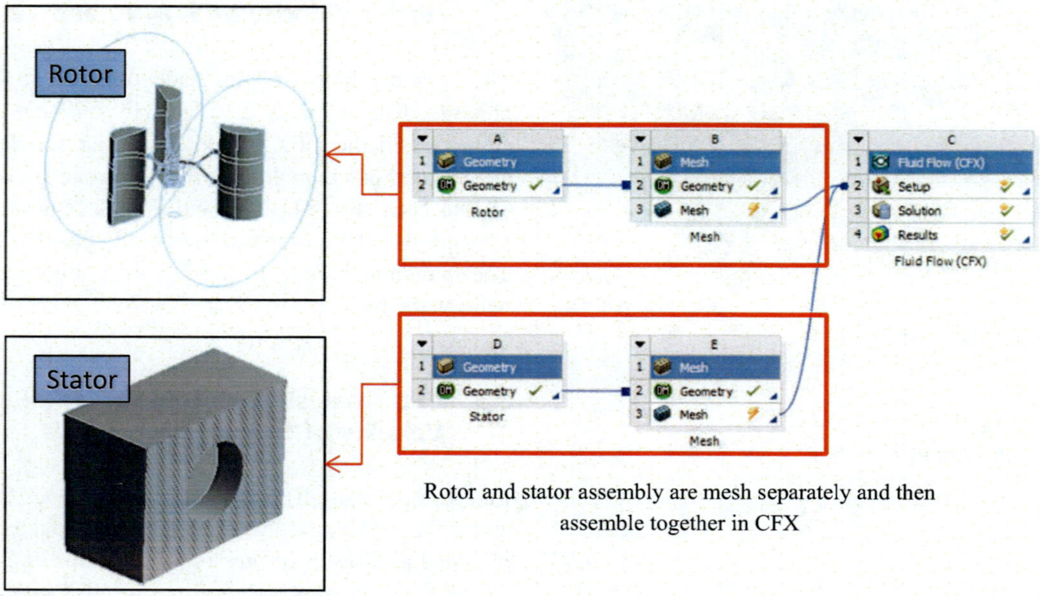

Fig. 5 Geometry details

Figure 21 shows the comparability of torque and velocity of Savonius VAWT without and with Dimples and Fins. From the graph it clears that, for both types of Savonius VAWT lowest torque was generated at 5 m/s air velocity while highest torque was achieved at air velocity of 10 m/s in both the cases. After that, if air velocity increased again torque generated capacity goes down.

Figure 22 shows the comparison of power and velocity of Savonius VAWT without and with Dimples and Fins at 5, 8, 10 and 12 m/s wind velocity. From the above evidence, at 10 m/s wind velocity power output produced is maximum.

4.2 Discussion

After analyzing the Savonius VAWT at velocities 5, 8, 10 and 12 m/s without and with Dimples and Fins, torque generated and hence output power both increase till the velocity of air at 10 m/s. If velocity of air increases again it shows torque and power output reduced in both the cases. It is clear from graphs 1 and 2, the highest torque was achieved and hence the output power at air velocity of 10 m/s when turbine blade incorporated with Dimples and Fins. Hence, turbine blade of Savonius VAWT with Dimples and Fins

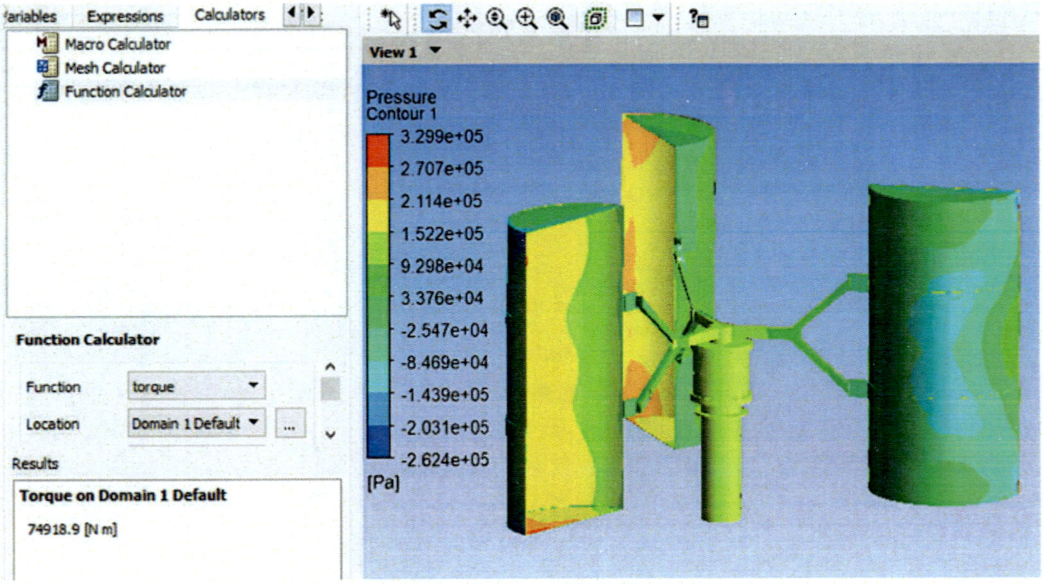

Fig. 6 Pressure contour on blade without Dimples and Fins at 5 m/s

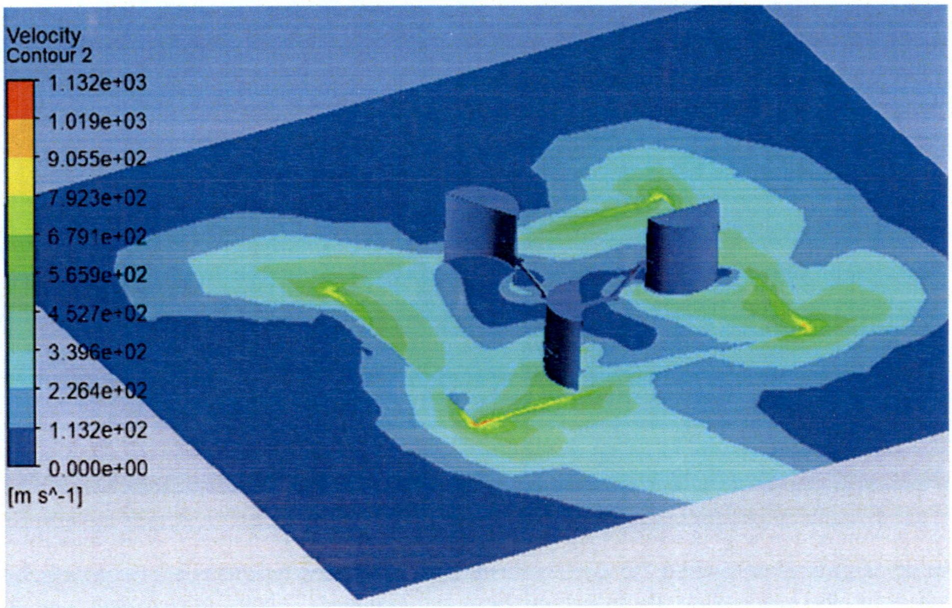

Fig. 7 Velocity contour on blade without Dimples and Fins at 8 m/s

gives excellent performance than turbine blade without Dimples and Fins.

4.3 Analysis for Torque and Power

Calculator available in CFX post is used to calculate the output torque on blade due to pressure distribution. Torque generated by the wind turbine blade is determined along the global z-axis as it is described axis of rotation.

For the blade without adding Dimples and fins, the torque induced by the turbine is found to be 116,293 N-m which can be converted to output power using the equation,

$$\text{Power(P)} = T \times \omega \quad \text{in watt} \tag{1}$$

where, T = Torque induced in N-meter and ω = Angular velocity in rad/sec

Using equation (i) output power is found to be 608.9 KW. Considering 20% loss when power is transmitted to the generator (Sarkar & Behera, 2012), the power of 487.12 KW

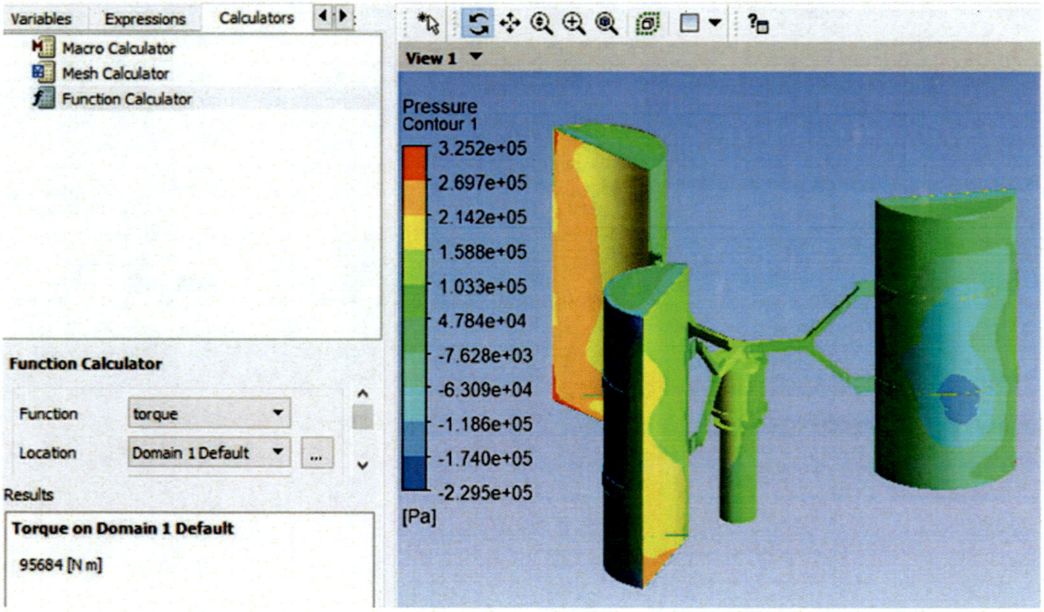

Fig. 8 Pressure contour on blade without Dimples and Fins at 8 m/s

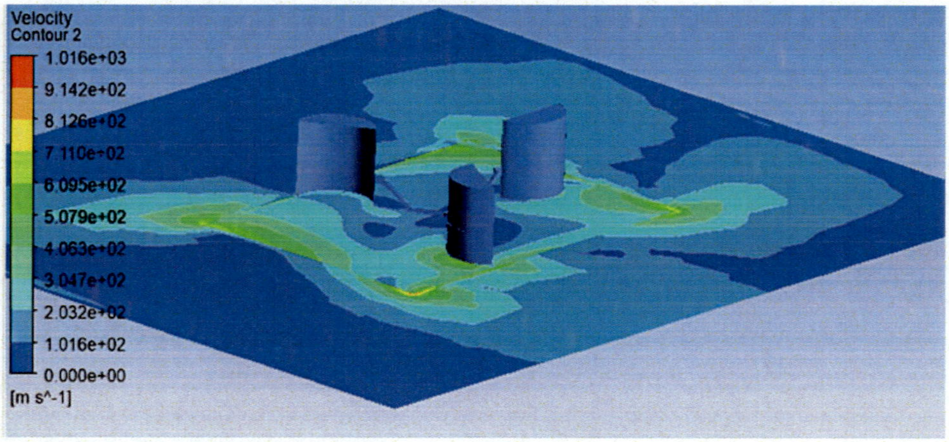

Fig. 9 Velocity contour on blade without Dimples and Fins at 10 m/s

is transmitted to the generator which is not far from the real value. Torque found to be 126,293 N-m for wind turbine when turbine blade with Dimples and Fins and produce 661.3 KW power. After considering loss, it gives 529.04 KW to the generator.

5 Conclusion

The study on Savonius VAWT invented with Dimples and Fins was done in order to diminish air drag to improve power and hence efficiency of the turbine. From the results, the following conclusions were reported. With the combination of Dimples and Fins, the Torque and Power are increased till air velocity of 10 m/sec. After increasing the air velocity, it is found that the Torque dropped and hence output power of turbine also decreased. With the combination of Dimples and Fins the efficiency of Savonius VAWT is better than turbine blade without Dimples and Fins. Hence, wind turbine designed at air velocity 10 m/sec with Dimples and Fins is observed to be best among all.

6 Future Scope

As per the overall discussion done in previous research papers and journals, research done is regarding Vertical Axis Wind Turbine blade of Savonius type with and without

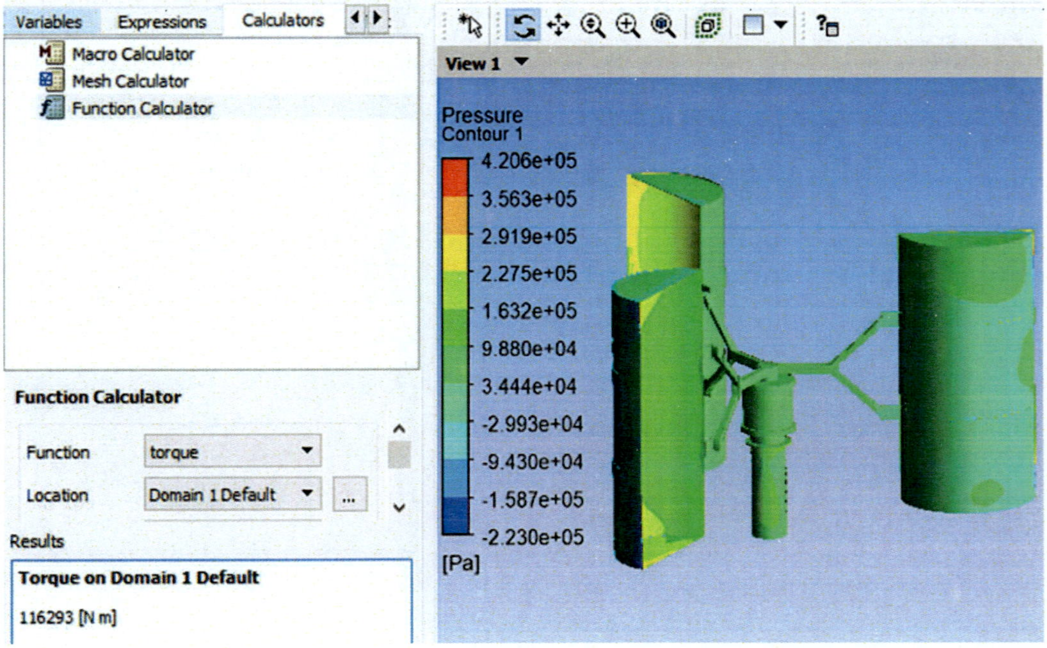

Fig. 10 Pressure contour on blade without Dimples and Fins at 10 m/s

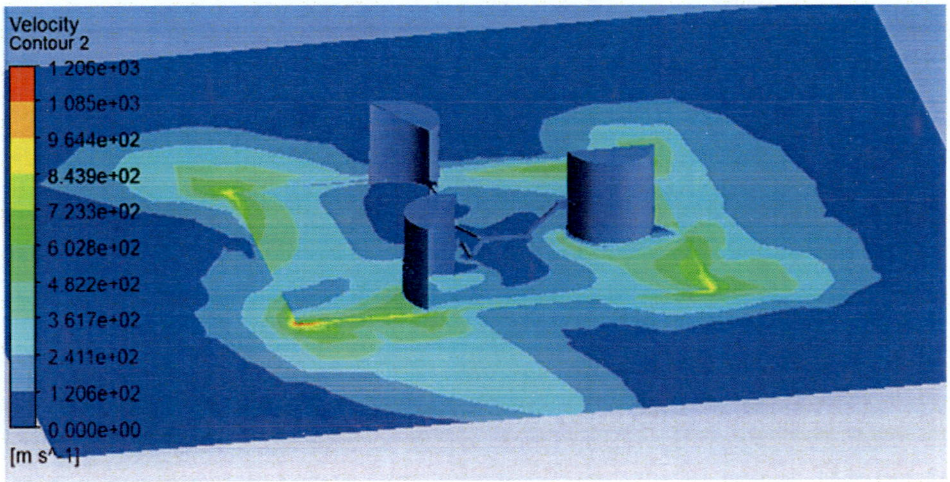

Fig. 11 Velocity contour on blade without Dimples and Fins at 12 m/s

Dimples and Fins. It has been shown that there is an improvement in efficiency and power of Savonius-type VAWT with Dimples and Fin. Some of the following modifications are suggested for future work for improvement in performance of Savonius-type VAWT. The present research is carried with three blades with single cup shape blade. It can be modified by three blades with two cup-shaped blades (like Pelton Wheel Turbine Blade) connected to each other with same surface area to improve the performance of the Savonius-type VAWT. In the present work, the turbine blades are designed with Dimples and Fins without obstacles, to improve the performance of VAWT it may be modified using a splitter at the end of the blade. The present research work if modified with the above changes, the efficiency and power output capacity may be increased by 1.5% more than the designed Savonius-type VAWT.

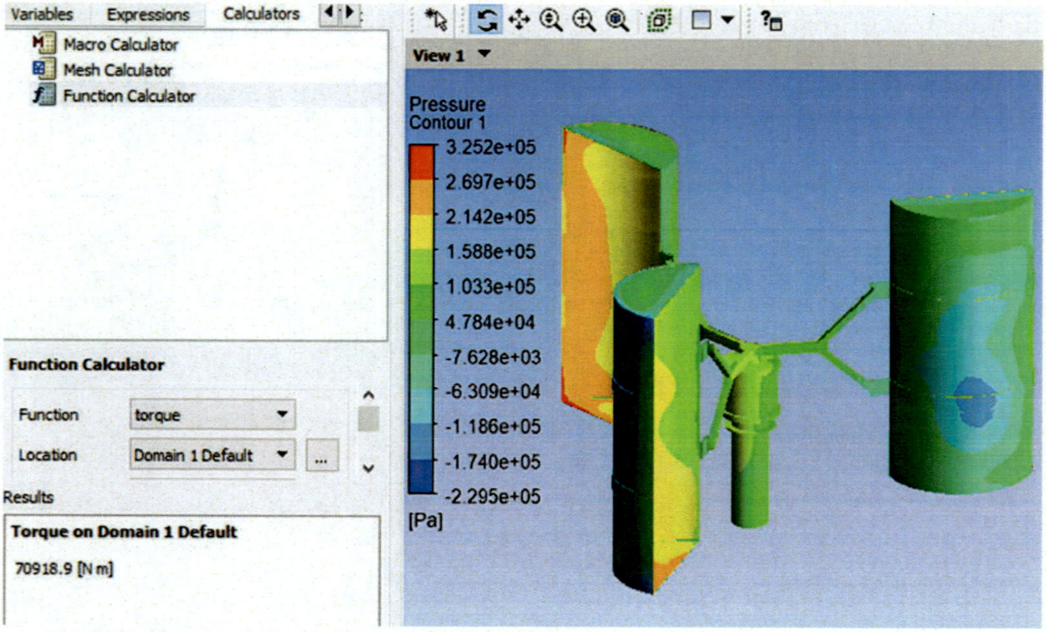

Fig. 12 Pressure contour on blade without Dimples and Fins at 12 m/s

Fig. 13 Velocity contour on blade with Dimples and Fins at 5 m/s

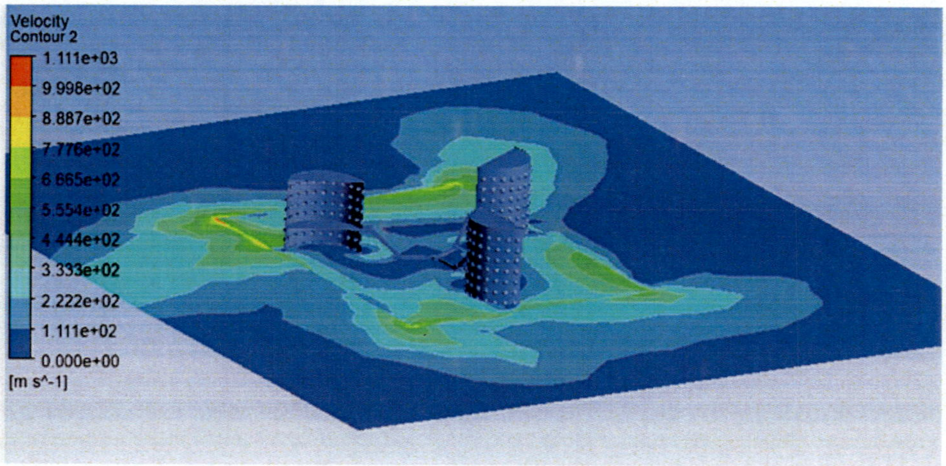

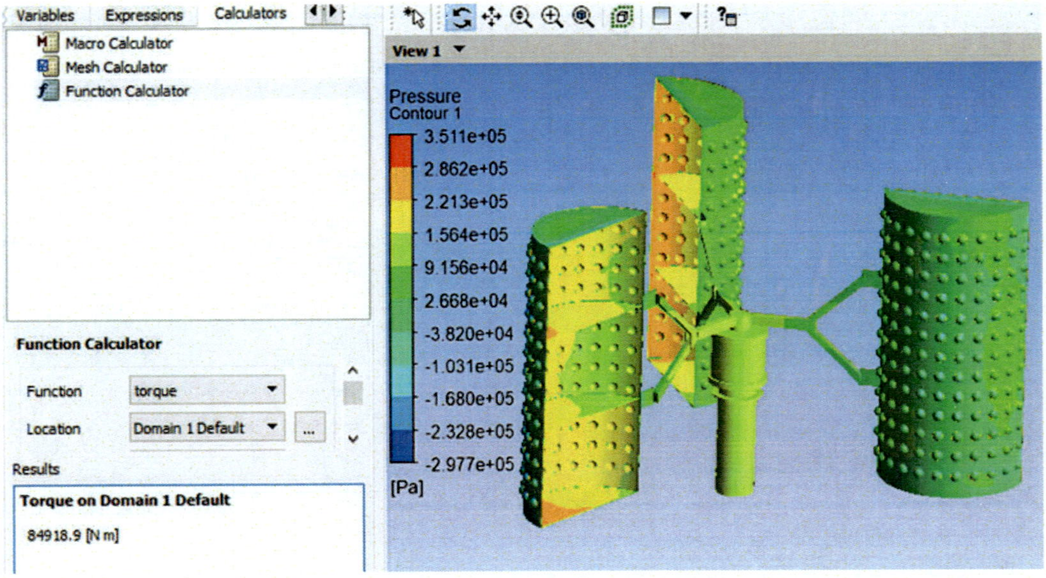

Fig. 14 Pressure contour on blade with Dimples and Fins at 5 m/s

Fig. 15 Velocity contour on blade with Dimples and Fins at 8 m/s

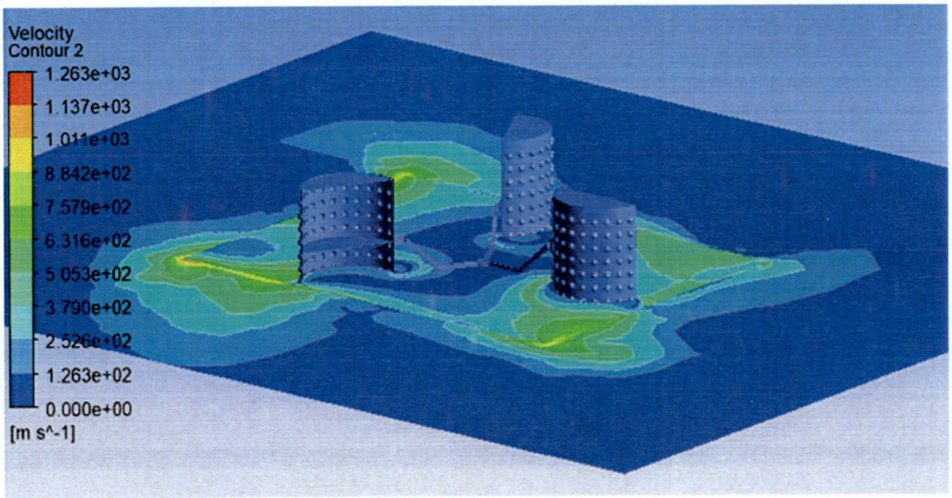

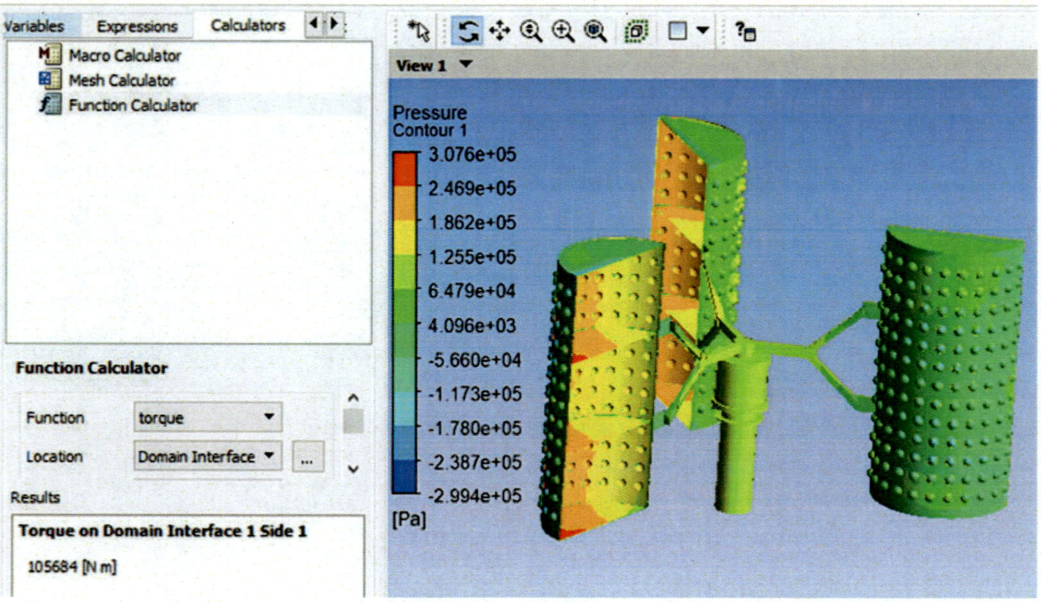

Fig. 16 Pressure contour on blade with Dimples and Fins at 8 m/s

Fig. 17 Velocity contour on blade with Dimples and Fins at 10 m/s

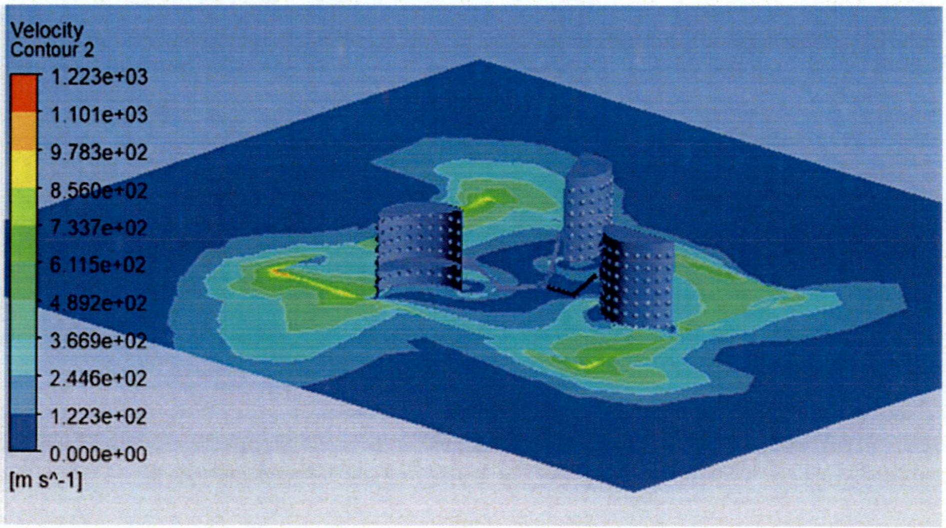

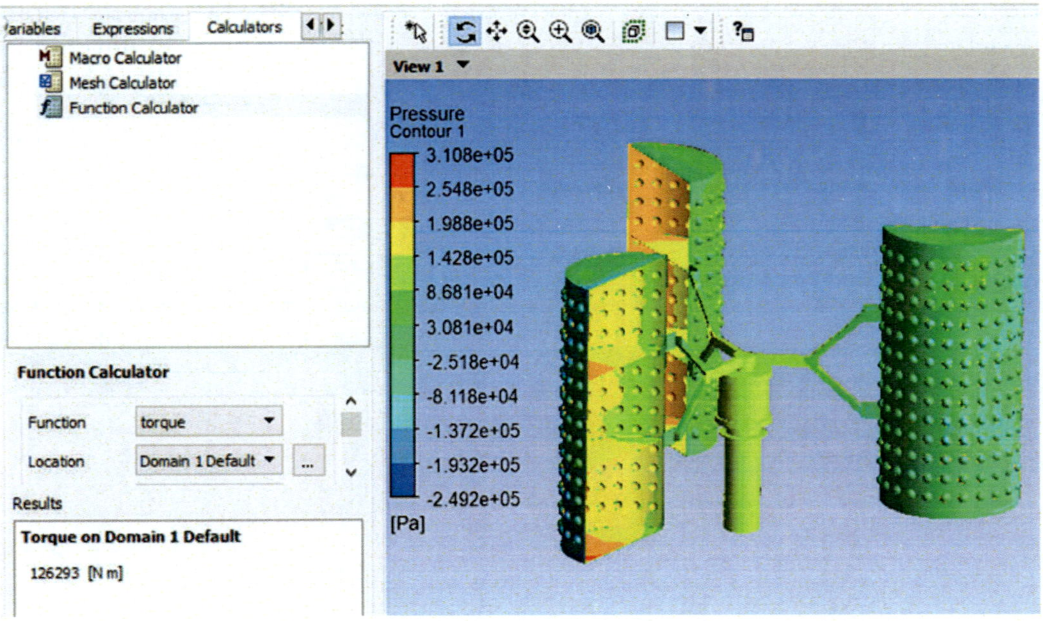

Fig. 18 Pressure contour on blade with Dimples and Fins at 10 m/s

Fig. 19 Velocity contour on blade with Dimples and Fins at 12 m/s

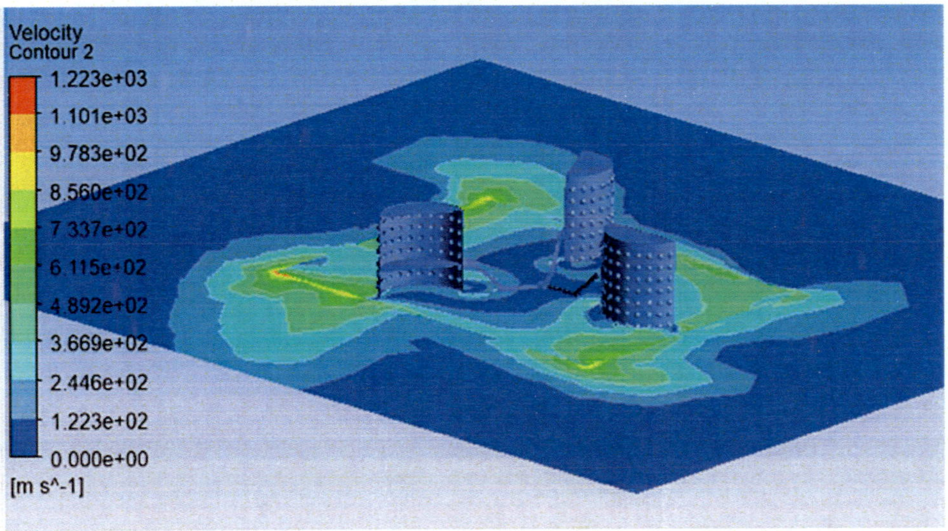

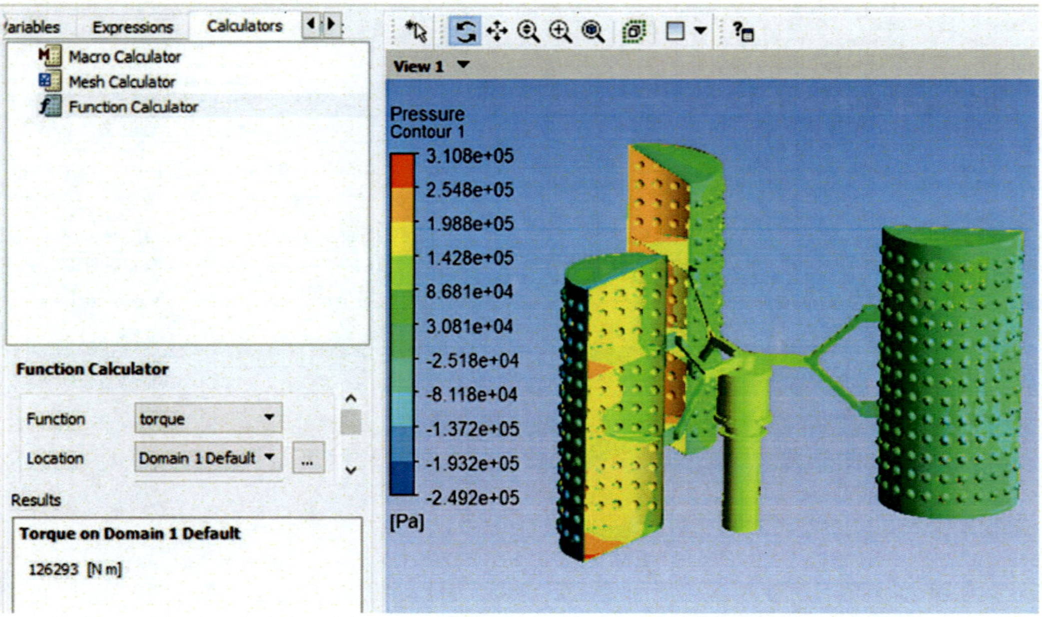

Fig. 20 Pressure contour on blade with Dimples and Fins at 12 m/s

Table 3 Output torque and power generated by Savonius VAWT without and with Dimples and Fins

Blade shape	case no	Velocity (m/sec)	Speed (RPM)	Torque output (N-m)	Speed (rad/sec)	Power (KW)
Without Dimples and Fins	1	5	25	74,919	2.6	196.1
	2	8	40	95,684	4.2	400.8
	3	10	50	116,293	5.2	608.9
	4	12	60	70,919	6.3	445.6
With Dimples and Fins	5	5	25	84,919	2.6	222.3
	6	8	40	105,684	4.2	442.7
	7	10	50	126,293	5.2	661.3
	8	12	60	80,919	6.3	508.4

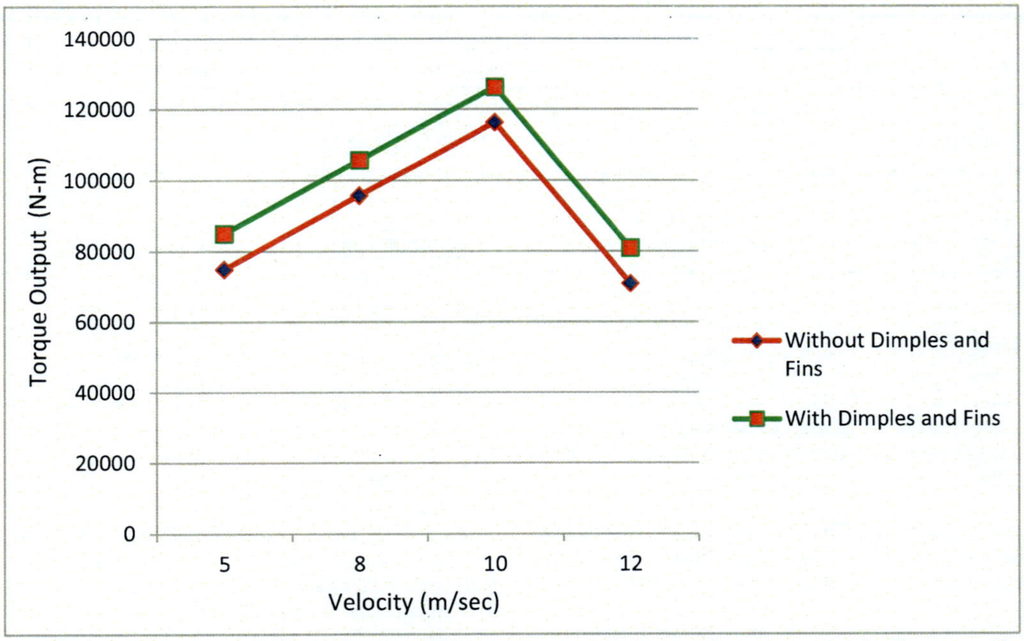

Fig. 21 Comparison of torque versus velocity with and without Dimples and Fins

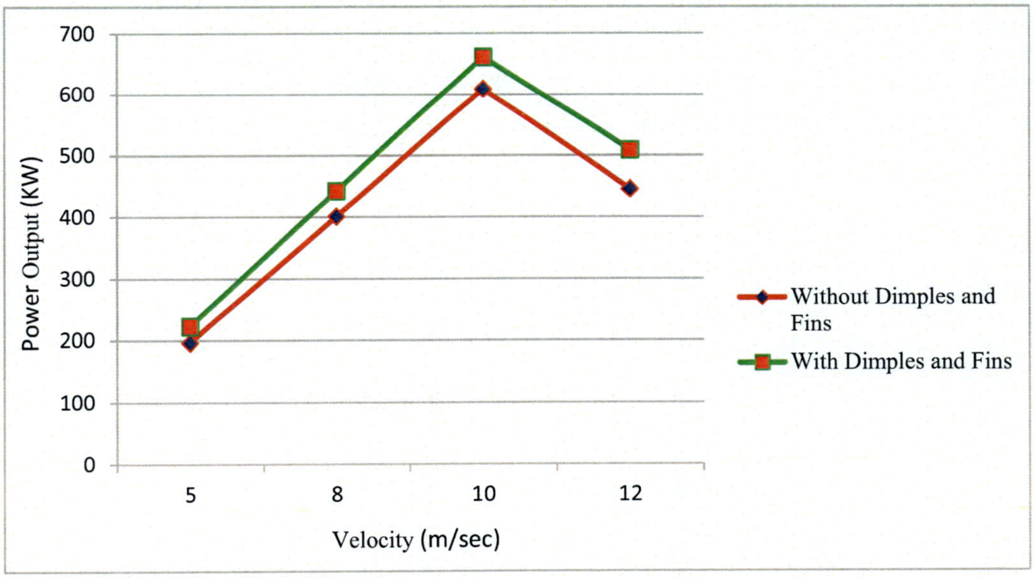

Fig. 22 Comparison of power versus velocity with and without Dimples and Fins

References

Aher, S. S., & Singh, N. P. (2021). Analysis for the improvement of efficiency and power of Savonius VAWT without Dimples and Fins. *Journal of Xidian University*, *15*(4), 482–487.

Aher, S. S., & Singh, N. P. (2021). Design and analysis of Savonius VAWT with Dimples and Fins. *IOSR Journal of Mechanical and Civil Engineering (IOSR-JMCE)* 18(6), 18–24.

Alam, F., Steiner, T., Chowdhury, H., & Subic, A. (2011). A study of golf ball aerodynamic drag. *Procedia Engineering 13*, 226–231. https://doi.org/10.1016/j.proeng.2011.05.077

Arun, K. K., Navaneeth, V. R., Sam Vimal Kumar, S., & Ajay, R. (2018). Analyzing the effect of Dimples on wind turbine efficiency using CFD. *International Journal of Applied Engineering Research*, *13*(6), 4484–4489.

Chowdhury, H., Loganathan, B., Wang, Y., Mustary, I., & Alam, F. (2016). A study of dimple characteristics on golf ball drag. *Procedia Engineering, 147*, 87–91. https://doi.org/10.1016/j.proeng.2016.06.194

Kalluvila, J. B. S., & Sreejith, B. (2018). Numerical and experimental study on a modified Savonius rotor with guide blades. *International Journal of Green Energy*. https://doi.org/10.1080/15435075.2018.1529574

Pamungkas, S. F., Wijayanto, D. S., Saputro, H., & Widiastuti, I. (2018). Performance 'S' type Savonius wind turbine with variation of fin addition on blade. *IOP Conference Series: Materials Science and Engineering*, *288*.

Prabowo, D., Wijayanto, D. S., & Widiastuti, I. (2018). Experimental study of incorporating fins on the rotor blades of Savonius wind turbine. *IOP Conference Series: Materials Science and Engineering*, *288*, 012046.

Rajaram Narayanana, M., Nallusamy, S., & Ragesh Sathiyan, M. (2018). Design and analysis of a wind turbine blade with Dimples to enhance the efficiency through CFD with ANSYS R16.0. *ICMMPM, MATEC Web of Conferences* (vol. 207).

Ridwan, I. S., & Setiyono. (2019). Performance of vertical axis Savonius wind turbines related to the fin number on the blade. *International Conference on Design, Energy, Materials and Manufacture, IOP Conf. Series: Materials Science and Engineering*, *539*.

Salih Meri, A. R., Hamidon Bin Salleh, B., Nemah, M. N., Al-Quraishi, B. A., & Asmuin, N. Z. B. (2019). Performance evaluation of Savonius wind turbine based on a new design of blade shape. *International Journal of Mechanical Engineering and Technology*, *10*(01), 837–846.

Sarkar, A., & Behera, D. K. (2012). Wind turbine blade efficiency and power calculation with electrical analogy.

Sedighi, H., Akbarzadeh, P., & Salavatipour, A. (2020). Aerodynamic performance enhancement of horizontal axis wind turbines by Dimples on blade: Numerical investigation. *Energy*.

Ting, L. (2003). Effects of Dimple size and depth on golf ball aerodynamic performance. In *Conference: ASME/JSME 4th joint fluids summer engineering conference, Volume 2: Symposia, Parts A, B, and C*. Paper No. FEDSM2003-45081 (pp. 811–817). https://doi.org/10.1115/FEDSM2003-45081

Utomo, I. S., Danardono, D., Tjahjana, D. P., & Hadi, S. (2018). Experimental studies of Savonius wind turbines with variations sizes and fin numbers towards performance. *AIP Conference Proceeding, 1931*, 030041–030047.

Zemamou, M., Aggour, M., & Toumi, A. (2017). Review of Savonius wind turbine design and performance. In *4th international conference on power and energy systems engineering, CPESE 2017*, Berlin, Germany (pp. 25–29).

Zheng, M., Li, Y., Tian, Y., Hu, J., Zhao, Y., & Yu, L. (2015). Effect of blade installation angle on power efficiency of resistance type VAWT by CFD study. *International Journal of Energy Environment*. https://doi.org/10.1007/s40095-014-0142-0

Effect of Different Rib Shapes on the Performance of Solar Thermal Collector

Rehan Fazal, Ravindra Mohan, Geetesh Goga, Santosh Kumar, Anil Singh Yadav, and Subhendu Chakroborty

Abstract

Solar Thermal Collectors have been growing in demand due to their efficiency of converting the solar energy into electrical energy without producing any form of waste or pollution. Thus, a number of research projects are being conducted in the field of enhancing the performance of the solar thermal collectors. Enhancing the heat dissipation through the collectors by adding ribs to its surface is an effective method to save money on energy. In this paper, a solar collector having total length of 605 mm is selected which is then divided into an inlet section of 120 mm, an outlet section of 225 mm and a test section in the middle of 260 mm. The test section is provided with ribs of 1.25 mm thickness and 5 mm height at 18 mm distance each. Three different shapes of the ribs, i.e. pentagonal, Circular and T-section have been analysed. The results show that the T-section ribs have been found more efficient with a maximum temperature of 392.99 K at the rib walls. Hence, T-section ribs have been suggested in this analysis with the best performance.

Keywords

Solar thermal collectors • Pentagonal • Circular • T-sectional ribs • Performance

R. Fazal · R. Mohan · A. S. Yadav
Department of Mechanical Engineering, IES College of Technology, Bhopal, Madhya Pradesh, India

G. Goga
Department of Mechanical Engineering, Bharat Group of Colleges, Sardulgarh, Mansa, Punjab, India

S. Kumar (✉)
Department of Mechanical Engineering, CGC-College of Engineering, CGC, Landran, Mohali, Punjab, India
e-mail: falsesantoshdgc@gmail.com

S. Chakroborty
Department of Basic Sciences, IES University, Bhopal, Madhya Pradesh, India

1 Introduction

Solar energy, or sun energy, has received increased attention among the renewable energy sources due to the wide range of uses it provides. With solar collector technology, this solar energy is transformed into thermal heat or electricity for usage. It has been used for heating purposes for thousands of years. Yet, as technology advanced, its range of uses and conversion efficiency rose. To present, a variety of applications have exploited solar thermal collector technologies. The fundamental idea behind a solar thermal collector is that part of the solar energy that strikes a surface is absorbed, which raises the surface temperature. Heat from the higher temperature is carried to the application site through a fluid-based medium. Nevertheless, the effectiveness of the solar collector is dependent on the surface's ability to absorb light, how much radiation is lost to the environment, and how much heat energy can be extracted from the collector (Hussain & Hasanuzzaman, 2022; Yang et al., 2015). The basic difference between both types of correctors is that in the case of non-concentrating collector, a large surface area is required to enhance the efficiency of the collector. Compared to flat collectors, concentrating solar collectors often offer greater temperatures. Reducing the heat loss area often improves the energy delivery temperature for concentrating collectors. Radiation is absorbed by the receiver, which transforms it into more energy (usually thermal energy). An absorber, insulation, and cover materials make up the device. The radiation is delivered to the receiver via an optical system or concentrator. Concentrators come in a variety of shapes, including cylindrical or surface-of-revolution, segmented, or continuous. Receivers can be flat, convex, or concave, as well as covered or uncovered. Concentrators can be used in conjunction with storage and produce greater temperatures (Hussain & Hasanuzzaman, 2022; Yang et al., 2015). Recently, several researches are being carried out to enhance the effectiveness of the solar thermal collectors (Yang et al., 2015). Various

M. Sumesh et al. (eds.), *2nd International Conference on Smart Sustainable Materials and Technologies (ICSSMT 2023)*, Advances in Science, Technology & Innovation, https://doi.org/10.1007/978-3-031-49826-8_21

researchers used sol–gel method and dip coating technique to create SiO_2 anti-reflective nonporous films. The greatest solar transmittance measurements (96% in the wider spectrum range of 250–2500 nm) are made (Ataee & Ameri, 2015; Broughton et al., 2018; Hussain & Hasanuzzaman, 2022; Kumar et al., 2017; Yang et al., 2015). Broughton et al., 2018 summarised the different ways to enhance the performance of thermal collectors: (i) enhancing the solar selectivity of the absorber, (ii) by decreasing the thermal losses from the solar thermal collector and (iii) enhancing the heat transfer coefficient b/w the working fluid and the absorber surface (Broughton et al., 2018). However, the main variables influencing the performance of thermal solar collectors are: absorber absorptivity, area, emissivity of glass cover, absorber plate temperature, tilt angle of collector, number of glass covers, etc. (Kumar et al., 2017). Darhigh and Shafieian (2016) used a solar water heating system with an evacuated tube heat pipe collector as the subject of a study to theoretically and empirically evaluate its performance. The performance of the collector was first analysed mathematically using a thermal and exergy analysis model, and then the given system was built and physically evaluated using real consumption data (Essa & Mostafa, 2017; Ghaderian et al., 2017; Goga, et al., xxxx; Kumar et al., 2022; Mohan et al., 2023). To study the performance of the collector, a mathematical model depending on thermal and exergy analysis was first presented. Thereafter, the system was built and physically tested in accordance with actual consumption pattern (Essa & Mostafa, 2017; Ghaderian et al., 2017; Goga, et al., xxxx; Kumar et al., 2022; Mohan et al., 2023). The findings collected demonstrated that 15 were considered to be the recommended number of collection pipes.

Essa & Mostafa, (2017) conducted a simulation of water flow in an evacuated tube solar collector numerically that was linked to a storage tank in three dimensions. It uses experimental data from several researches to confirm the temp. distribution inside the evacuated tube. Discussions were on variations in solar radiation intensity and incidence angle. Sobhansarbandi et al., (2017) (Kumar et al., 2022) developed a solar collector that makes use of Octadecane paraffin, a phase transition material with a melting point of 28 °C and long-term chemical stability PCM. Phase change materials (PCMs), due to PCMs' poor heat transfer rate, low thermal diffusivity, and low thermal conductivity, especially in powder form, they may not be efficient on their own. However, authors were able to overcome these drawbacks and create a novel and effective solar water heater by combining CNT layers with the high thermal diffusivity and thermal conductivity compared to phase change materials (Jhariya et al., 2019; Mahbubul et al., 2018). Mahbubul et al. 2018 investigated how well an evacuated tube solar collector system heated water while it was first operating (Ghaderian & Sidik, 2017; Henein & Abdel-Rehim, 2022). The collector is a component of an absorption cooling system with a 20 kW heating capability. The goal of the investigation was to determine how a water-single-walled carbon nanotube nanofluid affects the collector's ability to function. The outcomes show that when the collector is operated with water and 0.2 vol% nanofluid, respectively, up to 56.7 and 66% of efficiencies are recorded (Amin et al., 2012; Boulemtafes-Boukadoum & Benzaoui, 2014; Henein & Abdel-Rehim, 2022; Kumar et al., 2017; Rajput & Yang, 2018). Further, Mercan & Yurddaş, (2019) used computational fluid dynamics (CFD) to do a numerical study of water and water-based nanofluids in evacuated tube solar collectors (ETSCs). A comparison between an experimental result and two alternative numerical outcomes was used to validate the study. The analyses allowed for comparisons between the impacts of H_2O-based $Al_2O_3{-}H_2O$ and $CuO{-}H_2O$ nanofluids and water on heat transfer for various nanoparticle volume fractions, collector angles, mass flow rates, and numbers of evacuated tubes. These characteristics were used to determine the tank output temperatures and to assess the thermal and hydraulic statuses of the collector employing the Boussinesq approximation. The use of nanofluid in ETSC systems has been found to increase heat transmission, with $CuO{-}H_2O$ nanofluid showing the greatest improvement. The influence of nanofluids on solar evacuated tube solar collectors is presented by Olfian et al., (2020). Gholipour et al., (2021) used three different types of absorber tubes were used to increase the effectiveness of a solar collector with a vacuum tube. The absorber tubes contained a silica/water-ethylene glycol nanofluid. Instantaneous entrance and output temperatures, the surrounding temperature, and sun radiation were all measured as part of the experiment. In 2022, Henein and Abdel-Rehim (Henein & Abdel-Rehim, 2022) used an effective mode to sustainably improve the thermal effectiveness of the solar collector by using a hybrid nanofluid. The experimental research is conducted to ascertain the effect of using a hybrid nanofluid consisting of multi-walled carbon nanotubes and magnesium oxide (MgO/MWCNT) on the thermal behaviour of the evacuated tube solar collector. For a water base combination of MgO and Multi-walled carbon nanotubes (MWCNTs). Four distinct weight ratios of (80:20), (70:30), (60:40), and (50:50) are utilised, correspondingly (Abhay et al., 2018; Bhakre et al., 2021; Jhariya et al., 2019; Qian et al., 2017). The trials are carried out at varied volume flow rates between 1 and 3 L/min with a particle concentration of 0.02%. The findings indicate that when MWCNTs nanoparticle weight ratios and volume flow rates rise, energy and exergy efficiency improve (Afridi et al., 2023; Prasad et al., 2014). Various researchers (Han et al., 1978; James, 2021) reported that the thermal performance of solar collector depends upon various parameters such as rib shapes, rib height to hydraulic diameter ratios, rib angles, and rib pitch. Hence, the aim of current investigation is to examine the effect of different rib shapes on thermal performance of solar collector.

From the literature review, it has been found that there are lots of experimentation to increase the performance of the thermal collectors, but there are very few studies carried out to check the effect of different rib shapes on the performance of solar thermal collector. Based upon that the authors have chosen three different rib shapes to check the performance of solar thermal collector.

2 Material and Method

2.1 Materials Employed

An aluminium solar collector has been used in this experiment with ribs cut on it at regular intervals. Air was used as a fluid. The properties of these materials have been discussed in Table 1.

However, the method employed in this current study is mesh generation. During the study, SOLIDWORKS and ANSYS was used.

2.2 Governing Equation

The turbulent fluid flow through pentagon roughened solar air collector is governed by continuity, momentum (Navier–Stokes equations), and energy equations and are shown below: As the analysis domain is two dimensional so governing equations are:

$$\frac{\partial u}{\partial x} + \frac{\partial v}{\partial y} = 0$$

where, ∂ u and ∂ v is partial velocity in x and y direction, respectively.

Momentum eqn. in the direction of x and y axis:

$$\left(u\frac{\partial u}{\partial x} + v\frac{\partial u}{\partial y}\right) = -\frac{\partial p}{\partial x} + \frac{1}{Re}\left(\frac{\partial^2 u}{\partial x^2} + \frac{\partial^2 u}{\partial y^2}\right) + \frac{\partial}{\partial x}\left(-\rho\overline{u_1{}'u_x{}'}\right)$$

$$\left(u\frac{\partial v}{\partial x} + v\frac{\partial v}{\partial y}\right) = -\frac{\partial p}{\partial y} + \frac{1}{Re}\left(\frac{\partial^2 v}{\partial x^2} + \frac{\partial^2 v}{\partial y^2}\right) + \frac{\partial}{\partial y}\left(-\rho\overline{u_1{}'u_y{}'}\right)$$

Energy equation:

$$\frac{\partial}{\partial x_i}\left(\rho u_j T\right) = \frac{\partial}{\partial x_j}\left[(\Gamma + \Gamma_t)\frac{\partial T}{\partial x_j}\right]$$

where, Γ is "molecular thermal diffusivity" & t is the "turbulent thermal diffusivity" and depicted as:

$$\Gamma = \mu/Pr$$

$$\Gamma_t = \mu_t/Pr_t$$

$$-\rho\overline{u_1{}'u_{x,y}{}'} = \mu_t\left(\frac{\partial u_x}{\partial u_y} + \frac{\partial u_y}{\partial u_x}\right)$$

The average Nusselt number evaluated for roughened solar air collector is:

$$Nu_r = hD/k$$

where h is the convective heat transfer coefficient. The friction factor is presented below as a function of pressure drops across the length of the analysis section.

$$ff = \frac{(\Delta P/L_2)D}{2\rho v^2}$$

For smooth duct Nusselt number is computed as

$$\mathrm{Nu_s} = 0.0481 Re^{0.7354}$$

And for the friction factor in smooth duct is evaluated as

$$ff_s = 0.788 Re^{-0.3402}$$

where, Nu_s = Nusselt number for smooth duct, Re = Reynold no., ff_s = friction factor in smooth duct.

2.3 Design of the Solar Collector

A solar collector of total length 605 mm is selected which is then divided into an inlet section of 120 mm, an outlet section of 225 mm and a test section in the middle of 260 mm. The test section is provided with ribs of 1.25 mm thickness at 18 mm distance each. The height of the ribs will be varied for the analysis (Fig. 1).

Table 1 Material properties of AIR and ALUMINIUM

Properties	For AIR	For ALUMINIUM
Thermal conductivity ($Wm^{-1}K^{-1}$)	0.0242	202.4
Density (Kg/m^3)	1.225	2719
Specific heat (J/KgK)	1006.43	871
Viscosity	1.7894×10^{-5}	–

Table 2 No. of nodes and elements in each case

Rib height	Nodes	Elements
Pentagonal ribs	12,848	12,152
Circular ribs	40,335	38,232
T-sectional ribs	13,098	12,297

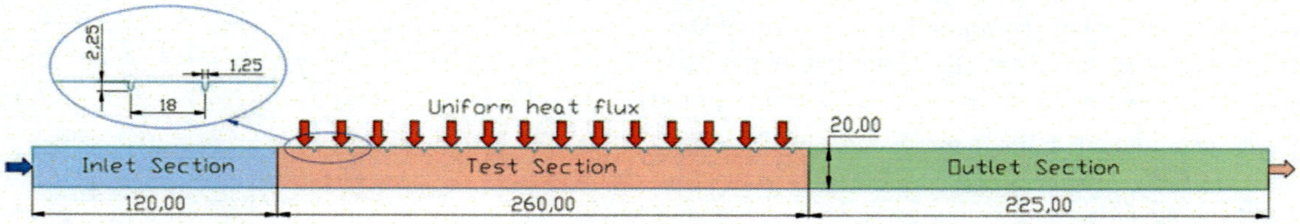

Fig. 1 2-D solar collector

2.4 Cases Considered

Three different shapes of the ribs with 5 mm height have been considered which are Pentagonal (Fig. 2a), Circular (Fig. 2b) and T-section (Fig. 2c).

Mesh Generation: Mesh generation is depicted in Fig. 3.

The total number of nodes and elements generated in each case has been provided in Table 2.

2.5 Boundary Condition

The boundary conditions have been mentioned below.: 2-D study is performed on solar collector, Pentagonal ribs are added at the absorbing plate section.

Energy ON, Turbulent fluid flow, K-e (RNG) model is selected, Velocity = 7 m/s, Inlet temperature is selected as atmospheric temperature which is 300 k, Pressure outlet is selected, Air is selected as flowing fluid. Aluminium is selected as wall material for absorbing plate (Fig. 4).

3 Results and Discussions

For CAD modelling, SOLIDWORKS software has been used in the present study. Workbench structure modeller is used to create Heat exchanger geometry and for further research in software model competent in Shell and tube-type heat exchanger in ANSYS. By allowing some wiggle room for plane symmetry, geometry becomes much easier to understand. Baffles support tubes with correct spacing in this Shell and Tube heat exchanger. Run a heat exchanger thermal simulation once the model has been created. The repercussions that had been overcome had been carefully examined in light of the whole situation. We arrive at the correct conclusions via CFD analysis.

3.1 Pressure with Different Rib Shape

The pressure contours for different rib shapes are shown in Fig. 5.

As can be observed through the pressure contours for different rib shapes above, the T-sectional ribs show the maximum variation of pressure, the range being -64.96 Pa to 99.99 Pa. The graph in Fig. 6 shows the variation of the pressures along the length of the solar collector for different rib shapes of 5 mm height.

3.2 Velocity with Different Rib Shape

The velocity contours with different rib shapes are shown in Fig. 7.

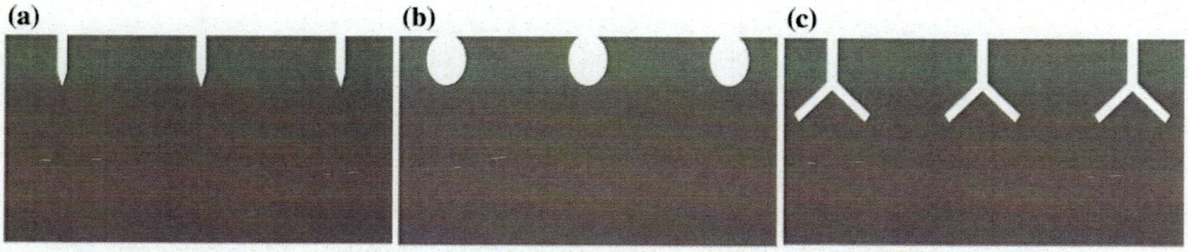

Fig. 2 **a** Pentagonal ribs, **b** Circular ribs, **c** T-sectional ribs

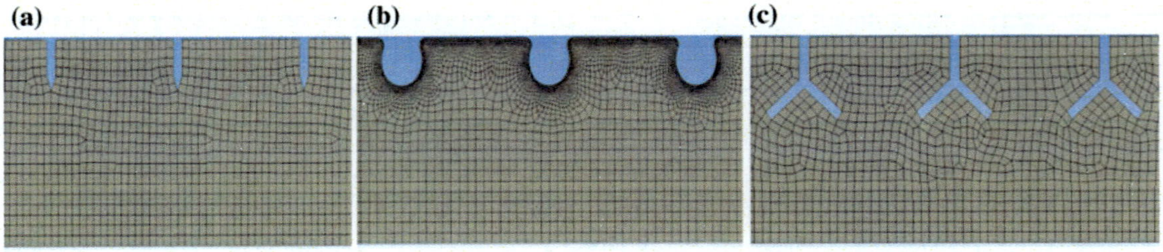

Fig. 3 **a** Pentagonal ribs, **b** Circular ribs, **c** T-sectional ribs

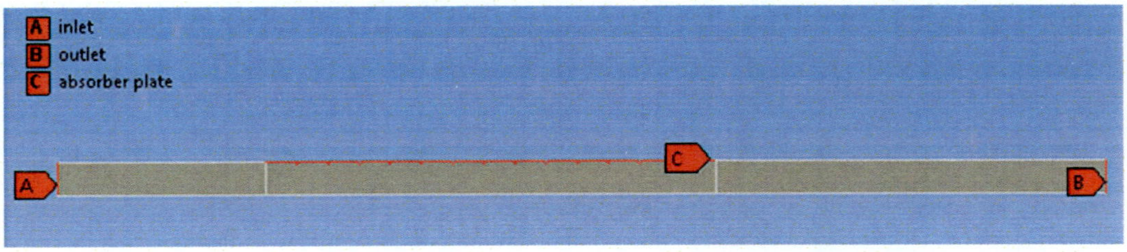

Fig. 4 Boundary conditions

Fig. 5 **a** Pentagonal ribs, **b** Circular ribs, **c** T-Sectional ribs

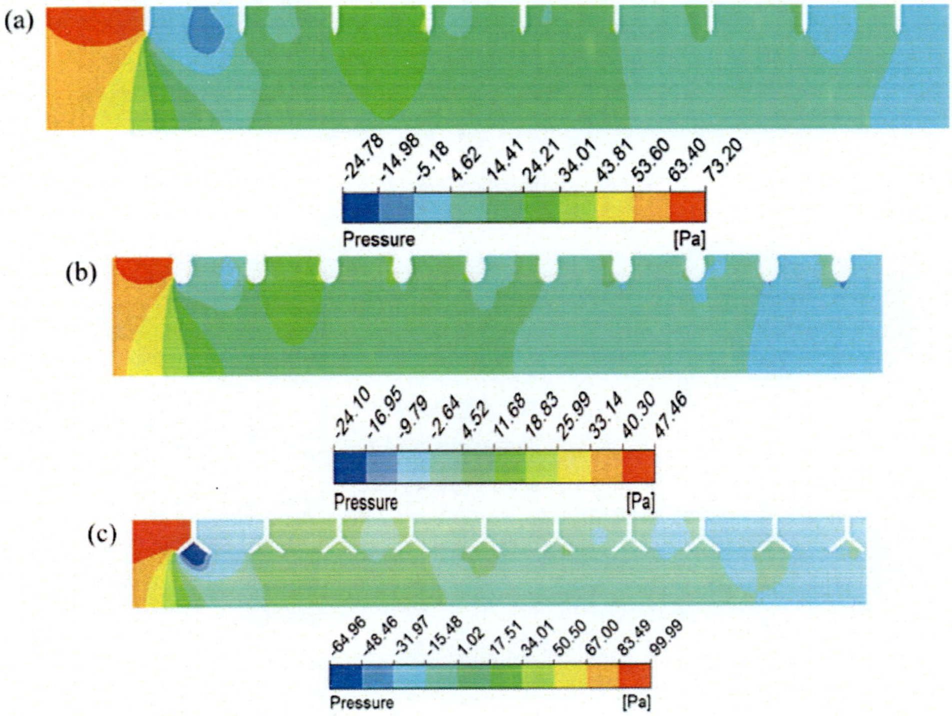

The velocity contours above also show that the maximum mainstream velocity is achieved in the case of the T-sectional ribs, i.e. 15.42 m/s. The velocity contours are shown in Fig. 8.

From Fig. 8, it is observed that the velocity profiles are almost similar in all the different rib shapes except the T-sectional rib. However, the T-section rib showed an abrupt velocity rise near the inlet end.

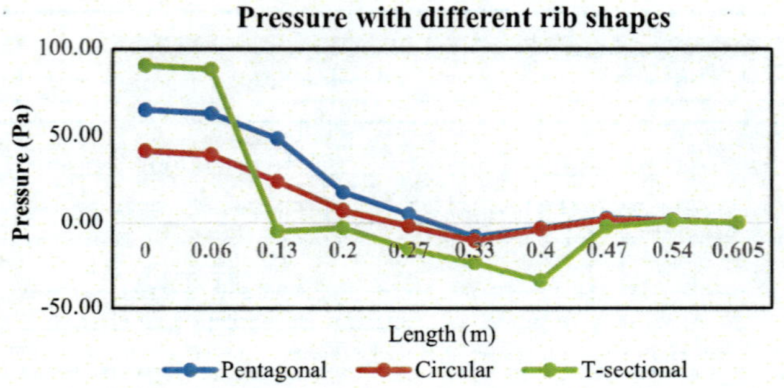

Fig. 6 Graphical comparison of maximum temperature for different rib shapes

3.3 Turbulence Kinetic Energy with Different Rib Shape

The variation of turbulence K.E. w.r.t distinct rib shapes is shown in Fig. 9.

The graph in Fig. 10 shows the variation of the same along the length of the solar collector.

3.4 Temperature with Different Rib Shape

The temperature contours w.r.t distinct rib shapes are represented in Fig. 11.

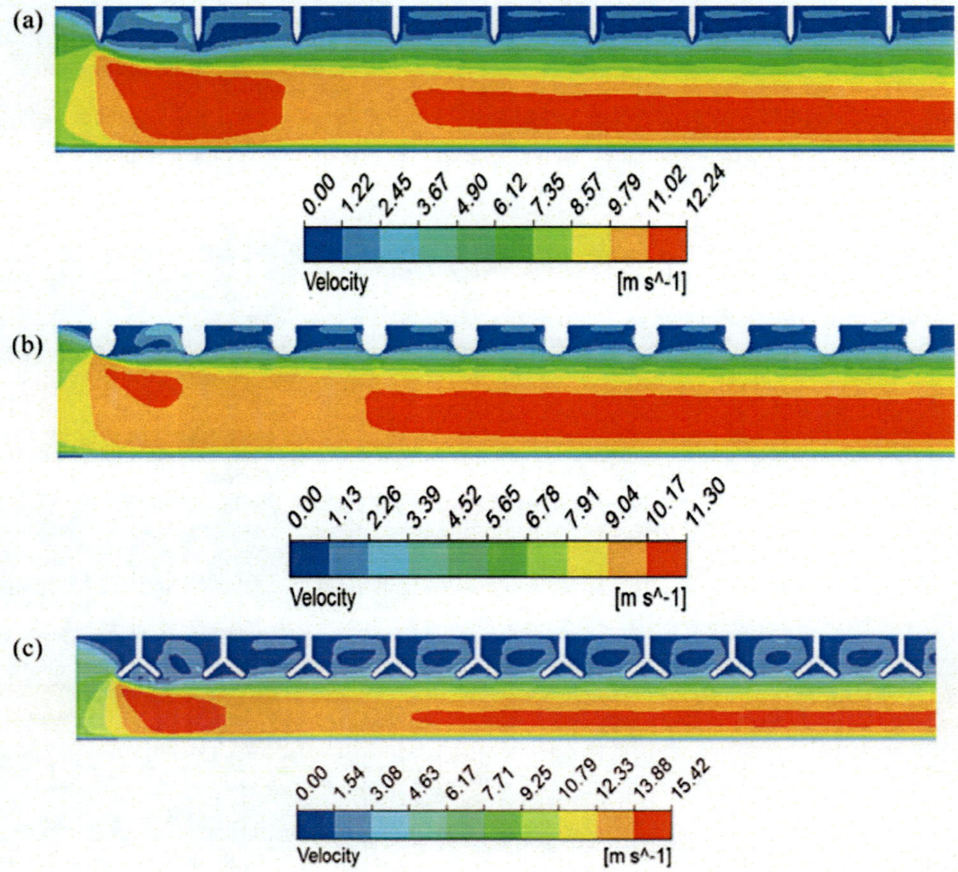

Fig. 7 **a** Pentagonal ribs, **b** Circular ribs, **c** T-Sectional ribs

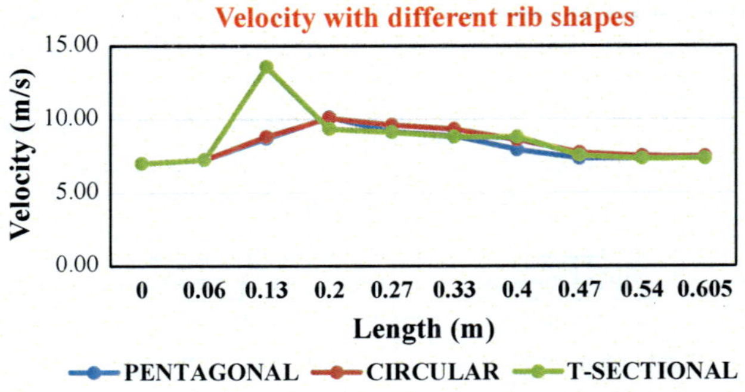

Fig. 8 Graphical comparison of velocities for different rib shapes

The temperature contours show that the maximum temperature is achieved in the case of T-sectional ribs, near the walls as depicted in Fig. 12.

The maximum temperature has been observed in the case of T-sectional ribs, i.e. 392.99 K. Thus, in Fig. 12, the T-sectional ribs with height 5 mm have been found to be the favourable choice for the solar collectors in this analysis. The higher performance of T-section rib may be attributed to larger recirculating zones that tend to give better heat transfer performance (Han et al., 1978).

Fig. 9 **a** Pentagonal ribs, **b** Circular ribs, **c** T-sectional ribs

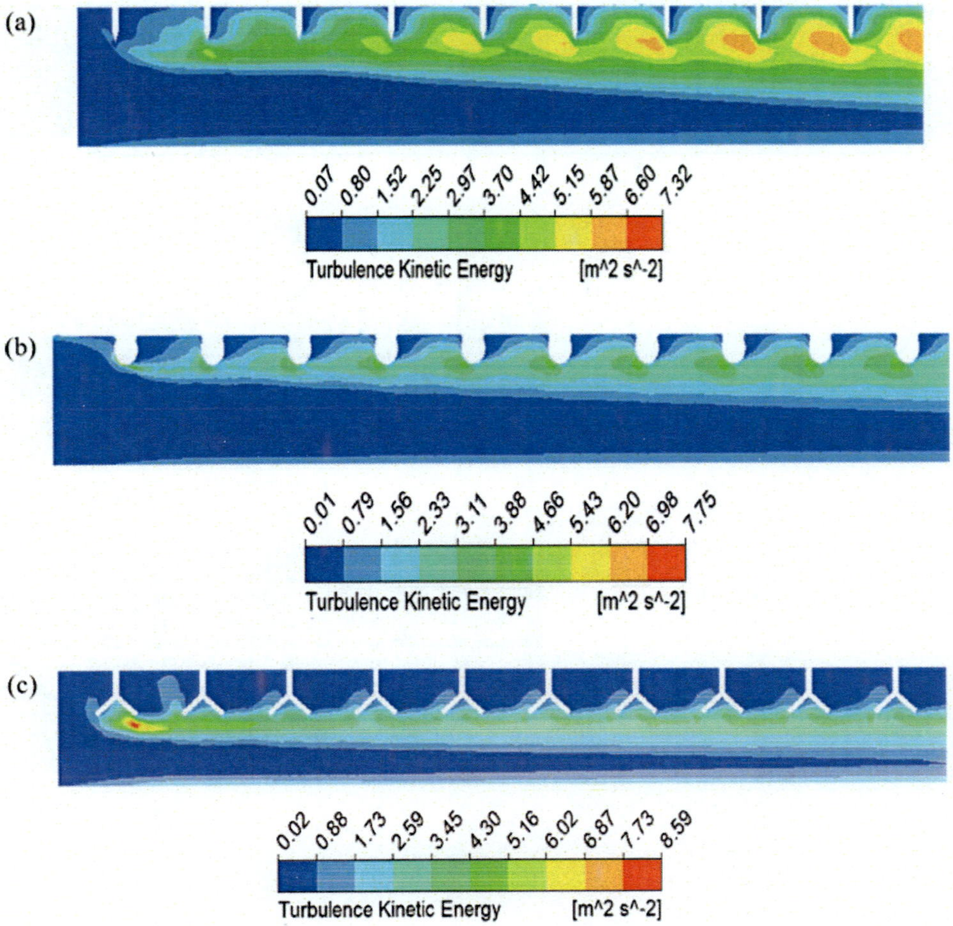

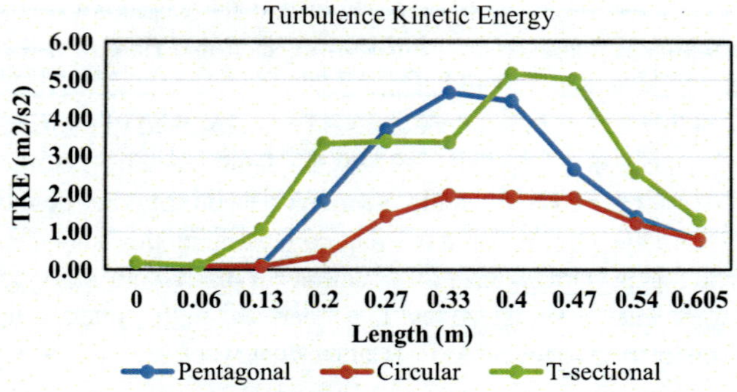

Fig. 10 Graphical comparison of the turbulence kinetic energies

4 Conclusion

A solar collector of total length 605 mm is selected which is then divided into an inlet section of 120 mm, an outlet section of 225 mm and a test section in the middle of 260 mm. The test section is provided with ribs of 1.25 mm thickness and 5 mm height at 18 mm distance each. Heat dissipation through the collectors is an important factor and hence ribs are used as one of the methods. The inlet pressure has been observed maximum in the case of T-sectional ribs (5 mm), i.e. 90.46 Pa, also with maximum pressure drop, -33.77 Pa. The velocity contours are observed to be almost

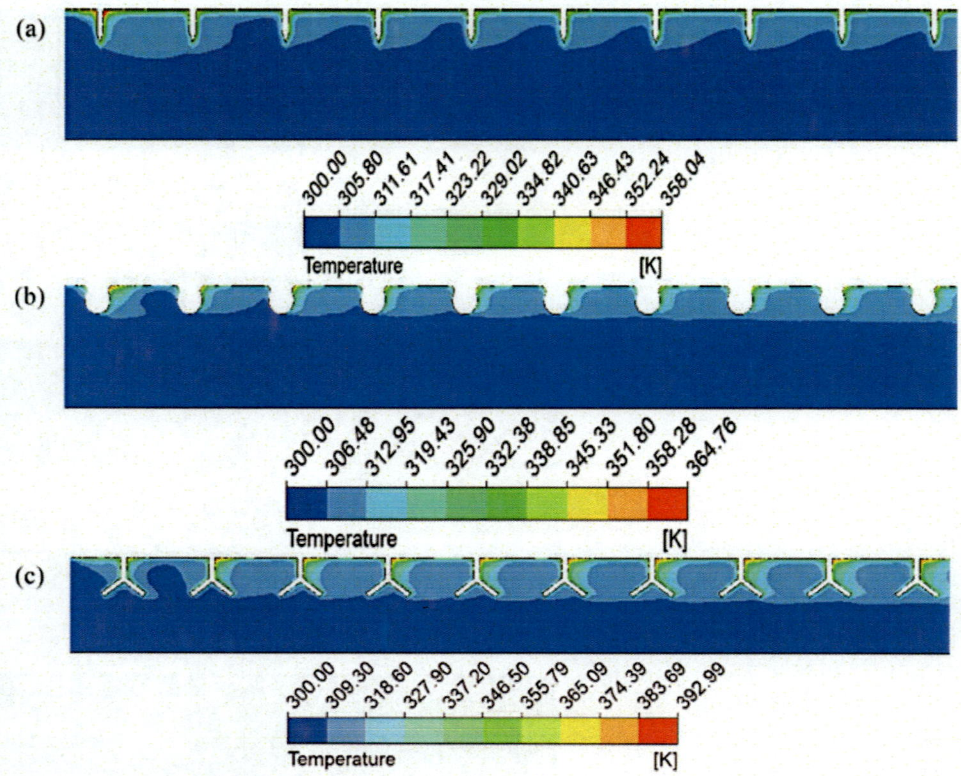

Fig. 11 **a** Pentagonal ribs, **b** Circular ribs, **c** T-sectional ribs

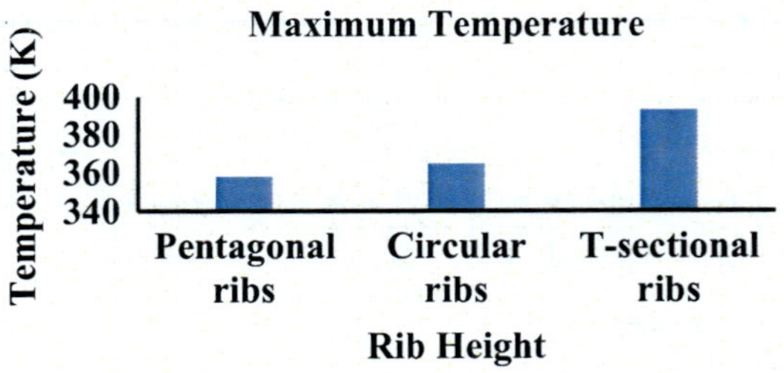

Fig. 12 Graphical comparison of maximum temperature for different rib shapes

similar in all the different rib shapes except the T-sectional rib showing an abrupt velocity rise near the inlet end, while the turbulence kinetic energy graph is also maximum for the T-sectional ribs. The maximum temperature has been observed in the case of T-sectional ribs, i.e. 392.99 K. Thus the T-sectional ribs with height 5 mm have been found to be the favourable choice for the solar collectors in this analysis.

References

Abhay, L., Chandramohan, V. P., & Raju, V. R. K. (2018). Numerical analysis on solar air collector provided with artificial square shaped roughness for indirect type solar dryer. *Journal of Cleaner Production, 190*, 353–367. https://doi.org/10.1016/j.jclepro.2018.04.130

Afridi, M. S., Mohan, R., Kumar, S., & Goga, G. (2023). Overview on Solar Pond. *International Journal of Enhanced Research in Science, Technology & Engineering, 12*(2), 34–41.

Amin, N. A. M., Belusko, M., Bruno, F., & Liu, M. (2012). Optimising PCM thermal storage systems for maximum energy storage effectiveness. *Solar Energy, 86*(9), 2263–2272. https://doi.org/10.1016/j.solener.2012.04.020

Ataee, S., & Ameri, M. (2015). Energy and exergy analysis of all-glass evacuated solar collector tubes with coaxial fluid conduit. *Solar Energy, 118*, 575–591.

Bhakre, S. S., Sawarkar, P. D., & Kalamkar, V. R. (2021). Performance evaluation of PV panel surfaces exposed to hydraulic cooling–a review. *Solar Energy, 224*(May), 1193–1209. https://doi.org/10.1016/j.solener.2021.06.083

Boulemtafes-Boukadoum, A., & Benzaoui, A. (2014). CFD based analysis of heat transfer enhancement in solar air heater provided with transverse rectangular ribs. *Energy Procedia, 50*, 761–772. https://doi.org/10.1016/j.egypro.2014.06.094

Broughton, J., Smet, V., Tummala, R. R., & Joshi, Y. K. (2018). Review of thermal packaging technologies for automotive power electronics for traction purposes. *ASME Journal of Electronic Packaging, 140*(4), 040801. https://doi.org/10.1115/1.4040828

Daghigh, R., & Shafieian, A. (2016). Theoretical and experimental analysis of thermal performance of a solar water heating system with evacuated tube heat pipe collector. *Applied Thermal Engineering, 103*, 1219–1227. https://doi.org/10.1016/j.applthermaleng.2016.05.034

Essa, M. A., & Mostafa, N. H. (2017). Theoretical and experimental study for temperature distribution and flow profile in all water evacuated tube solar collector considering solar radiation boundary condition. *Solar Energy, 142*, 267–277.

Ghaderian, J., & Sidik, N. A. C. (2017). An experimental investigation on the effect of Al2O3/distilled water nanofluid on the energy efficiency of evacuated tube solar collector. *International Journal of Heat and Mass Transfer, 108*, 972–987. https://doi.org/10.1016/j.ijheatmasstransfer.2016.12.101

Ghaderian, J., et al. (2017). Performance of copper oxide/distilled water nanofluid in evacuated tube solar collector (ETSC) water heater with internal coil under thermosyphon system circulations. *Applied Thermal Engineering, 121*, 520–536.

Gholipour, S., Afrand, M., & Kalbasi, R. (2021). Introducing two scenarios to enhance the vacuum U-tube solar collector efficiency by considering economic criterion. *Journal of the Taiwan Institute of Chemical Engineers, 124*, 228–237. https://doi.org/10.1016/j.jtice.2021.04.015

Goga, G., Kumar et al. Heat Transfer Enhancement in Solar Pond Using Nano Fluids. *Material Today proceeding*. https://doi.org/10.1016/j.matpr.2022.12.238.

Han, J. C., Glicksman, L. R., & Rohsenow, W. M. (1978). An investigation of heat transfer and friction for rib-roughened surfaces. *International Journal of Heat and Mass Transfer, 21*(8), 1143–1156. https://doi.org/10.1016/0017-9310(78)90113-8

Henei, S. M., Abdel-Rehim, A. A. (2022). The performance response of a heat pipe evacuated tube solar collector using MgO/MWCNT hybrid nanofluid as a working fluid. *Case studies of thermal engineering, 33*(March), 101957. https://doi.org/10.1016/j.csite.2022.101957.

Hussain, F., & Hasanuzzaman, M. (2022). Chapter 4: Solar thermal collector. In *Technologies for Solar Thermal Energy*. https://doi.org/10.1016/b978-0-12-823959-9.00008-8

Jhariya, O., Verma, A., & Selokar, G.R. (2019). Efficiency prediction of solar air heater with the using of difference size of ribs By. *IJARIIE-ISSN(O)*, (6), 515–521.

Kumar, A., Lal, S., & harender. (2017). Thermodynamic analysis of factors affecting the performance of solar collectors. In *Advances in Power Generation from Renewable Energy Sources (APGRES 2017) December 22–23* (pp. 1–6). 2017 at GEC Banswara. www.apgres.in.

Kumar, R., Kumar, M., Chauhan, J. S., et al. (2022). Overview on metamaterial: History, types and applications. *Material Today Proceedings, 56*(5), 3016–3024. https://doi.org/10.1016/j.matpr.2021.11.423

Mahbubul, I. M., Khan, M. M. A., Ibrahim, N. I., Ali, H. M., Al-Sulaiman, F. A., & Saidur, R. (2018). Carbon nanotube nanofluid in enhancing the efficiency of evacuated tube solar collector. *Renewable Energy, 121*, 36–44. https://doi.org/10.1016/j.renene.2018.01.006

Mercan, M., & Yurddaş, A. (2019). Numerical analysis of evacuated tube solar collectors using nanofluids. *Solar Energy, 191*(July), 167–179. https://doi.org/10.1016/j.solener.2019.08.074

Mohan, R., Saxena, N. V., Kumar, S., et al. (2023). Performance optimization and numerical analysis of boiler at husk fuel based thermal power plant. In *E3S Web of Conferences* (vol. 405, p. 02010, pp. 1–12). https://doi.org/10.1051/e3sconf/202340502010

Olfian, H., Ajarostaghi, S. S. M., & Ebrahimnataj, M. (2020). Development on evacuated tube solar collectors: A review of the last decade results of using nanofluids. *Solar Energy, 211*(July), 265–282. https://doi.org/10.1016/j.solener.2020.09.056

Prasad, P. V. D., Gupta, A. V. S. S. K. S., & Deepak, K. (2015). Investigation of trapezoidal-cut twisted tape insert in a double pipe U-Tube heat exchanger using Al2O3/Water Nanofluid. *Procedia Materials Science, 10*(2014), 50–63.

Qian, T., Li, J., Feng, W., & Nian, H. (2017). Enhanced thermal conductivity of form-stable phase change composite with single-walled carbon nanotubes for thermal energy storage. *Science and Reports, 7*(February), 1–10. https://doi.org/10.1038/srep44710

Rajput, U. J., & Yang, J. (2018). Comparison of heat sink and water type PV/T collector for polycrystalline photovoltaic panel cooling. *Renewable Energy, 116*, 479–491. https://doi.org/10.1016/j.renene.2017.09.090

Tyacke, J. C., Dai, Y., Tucker, P. G. (2021). Impact of rib shape on heat transfer using LES. *Applied Mathematical Modelling*, *97*, 244–267. https://doi.org/10.1016/j.apm.2021.03.053

Yang, H. L., Hao, L., Wang, J. N., Zhang, Z. N., Liu, X. P., & Jiang, L. J. (2015). Self-cleaning and antireflective films for all-glass evacuated tube solar collectors. *Energy Procedia, 69*, 226–232. https://doi.org/10.1016/j.egypro.2015.03.026

Exploring the Complexities and Challenges of Plastic Recycling: A Comprehensive Research Review

Sunil Sable, Mitesh Ikar, and Priyanka Dudheinamdar

Abstract

This chapter discusses the challenges related to recycling one of the most commonly used plastics worldwide—polyethylene (PE). Although PE is widely used for packaging materials due to its low cost and durability, it is difficult to recycle due to its physical and chemical properties. One of the primary challenges associated with recycling PE is identifying and separating different types of PE, which have varying properties. This makes it challenging to obtain high-quality recycled materials that meet industry standards. Furthermore, contaminants such as food residue and other materials can adversely affect the quality of recycled PE. The chapter also highlights technical difficulties involved in the recycling process, such as the need for specialized equipment and the high energy consumption required for melting and reforming the plastic. Economic challenges such as the high cost of the recycling infrastructure and the low value of recycled materials compared to virgin plastic are also addressed. The chapter stresses the significance of investing in research and development to create new technologies and methods that can overcome these challenges and make PE recycling more feasible, both economically and environmentally. Additionally, reducing plastic waste at source by adopting measures such as minimizing packaging and encouraging the use of alternative materials is equally important.

Keywords

Circular economy • Contamination • Polyethylene • Recycling • Sustainability

1 Introduction

It is vital to recycle plastic waste, in order to conserve resources and minimize environmental impacts. Important guidelines are required for: (1) sorting, collection, transportation, and processing of plastic waste; (2) identification of different types of plastic resins and their recycling methods; (3) quality control procedures to ensure the safety and purity of recycled plastic; and (4) environmental and economic benefits of using recycled plastic in the manufacturing of new products (ACRR, 2004). Feedstock recycling is one method for the recycling of plastic waste, which involves the conversion of plastics into chemicals and fuels. Feedstock recycling technologies include pyrolysis, gasification, and liquefaction. Feedstock recycling has the potential to recover a significant amount of energy from plastic waste, which can help to reduce greenhouse gas emissions. The challenges associated with feedstock recycling involve the lack of infrastructure and the need for regulatory support. Feedstock recycling can be an important part of the overall strategy for managing plastic waste in Europe, but it requires investment and support from both the public and private sectors (Aguado et al., 2007). Synthetic polymers, including plastics, are not easily biodegradable due to their high molecular weight and lack of natural enzymatic breakdown pathways. Biodegradation of plastics can occur only under specific conditions such as exposure to UV light, moisture, and heat. Plastics that are biodegradable under specific conditions can still pose a threat to the environment if they are not disposed of properly. The accumulation of plastic waste in the environment can lead to serious environmental problems, including pollution of oceans,

S. Sable · M. Ikar (✉)
Department of Chemical Engineering, Vishwakarma Institute of Technology, Pune, Maharashtra 411037, India
e-mail: mitesh.ikar21@vit.edu

S. Sable
e-mail: sunil.sable@vit.edu

P. Dudheinamdar
Sinhgad College of Engineering, Pune, Maharashtra 411037, India

M. Sumesh et al. (eds.), *2nd International Conference on Smart Sustainable Materials and Technologies (ICSSMT 2023)*, Advances in Science, Technology & Innovation, https://doi.org/10.1007/978-3-031-49826-8_22

harm to wildlife, and disruption of natural ecosystems. The most effective way to manage plastic waste is through waste reduction, recycling, and proper disposal. There is a real need for responsible plastic waste management to prevent the negative environmental impacts of plastic waste, and for developing more sustainable alternatives to conventional plastics, such as biodegradable and compostable materials (Andrady, 1994). Plastic production has increased dramatically since the mid-twentieth century and shows no sign of slowing down. This has resulted in significant environmental challenges, including waste generation, litter, and pollution. Plastics have a wide range of uses, but their durability also means that they persist in the environment for long periods of time. This can result in harm to wildlife, ecological damage, and contamination of food chains. While plastic recycling is an important strategy for reducing waste and conserving resources, it is not a perfect solution. Challenges include contamination, sorting, and difficulties in recycling certain types of plastics. To address these challenges, a combination of strategies is needed, including reducing plastic consumption, improving waste management systems, and developing new recycling technologies. A circular economy model is needed for plastics, in which waste is minimized, recycling is maximized, and new plastic production is reduced. This will require a collaborative effort between industry, government, and consumers (Andrady & Andrady, 2003).

Figure 1 shows worldwide plastics consumption from 1990 to 2018. The demand for plastic has risen sharply during this period. Plastics have numerous applications and have contributed to the development of several industries, such as packaging, healthcare, construction, and transportation. Plastics are lightweight and durable, which makes them suitable for use in various applications. Plastics have also contributed to energy savings, particularly in the transportation sector, where the use of lightweight plastics has led to improved fuel efficiency. Plastics have improved food safety by providing packaging that protects food from contamination and extends its shelf life. There are challenges associated with plastic waste management and a need for a comprehensive approach that involves reducing plastic waste generation, improving recycling technologies, and promoting the use of biodegradable plastics (Andrady & Neal, 2009). In 2002, the consumption of plastic in Europe was 44.7 million tons, with packaging being the largest application (38%). In the same year, 22.5 million tons of plastic waste were generated, out of which 19% was mechanically recycled, 17% was incinerated, and 64% was landfilled. This report also emphasized the importance of improving the collection and sorting of plastic waste to increase the recycling rate. Harmonized waste management policies and practices across Europe to promote a circular economy for plastics are much needed. It is worth remembering the value of plastics as a resource, and the benefits of plastics in various applications, including reducing greenhouse gas emissions in transportation and extending the shelf-life of food products. However, the negative environmental impacts of plastic waste, such as littering and marine pollution should not be ignored. A more sustainable approach to plastic use and waste management is necessary, which involves designing products for recyclability, promoting resource efficiency, and increasing the use of recycled plastics in new products (APME, 2004).

The life cycle assessment (LCA) of a plastic packaging recycling system is required to evaluate the environmental impact of recycling plastic packaging waste using a mechanical recycling system. The environmental impact of recycling plastic packaging is still lower than the impact of

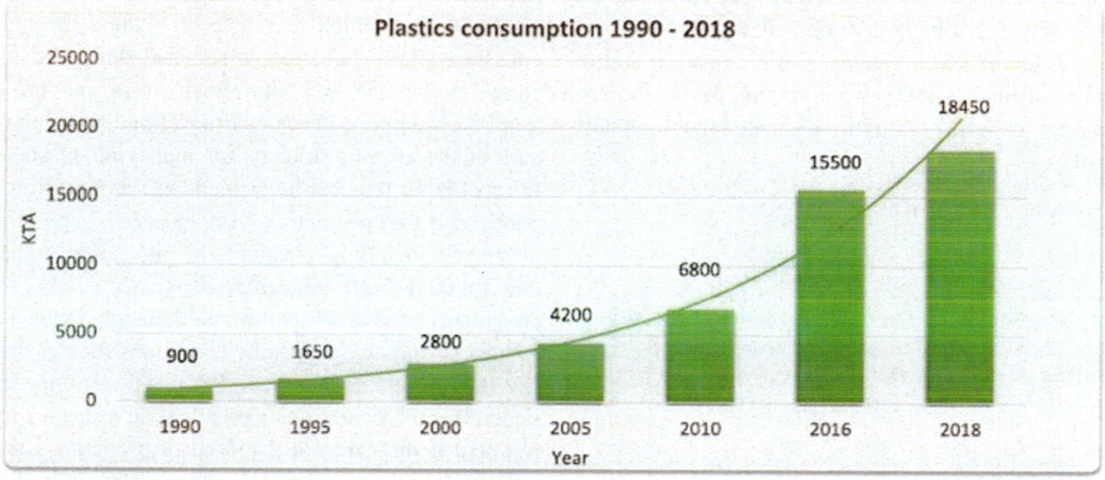

Fig. 1 Plastics consumption 1990–2018

producing new plastic from virgin materials. The use of recycled plastic in the production of new packaging results in lower energy consumption and greenhouse gas emissions compared to virgin material production. Nonetheless, there is a need for proper collection and sorting of plastic waste to ensure high-quality recycled material. This suggests that recycling plastic packaging waste can have significant environmental benefits if it is properly collected, sorted, and processed (Arena et al., 2003). The current status of and future perspectives on recycling polymeric materials used for food packaging must also be examined, including the types of polymeric materials used in food packaging and their properties, as well as the different methods used for recycling, such as mechanical recycling, chemical recycling, and energy recovery. Recycling of polymeric materials used for food packaging is a challenging task due to the complex nature of the materials, as they often contain multiple layers and different types of plastics. However, the potential benefits of recycling, such as reducing waste and conserving natural resources, and the importance of developing more efficient and cost-effective recycling technologies make it a worthwhile venture. A combination of different recycling methods may be necessary to achieve higher levels of recycling of polymeric materials used for food packaging in the future (Arvanitoyannis & Bosnea, 2001). In addition, it is important to establish a comprehensive system of standards related to the recycling and use of recycled plastics. This includes the need to identify stakeholders, establish objectives, and develop a framework for the development of standards. This provides guidance on the key considerations that should be taken into account when developing standards related to recycled plastics. These considerations include the quality of the recycled material, the processing methods used, and the end-use application of the material. The standard emphasizes the importance of incorporating input from all relevant stakeholders in the development of standards related to recycled plastics, including industry representatives, government agencies, and environmental organizations. The standard provides guidance on the testing and evaluation of recycled plastics, including the need to establish appropriate test methods and performance criteria for different types of applications. It also emphasizes the importance of addressing safety and health concerns related to the use of recycled plastics, including the need to ensure that recycled materials do not contain harmful contaminants. ASTM Standard D5033 provides important guidance on the development of standards related to the recycling of plastics and the use of recycled plastics. By following the recommendations outlined in this standard, stakeholders can help ensure the safe and effective use of recycled materials in a wide range of applications (Standard & D5033, 2000).

Figure 2 shows the global demand in 2018–2019 for major commodity plastic materials and it can be seen that

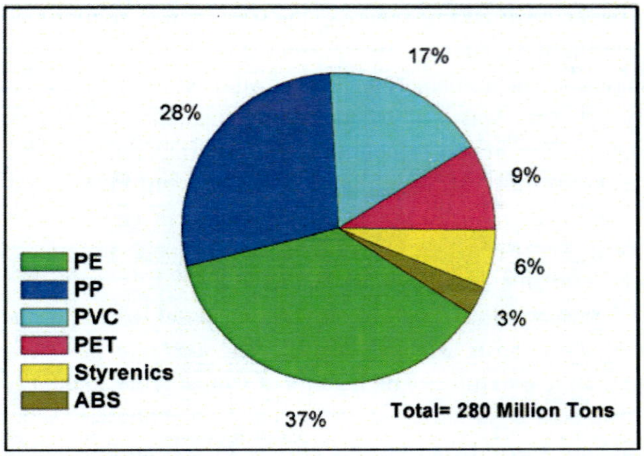

Fig. 2 Global plastic demand in 2018–2019

polyethylene (PE), and polypropylene (PP), and Polyvinyl chloride (PVC) contributed the most to the use of plastic. Plastic debris has become ubiquitous in the global environment and is present in virtually all marine and terrestrial habitats. The amount of plastic debris in the environment has been increasing over the past several decades and is expected to continue to increase in the future. Plastic debris poses a range of threats to wildlife, including ingestion, entanglement, and habitat modification. Plastic debris can also have indirect impacts on wildlife, such as by acting as a vector for transporting invasive species. Plastic debris can persist in the environment for hundreds or even thousands of years and can undergo fragmentation into smaller and smaller pieces. The accumulation and fragmentation of plastic debris can have cascading effects on ecosystem structure and function, including altering nutrient cycling and affecting food webs. Addressing the problem of plastic pollution will require a range of solutions, including improved waste management practices, increased recycling rates, and the development of more sustainable materials. There is an urgent need to address the issue of plastic pollution in order to protect the health of ecosystems and the species that depend on them (Barnes et al., 2009). Refillable beverage containers have been successful in reducing waste in Western Europe. In countries such as Germany, Switzerland, and the Netherlands, refillable containers make up a significant portion of the beverage market. Refillable containers are often seen as more environmentally friendly than single-use containers because they can be reused multiple times, reducing the amount of waste generated. The success of refillable container systems depends on a variety of factors, including the availability of collection and processing infrastructure, consumer behaviour, and government policies. Some challenges to the widespread adoption of refillable containers include the higher upfront costs of establishing a refillable system, the need for a deposit system to incentivize returns, and the logistics of transporting and cleaning the containers. Despite these challenges, the report suggests that refillable containers could be a viable option for reducing waste in other regions, including the United States. The report recommends exploring policies that incentivize the use of refillable containers and investing in the necessary infrastructure to support them (Institute for Local Self-Reliance, 2002 Western Europe's experience with refillable beverage containers. 2002).

The landfill tax is a tax on waste disposal, based on the weight of waste sent to landfill. The tax rate has been increasing over time, with the aim of reducing the amount of waste sent to landfill and encouraging recycling and other forms of waste management. The revenue from the landfill tax is used to fund environmental projects and initiatives, such as recycling and waste reduction programs. The landfill tax has been successful in reducing the amount of waste sent to landfill in the UK, but there are concerns that it may lead to illegal dumping and fly-tipping in some areas. The UK government has set targets for reducing the amount of waste sent to landfill and increasing recycling rates, and the landfill tax is seen as a key tool in achieving these targets. Information on the landfill tax in the UK and its role in promoting recycling and reducing waste sent to landfill is available from the UK Department for Environment, Food, and Rural Affairs (DEFRA, 2007). In the recycling industry, selective froth flotation can effectively separate PVC from PVC/PET (polyethylene terephthalate) mixtures. The separation efficiency is influenced by various factors such as temperature, pH, frother and collector dosage, and particle size. Research results from one study (Drelich et al., 1998) could be useful for the plastic recycling industry to improve the separation of PVC from PVC/PET mixtures A European Environment Agency (EEA) report discusses the importance of improving the management of municipal waste in order to reduce greenhouse gas emissions. It highlights the need for implementing waste-prevention measures, increasing recycling rates, and promoting the use of renewable energy from waste. The report also presents several case studies from different European countries, demonstrating the positive impact of improved waste management on reducing greenhouse gas emissions. The conclusions emphasize the importance of taking action on municipal waste management as a key strategy to combat climate change (EEA, 2008). Plastic recycling is essential for reducing waste and conserving natural resources, and it has the potential to provide economic benefits and mitigate environmental impacts. The recycling of plastic is complicated due to the diversity of plastic materials and their varying properties, which require different processing methods. Mechanical recycling, which involves the shredding and melting of plastic to produce new products, is the most common form of plastic recycling. Chemical recycling, which involves breaking down plastics into their constituent molecules to create new products, is a promising area of research and development. The economics of plastic recycling are influenced by factors such as the price of virgin plastic, the cost of collection and sorting, and the availability of markets for recycled products. The environmental benefits of plastic recycling include reducing greenhouse gas emissions, conserving energy and resources, and reducing the amount of plastic waste in landfills and the environment. To improve the efficiency and effectiveness of plastic recycling, it is important to increase public awareness and participation, promote the use of recycled plastic in new products, and develop better technologies and infrastructure for collection, sorting, and processing of plastic waste (Fisher & Andrady, 2003).

Figure 3 shows worldwide production of bioplastics including PP, PE, PET, PVC and others, with significantly high production of PE and PP. A model explores the relationship between plastic recycling and waste generation. The model suggests that recycling can reduce the amount of waste generated, but the extent of the reduction depends on factors such as the rate of recycling and the demand for recycled materials. It is important to consider the entire life cycle of a product when evaluating the effectiveness of recycling. For example, if recycling leads to an increase in the production of plastic products, then the overall environmental impact may be greater than if recycling did not occur. Recycling should be considered as part of an overall waste management strategy that includes reduction in the use of virgin materials and energy recovery from waste. The model suggests that recycling may be more effective in reducing waste in the long term rather than in the short term. This is because the benefits of recycling may not be immediately apparent, and it may take time for changes in behaviour and technology to occur. Fletcher and Mackay consider the importance of the broader context of recycling and waste management when evaluating the effectiveness of

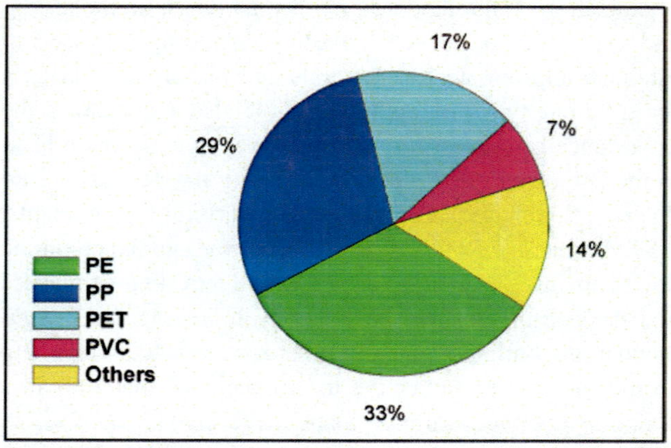

Fig. 3 Worldwide production of bioplastics

recycling (Fletcher & Mackay, 1996). They discuss various strategies for manufacturing, including the conventional "take-make-dispose" model; the "industrial ecology" model that emphasizes reuse and recycling of materials; and the "biomimicry" model that seeks to emulate nature's closed-loop systems, concluding that the conventional model of manufacturing is unsustainable, and the industrial ecology and biomimicry models offer promising alternatives. They suggest that adopting these models would require fundamental changes in the way products are designed, produced, and consumed, and emphasize the potential benefits of these models, including reduced waste and pollution, increased efficiency and cost savings, and improved environmental and social outcomes (Frosch & Gallopoulos, 1989).

Garforth et al. researched feedstock recycling of polymer wastes. Feedstock recycling is a process that involves the breaking down of polymers into their constituent monomers or other intermediate products. The authors describe the different types of feedstock recycling methods, including thermal and catalytic processes, and discuss their advantages and limitations, as well as the potential of feedstock recycling in the context of the increasing demand for polymer materials and the growing problem of plastic waste. Feedstock recycling has the potential to make a significant contribution to the sustainable use of polymers by allowing waste materials to be reused in the production of new materials. The technology is still in its early stages of development and further research is needed to optimize the process and make it economically viable (Garforth et al., 2004). Gilpin et al. discuss the production, distribution, and fate of polychlorinated dibenzo-p-dioxins (PCDDs), dibenzofurans (PCDFs), and related organohalogens in the environment. They note that PCDDs and PCDFs are unintentionally produced by human activities such as combustion and industrial processes. These compounds are persistent organic pollutants that can travel long distances and bioaccumulate in the food chain. Human exposure to PCDDs and PCDFs has been linked to a variety of health effects, including cancer, developmental abnormalities, and immune system dysfunction. Efforts have been made to reduce the production and release of these compounds into the environment, including the Stockholm Convention on Persistent Organic Pollutants (Gilpin et al., 2003). The increasing amount of plastic debris in marine environments has significant environmental impact. The impact of plastic debris includes entanglement, ingestion, smothering, and hitchhiking of organisms. Plastic debris can also act as a raft for invasive species to reach new areas and cause ecological harm. There is a need to reduce plastic waste and prevent it from entering marine environments to protect marine organisms and ecosystems. Improved waste management, increased recycling, and reduced use of single-use plastics are some of the measures that can be taken to address the problem of plastic debris in the oceans (Gregory, 2009).

2 Challenges Associated with Recycling of Plastic

A chemical recycling method called "upcycling" can convert waste polyethylene into valuable chemicals, such as olefins and aromatics, which can be used in the production of new plastics and other materials. This approach could help reduce the environmental impact of plastic waste while also creating new economic opportunities (Rahimi & García, 2017). The challenges associated with the mechanical recycling of polyethylene, including issues with contamination, the need for sorting and cleaning, and limitations in the ability to recycle mixed plastics are discussed in Datta and Kopczynska (2016). The challenges of recycling plastics in the US, including polyethylene, such as inadequate infrastructure, limited markets for recycled plastics, and difficulties in achieving high levels of recycling are covered in Epa and EPA530-F-18-004 (2018). Information is provided regarding greenhouse gas emissions and energy use, including the need for energy-intensive processes to transform plastic waste into new products in Epa (2019). The need for high temperatures and pressures to break down the polymer, as well as the potential for the formation of toxic by-products during the process is further discussed in Al-Salem et al. (2009). There are challenges in recycling polyethylene through pyrolysis, which involves heating the plastic in the absence of oxygen to break it down into smaller molecules that can be used to create new products. While pyrolysis has the potential to be an effective method of recycling polyethylene, there are challenges associated with controlling the process to ensure consistent product quality, as well as concerns about the environmental impact of the energy-intensive process (Ignatyev et al., 2014). In addition, there are challenges with recycling of post-consumer polyethylene waste, including the presence of contaminants such as food residues, chemicals, and non-polyethylene plastics, which can affect the properties and quality of the recycled material. The difficulty in separating different types of polyethylene for recycling purposes, as well as the challenges in achieving cost-effective and energy-efficient recycling processes are discussed in Karthikeyan et al. (2012). There are high costs associated with the recycling process, low demand for recycled products, and low public awareness of the importance of recycling. The challenges associated with the collection and sorting of post-consumer waste, as well as the variability in the properties of different types of polyethylene are examined in Wong et al. (2015).

Mohanraj et al. consider the challenges associated with recycling of polyethylene waste in the context of a circular economy, including the need for efficient waste management systems to collect and sort post-consumer waste, as well as the development of cost-effective and energy-efficient recycling technologies. They note the importance of creating a market for recycled polyethylene products to increase demand and promote the circular economy (Mohanraj et al., 2017). Challenges associated with recycling of polyethylene waste in the context of mining and mineral processing. Froelich et al. review the need for efficient waste management systems to reduce the environmental impact of mining activities, as well as the development of cost-effective and energy-efficient recycling technologies. The authors consider the importance of improving the quality of recycled polyethylene to increase its value as a raw material in mineral processing (Froelich et al., 2007). Ahmad discusses the challenges associated with recycling of polyethylene waste in the context of environmental sustainability. There is a need for efficient waste management systems to reduce the environmental impact of plastic waste, as well as the development of cost-effective and energy-efficient recycling technologies. Ahmad observes the importance of increasing public awareness of the benefits of recycling and the need for government policies to promote recycling practices (Ahmad, 2004). There are various challenges associated with the recycling of low-density polyethylene (LDPE) by pyrolysis. The challenges include the difficulty in achieving complete decomposition of LDPE due to the high molecular weight and crystallinity of the polymer, the formation of char during the pyrolysis process, and the limited understanding of the reaction mechanism and are discussed in George and Kurian (2014). Challenges associated with recycling of high-density polyethylene (HDPE) into value-added products include the lack of understanding of the fundamental reaction mechanisms involved in the recycling process, the need for efficient catalysts, and the potential for the formation of undesirable by-products during the process (Akato et al., 2018). Challenges associated with the recycling of mixed polyethylene (PE) wastes include the difficulty in separating different types of PE wastes from each other, the need for efficient and cost-effective separation methods, and the variability in the properties of different types of PE wastes (Geyer et al., 2016).

Challenges associated with the recycling of LDPE into value-added products include the difficulty in obtaining a high yield of desirable products, the need for efficient catalysts, and the potential for the formation of undesirable by-products during the recycling process (López et al., 2014). Challenges associated with the recycling of LDPE using thermal cracking include the low yield of desirable products, the potential for the formation of undesirable by-products, and the difficulty in controlling the reaction conditions to obtain a high yield of desirable products (Al-Sabagh et al., 2016). The recycling of polyethylene is through thermal cracking and subsequent upgrading of the pyrolysis oil. One of the main challenges associated with the recycling of polyethylene is the presence of contaminants, such as PVC and polyurethane, which can cause catalyst deactivation during the pyrolysis process. Additionally, the presence of additives and stabilizers used in the production of polyethylene can also negatively affect the quality of the pyrolysis oil (Kenny et al., 2008). Catalytic pyrolysis of polyethylene produces high-value chemicals. One of the main challenges associated with the recycling of polyethylene is the development of cost-effective and environmentally friendly catalysts that can efficiently convert polyethylene into useful products. Additionally, the presence of impurities in the feedstock, such as PVC and PET, can negatively affect the yield and selectivity of the desired products (Du et al., 2016).

Figure 4 shows various challenges in plastic recycling processes, mainly the high cost and less efficient processing. There are some challenges associated with the feedstock recycling of plastic wastes, including polyethylene. One of the main challenges is the development of efficient and cost-effective technologies that can handle the large volumes of plastic waste generated. Additionally, the recycling process should be designed to minimize the production of hazardous by-products and emissions, such as dioxins and furans (Aguado & Serrano, 1999). In the catalytic cracking of polyethylene to produce gasoline-range hydrocarbons, one of the main challenges associated with the recycling of polyethylene is the development of efficient catalysts that can selectively convert polyethylene into desired products with high yield and selectivity. Additionally, the presence of contaminants, such as PVC and polystyrene, can cause catalyst deactivation and negatively affect the quality of the final products (Yoshioka et al., 1998). Within the different methods for the recycling of polyethylene, including thermal degradation, mechanical recycling, and chemical recycling, one of the main challenges associated with the recycling of polyethylene is the development of cost-effective and environmentally friendly technologies that can efficiently convert polyethylene into useful products. Additionally, the recycling process should be designed to minimize the production of hazardous by-products and emissions, such as dioxins and furans (Paszun & Spychaj, 1997). Yoshioka et al. investigated the pyrolysis of high-density polyethylene (HDPE) and found that the presence of oxygen in the pyrolysis atmosphere resulted in the formation of oxygen-containing compounds, which could affect the quality of the recycled product. The rate of polymer degradation decreased at higher heating rates, which could limit the efficiency of the recycling process (Yoshioka et al., 2001). Polk et al. investigated the effects of various processing parameters on the recycling

Fig. 4 Challenges with recycling of plastic

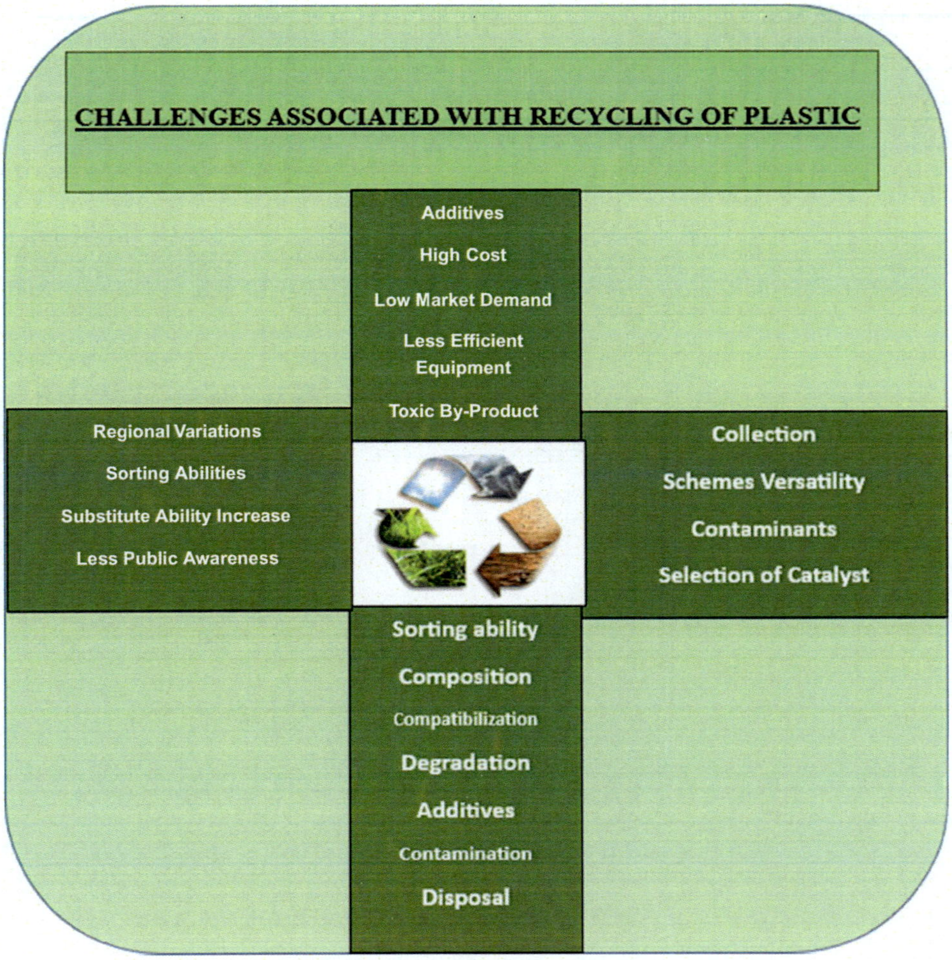

of mixed polyethylene wastes and found that the presence of contaminants, such as paper and metal, could reduce the quality of the recycled product. Additionally, the use of higher temperatures and longer residence times could lead to the formation of undesirable by-products, such as tar and char (Polk et al., 1999). Mechanical properties of recycled LDPE films and found that the properties deteriorated with increasing levels of contamination. The presence of contaminants could affect the melting behaviour of the polymer during processing, leading to defects in the final product (Campanelli et al., 1993).

Campanelli et al. looked into the effects of different processing conditions on the recycling of HDPE. The use of higher temperatures and longer residence times led to a decrease in the molecular weight of the polymer, which could affect the properties of the recycled product. The presence of contaminants, such as PVC and other plastics, could cause discoloration and reduce the quality of the recycled material (Genta et al., 2005). In a study by Goto et al. of the effects of electron beam irradiation on the recycling of LDPE, the authors found that the irradiation caused crosslinking of the polymer, which could affect the properties of the recycled material. They also noted that the presence of other types of plastics could lead to degradation of the LDPE during recycling, which could affect the quality of the final product (Goto et al., 2002). In the chemical recycling of low-density polyethylene (LDPE) using an ionic liquid as a solvent and a co-solvent, the challenge associated with the recycling of LDPE is its high thermal stability, which makes it difficult to degrade. The catalyst used in the recycling process should be carefully selected to ensure high efficiency and selectivity (Tang et al., 2019). The effect of various parameters such as temperature, time, and catalyst concentration on the thermal degradation of LDPE, have been studied and the main challenge associated with the recycling of LDPE is its high melting point, which requires high energy input to melt and degrade the polymer (Teotia et al., 2017). Soni et al. researched the effect of various parameters such as temperature and time on the thermal degradation of LDPE. They found the challenge associated with the recycling of LDPE is its high molecular weight, which makes it difficult to degrade the polymer using

Fig. 5 Plastics overview

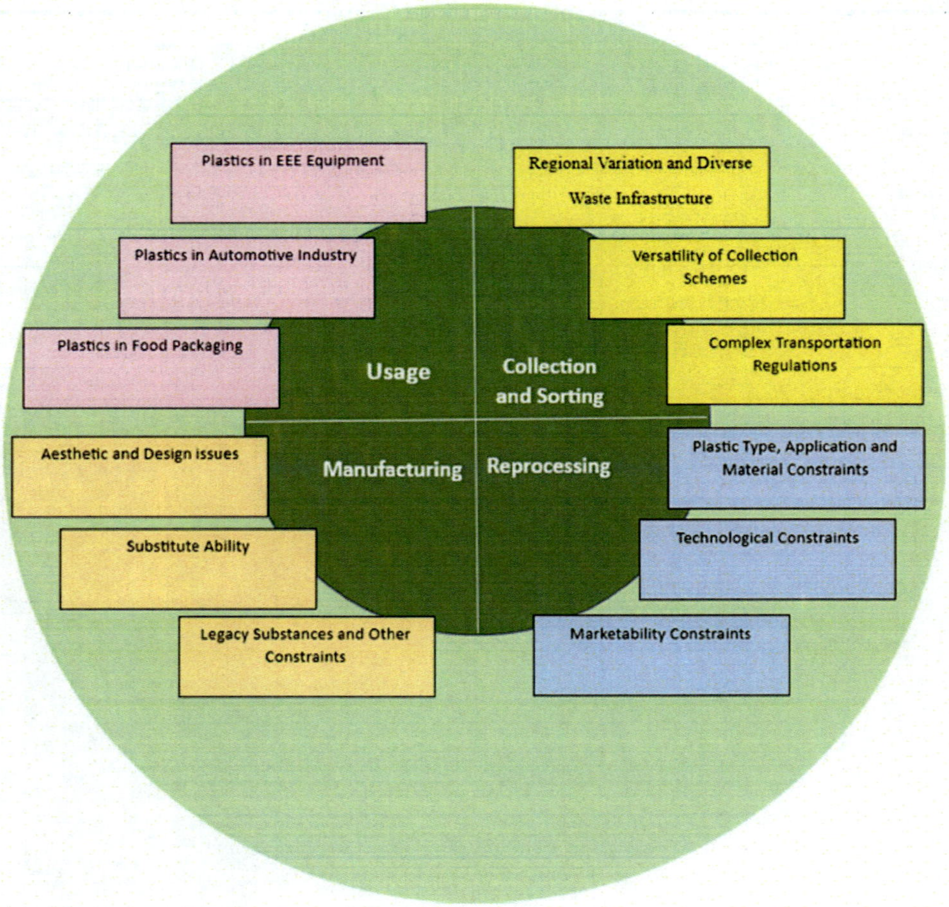

conventional thermal methods (Soni et al., 2010). Furthermore, the effect of various parameters such as temperature, time, and catalyst concentration on the thermal degradation of LDPE and the main challenge associated with the recycling of LDPE is its high crystallinity, which reduces the efficiency of the recycling process and results in low-quality products (Hoang & Dang, 2013).

Figure 5 shows plastic usage and challenges in collection and sorting, reprocessing and manufacturing. Tawfik et al. explored the effect of various parameters such as temperature and time on the thermal degradation of LDPE. The challenge associated with the recycling of LDPE is its high molecular weight, which results in low-quality products and reduces the efficiency of the recycling process. The authors found that the presence of impurities and contaminants in the polymer can affect the quality of the recycled products (Tawfik & Eskander, 2010). Furthermore, The effect of processing parameters, such as temperature and residence time, on the properties of recycled high-density polyethylene (HDPE) has also been investigated, and the authors found that the mechanical properties of recycled HDPE were inferior to those of the virgin material, and that the degree of degradation increased with increasing processing temperature and residence time (Tawfik et al., 2011). The degradation behaviour of low-density polyethylene (LDPE) during multiple recycling cycles has been studied and it was found that the mechanical properties of recycled LDPE gradually decreased with each recycling cycle, and that the degradation was caused by chain scission and oxidation (Langer et al., 2015).

In a study of the thermal degradation behaviour of LDPE during recycling, the recycled LDPE exhibited lower thermal stability and a higher degradation rate than the virgin material, and the degree of degradation increased with increasing recycling cycles (Suh et al., 2000). The degradation behaviour of polyethylene terephthalate (PET) during the recycling process was investigated and the presence of impurities, such as PVC and polypropylene (PP), in the PET waste stream, had a significant negative effect on the mechanical and thermal properties of the recycled PET (Liu et al., 2018). Troev et al. report the effect of processing parameters, such as temperature and pressure, on the properties of recycled LDPE. The authors found that the mechanical properties of recycled LDPE were inferior to those of the virgin material, and that the degree of degradation increased with increasing processing temperature and pressure. The recycled LDPE also contained a higher level of impurities, such as metals and residual monomers, than the virgin material (Troev et al., 2003). Wang et al. looked into the challenges associated with the recycling of low-density polyethylene (LDPE) through the process of pyrolysis. The properties of the resulting pyrolysis oil varied greatly depending on the pyrolysis temperature, with lower temperatures leading to oil with a higher heating value and higher concentrations of aromatic compounds. The pyrolysis process resulted in the formation of a significant amount of char, which could potentially limit the effectiveness of this method for LDPE recycling (Wang et al., 2012). The challenges associated with the recycling of high-density polyethylene (HDPE) through the process of pyrolysis have also been investigated and it was found that the presence of impurities in the HDPE feedstock, such as pigments and fillers, can have a significant impact on the properties of the resulting pyrolysis oil. The presence of oxygen in the pyrolysis reactor can lead to the formation of undesirable by-products, such as carbon monoxide and carbon dioxide (Wang et al., 2015).

Another study explored the challenges associated with the recycling of LDPE through the process of catalytic cracking. Results showed that the catalyst used in the process could significantly affect the properties of the resulting products, with a zeolite catalyst resulting in a higher yield of liquid products and a greater concentration of aromatic compounds. The presence of oxygen in the feedstock could lead to the formation of undesirable by-products, such as carbon monoxide and carbon dioxide (Wang et al., 2009). Yue et al. reported on the challenges associated with the recycling of LDPE through the process of reactive extrusion. The properties of the resulting recycled material were highly dependent on the extrusion conditions, with higher temperatures and shear rates resulting in greater molecular weight degradation and a decrease in the mechanical properties of the material. Results showed that the presence of impurities in the feedstock could lead to the formation of undesirable by-products and a decrease in the quality of the resulting material (Yue et al., 2011). Challenges associated with the recycling of LDPE through the process of melt blending with a compatibilizer were explored and the effectiveness of the compatibilizer was found to be highly dependent on its chemical composition, with a maleic anhydride-grafted polyethylene (MAPE) compatibilizer resulting in a greater improvement in the mechanical properties of the resulting material than a styrene-ethylene-butylene-styrene (SEBS) compatibilizer. The presence of impurities in the feedstock could limit the effectiveness of this method for LDPE recycling (Yue et al., 2013). In another report, the main challenge associated with the recycling of polyethylene was the poor compatibility of different types of polyethylene (PE) in the recycling process. Different types of PE exhibited different melting behaviours, which led to difficulties in obtaining a homogeneous melt. Additionally, impurities in the recycled material also affected the properties of the final product (Liu et al., 2019).

Mechanical recycling of high-density polyethylene (HDPE) waste was investigated and the study found that the main challenge was the decrease in the physical properties of the recycled material compared to the original material. The authors attributed this to the formation of defects during the recycling process, such as voids, cracks, and inclusions (Mecit & Akar, 2001). Desai et al. investigated the thermal degradation of LDPE and found that one of the main challenges associated with recycling was the generation of gaseous by-products, such as carbon dioxide, carbon monoxide, and hydrocarbons. These by-products not only affect the quality of the recycled material but also contribute to environmental pollution (Desai et al., 2005). Leung et al. reported on the energy consumption and greenhouse gas emissions associated with the recycling of plastics, including polyethylene. The main challenge was the high energy consumption and associated emissions from the recycling process, particularly from the use of thermal energy for melting and processing the recycled material (Leung et al., 2010).

3 Results from Recent Studies

3.1 Identifying Different Types of PE

Identifying and separating different types of PE can be challenging. However, several techniques and methods are commonly employed to overcome this challenge. These include visual identification based on resin codes, density-based separation techniques such as float-sink separation, infrared spectroscopy, and spectroscopic techniques like Raman spectroscopy. Additionally, advanced technologies such as near-infrared (NIR) sorting and automated sorting systems are being developed to enhance the identification and separation of PE materials.

Ways to reduce energy consumption

Optimizing the recycling process parameters such as temperature, heating time, and cooling rate can help minimize energy requirements. Fine-tuning these parameters based on the specific properties of the PE being recycled can lead to more efficient energy utilization. Exploring and adopting innovative technologies such as microwave heating, induction heating, or ultrasonic-assisted recycling can offer more energy-efficient alternatives to traditional heating methods. Implementing energy-recovery systems within recycling facilities can capture and reuse waste heat generated during the recycling process. This can contribute to overall energy savings and improve the energy efficiency of the recycling operation. Using PE materials with lower melting points or those that require less energy-intensive processing can contribute to energy savings during recycling. This can be achieved through the use of alternative PE formulations or the development of PE blends that exhibit favourable recycling properties.

Ways to reduce plastic waste

Encouraging widespread recycling programs and promoting the adoption of a circular economy model can significantly reduce plastic waste. This involves designing products and packaging with recycling in mind, improving collection and sorting systems, and increasing the use of recycled materials in manufacturing processes. Single-use plastics contribute significantly to plastic waste. Implementing measures to reduce the consumption of single-use plastics, such as banning or taxing certain items, promoting reusable alternatives, and raising awareness about the environmental impact of disposable plastics, can make a substantial difference. Reducing plastic waste can be achieved through product redesign and innovation. This includes exploring alternative materials that are more environmentally friendly, developing packaging designs that use less plastic or are easily recyclable, and creating durable and long-lasting products that reduce the need for frequent replacements. Educating consumers about the importance of reducing plastic waste and providing information on sustainable consumption habits can drive behavioural changes. Promoting awareness campaigns, organizing community initiatives, and collaborating with schools, businesses, and organizations can help spread the message and encourage responsible plastic use. Implementing effective policies and regulations at the governmental level is essential in reducing plastic waste. This can include imposing bans or restrictions on certain types of plastic products, implementing extended producer responsibility programs, and setting recycling targets and standards.

Separating different types of PE

Density is often utilized as a differentiating factor for PE separation. Thus, PE variants with different densities, such as high-density polyethylene (HDPE) and low-density polyethylene (LDPE), can be separated using techniques like float-sink separation or density gradient centrifugation. These methods exploit the density differences to separate the PE types based on their buoyancy or sedimentation rates. Thermal methods take advantage of the different melting points of various PE types. By subjecting the mixed PE samples to controlled heating, it is possible to selectively melt or soften specific types of PE while leaving others relatively unaffected. The molten or softened PE can then be separated using techniques like filtration or solid-state extrusion. Spectroscopic techniques such as infrared spectroscopy (IR) or Raman spectroscopy can be employed to analyse the chemical composition and molecular structure of the PE samples. Different PE types exhibit distinct spectroscopic signatures, allowing for their identification and separation based on their unique spectral characteristics. Some PE variants are soluble in specific solvents while others are not. By utilizing selective solvents, it is possible to dissolve certain types of PE while leaving others intact. The dissolved PE can be separated from the solution using precipitation or filtration methods, enabling the separation of different PE types. Mechanical sorting techniques rely on physical properties such as shape, size, or surface characteristics to separate different types of PE. This can be achieved through methods like sieving, air classification, or electrostatic separation, which exploit the differences in physical attributes to segregate the PE types.

Major challenges in plastic recycling

One of the primary challenges is contamination, where plastic waste is mixed with non-recyclable materials, such as food residues, dirt, or other types of plastic. Contamination reduces the quality and value of recycled plastic and makes it more difficult to process. Another challenge is the effective sorting and separation of different types of plastic. Plastics come in various forms, compositions, and colours, making it challenging to identify and separate them accurately. Automated sorting technologies are being developed, but the process remains complex and costly. Insufficient recycling infrastructure is a significant hurdle. Many areas lack the necessary facilities and resources to collect, sort, and process plastic waste effectively. Limited access to recycling centres and collection systems hinders the recycling process. The economic feasibility of plastic recycling is a significant concern. Recycled plastics often have lower market value compared to virgin plastics. The costs associated with collection, sorting, cleaning, and processing may outweigh the financial benefits, making it challenging for recycling operations to be economically sustainable. The demand for recycled plastic products is essential for creating a circular economy. However, the market demand for recycled plastics can be limited compared to virgin plastics. Increasing the demand and promoting the use of recycled plastic materials is crucial to support the growth of plastic recycling.

Major drawbacks to recycling plastic

Contamination is a significant drawback in plastic recycling. Plastic waste often gets contaminated with other materials, such as food residues, dirt, or different types of plastic.

Contamination affects the quality and purity of recycled plastic, making it less desirable for reuse in manufacturing processes. It can also lead to operational challenges in recycling facilities and increase processing costs. Economic viability is another significant drawback to recycling plastic. The costs associated with collecting, sorting, cleaning, and processing plastic waste can be high. Additionally, the market value of recycled plastics is often lower compared to virgin plastics, making it financially challenging for recycling operations to compete. The lack of a strong demand for recycled plastic products further affects the economic feasibility of plastic recycling.

Three negative effects of plastic on the environment

Plastic pollution is a significant environmental issue. When plastic waste is not properly managed, it can end up in oceans, rivers, and other natural habitats, causing harm to marine life, birds, and other animals. Plastic debris can be ingested by animals, leading to injuries, digestive issues, and even death. Moreover, plastic pollution disrupts ecosystems and degrades the aesthetic value of natural environments. Plastic takes an incredibly long time to decompose, often several hundred years or more. As a result, plastic waste accumulates in landfill, taking up valuable space and contributing to the burden on waste management systems. Landfill sites are not a sustainable long-term solution, and the accumulation of plastic waste poses risks such as leaching of harmful chemicals into the soil and groundwater. The production and disposal of plastic contribute to greenhouse-gas emissions, exacerbating climate change. Plastic is primarily derived from fossil fuels, and the extraction, refining, and manufacturing processes involved in plastic production release greenhouse gases. Additionally, when plastic is incinerated or breaks down in landfill, it releases carbon dioxide and other harmful emissions, further contributing to climate change.

Why can't plastic be recycled multiple times

Plastic can be recycled multiple times, but there are limitations to the number of times it can be recycled for several reasons. Each time plastic is recycled, its quality tends to degrade. The recycling process involves melting and reprocessing the plastic, which can cause a loss of certain properties and result in a lower-quality material compared to the original plastic. The repeated recycling cycles can lead to a decrease in the mechanical strength, flexibility, and overall performance of the plastic. Contamination is also a significant challenge in plastic recycling. During the recycling process, it is difficult to completely remove all impurities and contaminants present in the plastic waste, such as dirt, food residue, or other materials. These contaminants can negatively affect the quality and properties of the recycled plastic, making it less suitable for certain applications. As the number of recycling cycles increases, the accumulation of contaminants becomes more challenging to manage, further reducing the quality of the recycled plastic. Some plastic polymers, particularly those in certain plastic types like polyethylene terephthalate (PET), can undergo chemical degradation during recycling. The high temperatures involved in the recycling process can cause the polymer chains to break down, leading to a decrease in molecular weight and deteriorating the properties of the plastic. This degradation becomes more pronounced with each recycling cycle, limiting the number of times the plastic can be effectively recycled. The ability to recycle plastic multiple times also depends on the availability of appropriate recycling infrastructure and technologies. Some plastic types are easier to recycle than others, and the infrastructure and processes needed to recycle certain plastics may be limited or not yet fully developed. This can restrict the number of recycling cycles and the overall efficiency of the recycling process.

Shocking facts about plastic pollution

Over 380 million tons of plastic are produced worldwide each year, and this number continues to rise. The scale of plastic production contributes to the abundance of plastic waste that ends up polluting the environment. Single-use plastics, such as plastic bags, straws, bottles, and food packaging, contribute significantly to plastic pollution. It is estimated that around 50% of plastic produced is for single-use purposes, and a large portion of it ends up as litter in landfill, oceans, and other natural habitats. Every year, millions of tons of plastic waste enter the oceans, causing severe harm to marine life. It is estimated that there are more than 150 million tons of plastic already circulating in the marine environment. Marine animals often mistake plastic debris for food, leading to ingestion and entanglement, which can be fatal. Microplastics, tiny plastic particles less than 5 mm in size, have infiltrated various ecosystems, including rivers, oceans, and even the air we breathe. These microplastics can be consumed by marine organisms and enter the food chain, ultimately reaching humans. Recent studies have found microplastics in seafood, drinking water, and even in the air we breathe. Plastic takes hundreds of years to decompose naturally. This means that the plastic waste we generate today will persist in the environment for generations to come, continuously contributing to pollution and harming ecosystems. Plastic pollution poses a significant threat to wildlife. Marine animals can become entangled in plastic debris, leading to injuries, suffocation, and death. Additionally, the ingestion of plastic can cause digestive issues, blockages, and toxicity, leading to population decline and ecological imbalances. Plastic pollution not only affects ecosystems but also poses risks to human health. Some plastics contain harmful chemicals that can leach into the environment and potentially enter the human body through ingestion or exposure. These chemicals have been linked to various health issues, including endocrine disruption, developmental problems, and certain cancers.

Greatest challenges with plastics

The durability of plastic contributes to its widespread pollution. Plastic waste, especially single-use plastics, often ends up in landfill, rivers, lakes, and oceans, causing environmental pollution. The accumulation of plastic waste in these ecosystems poses a threat to wildlife, marine life, and overall ecosystem health. The sheer volume of plastic waste generated poses a significant challenge for waste management systems. Proper disposal and recycling of plastic waste require effective infrastructure and processes, which may be lacking in many regions. Inadequate waste management can result in plastic pollution, as well as contribute to the depletion of natural resources and the release of harmful chemicals from plastic waste. While recycling is an essential part of reducing plastic waste, it is not without its challenges. Many types of plastics are difficult to recycle due to their complex composition, different melting points, and the presence of contaminants. The lack of standardized recycling processes and technologies for all types of plastic creates hurdles in achieving high recycling rates. The breakdown of larger plastic items into smaller fragments, known as microplastics, is a growing concern. Microplastics can be found in various environments, including oceans, rivers, soil, and even the air we breathe. These tiny particles can enter the food chain, potentially impacting ecosystems and human health. Most plastics are derived from fossil fuels, primarily petroleum. This creates a challenge in terms of sustainability and environmental impact. The extraction and processing of fossil fuels for plastic production contribute to carbon emissions and exacerbate climate change.

Economic disadvantages of recycling

Establishing recycling infrastructure, such as collection systems, sorting facilities, and recycling plants, requires significant capital investment. The costs associated with purchasing and maintaining equipment, hiring skilled labour, and ensuring compliance with regulations can be substantial. Recycling operations require ongoing expenses for collecting, sorting, processing, and transporting recyclable materials. These costs can include labour, energy, maintenance, and transportation, which can be higher compared to traditional waste disposal methods. The economic viability of recycling heavily depends on the market demand and the value of recycled materials. Prices for recycled materials can fluctuate due to various factors such as supply and demand dynamics, global market conditions, and changes in government policies. In some cases, the value of recycled materials may be lower than the cost of collection and processing, making recycling financially challenging. The demand for recycled materials is not always consistent or sufficient to absorb the entire supply. Some recycled materials may have limited applications or face competition from cheaper virgin materials. Lack of market demand can result in stockpiling of recycled materials or their disposal as waste, leading to additional costs for recycling facilities. Inefficiencies in recycling processes can contribute to higher costs. Outdated technology, inadequate sorting systems, and inefficient transportation can increase operational expenses and decrease the overall economic feasibility of recycling.

Problems caused by recycling plastic

One of the major problems in plastic recycling is contamination. Plastic waste can be contaminated with various substances, such as food residue, oils, chemicals, and non-recyclable materials. Contamination hinders the recycling process and reduces the quality of recycled plastic, making it less valuable and limiting its usability. Plastic recycling involves sorting different types of plastic based on their resin codes or polymer types. However, identifying and separating various plastic types can be challenging due to the lack of standardized labelling and the presence of similar-looking plastics. This complexity adds to the cost and time required for sorting, impacting the efficiency of recycling operations. In some cases, plastic recycling involves downcycling, which means that the recycled plastic is converted into lower-value products. This occurs when the quality or properties of the recycled plastic degrade during the recycling process, limiting its reuse in high-value applications. Downcycling reduces the overall economic and environmental benefits of recycling. Plastic recycling processes require energy for collection, sorting, cleaning, shredding, melting, and reforming the plastic. These processes can consume significant amounts of energy, contributing to carbon emissions and environmental impact. The energy-intensive nature of plastic recycling can offset some of the environmental benefits gained from recycling. Insufficient recycling infrastructure is a significant challenge in many regions. Lack of recycling facilities, collection systems, and adequate resources for processing and handling plastic waste hinder effective recycling efforts. Limited infrastructure leads to lower recycling rates and a higher likelihood of plastic waste ending up in landfill or the natural environment.

Economic factors of recycling

Establishing and maintaining recycling infrastructure, including collection systems, sorting facilities, and recycling plants, requires significant investment. The cost of setting up recycling facilities, purchasing equipment, and implementing efficient waste management systems can be substantial. The availability of adequate funding and financial resources is crucial for the development and expansion of recycling infrastructure. The economic viability of recycling depends on the demand for and value of recycled materials. The market demand for recycled materials can fluctuate based on factors such as supply-and-demand dynamics, consumer preferences, and market trends. The value of recycled materials may be influenced by their quality, purity, and suitability for various applications. Higher demand and better market value for recycled materials contribute to the economic feasibility of recycling programs. Recycling involves the collection, transportation, and processing of recyclable materials. The cost of transporting materials from collection points to recycling facilities can be significant, especially when dealing with large volumes of waste over long distances. Efficient transportation and logistics systems help minimize costs and optimize the recycling process. Recycling can generate economic benefits and contribute to job creation. Recycling industries create employment opportunities in various sectors, including waste collection, sorting, processing, and manufacturing of recycled products. A thriving recycling sector can stimulate local economies and promote sustainable development. Recycling offers potential cost savings and resource conservation benefits. By reusing and recycling materials, businesses and industries can reduce their reliance on virgin resources, which can be expensive to extract and process. Recycling also helps conserve energy, reduce greenhouse gas emissions, and mitigate environmental impacts associated with the production of new materials. Extended Producer Responsibility (EPR) programs shift the responsibility of managing and financing recycling from governments to producers. In such programs, manufacturers are financially responsible for the end-of-life management of their products, including recycling or proper disposal. EPR programs can incentivize manufacturers to design products that are more recyclable and support the overall economic viability of recycling systems.

4 Conclusion

The recycling of polyethylene (PE) faces numerous challenges due to its physical and chemical properties. These challenges include difficulties in identifying and separating different types of PE, contamination issues, and the high energy consumption required during the recycling process. Despite these challenges, recycling of PE is crucial for reducing plastic waste and achieving sustainability goals. The chapter suggests the need for continuous research and investment in new technologies and methods to overcome these challenges and make the recycling of PE more economically and environmentally feasible. Additionally, reducing plastic waste at the source through measures like reducing packaging and promoting alternative materials is crucial. To address these challenges, collaboration between industry, government, and consumers is necessary to promote innovative solutions and a circular economy that prioritizes sustainability.

Acknowledgements Author would like to thank Dr Sunil Sable from Vishwakarma Institute of Technology for their constant support and guidance. And HOD Chemical Engineering Department, Vishwakarma Institute of Technology: - Prof. (Dr.) Manik Deosarkar.

Author Contributions Statement Acquisition of data: Mitesh Ikar; Writing of the manuscript: Sunil Sable and Mitesh Ikar; Supervision: Sunil Sable.

References

ACRR. (2004). *Good practices guide on waste plastics recycling*. Association of Cities and Regions for Recycling.

Aguado, J., & Serrano, D. P. (1999). J. H. Clark (Ed.) *Feedstock recycling of plastic wastes* (p. 1). The Royal Society of Chemistry.

Aguado, J., Serrano, D. P., & San Miguel, G. (2007). European trends in the feedstock recycling of plastic wastes. *Global NEST Journal, 9*, 12–19.

Ahmad, S. R. (2004). *Environmental Technology, 25*, 1143.

Akato, K. M., Nguyen, N. A., Bonnesen, P. V., Harper, D. P., & Naskar, A. K. (2018). *ACS Omega, 3*, 10709.

Al-Sabagh, A. M., Yehia, F. Z., Eshaq, G., Rabie, A. M., & ElMetwally, A. E. (2016). *Egyptian Journal of Petroleum, 25*, 53.

Al-Salem, S. M., Lettieri, P., & Baeyens, J. (2009). *Waste Management, 29*, 2625.

Andrady, A. (1994). Assessment of environmental biodegradation of synthetic polymers. *Polymer Reviews, 34*, 25–76. https://doi.org/10.1080/15321799408009632

Andrady, A. (2003). An environmental primer. In A. Andrady (Ed.), *Plastics and the environment* (pp. 3–76). Wiley Interscience.

Andrady, A. L., & Neal, M. A. (2009). Applications and societal benefits of plastics. *Philosophical Transactions of the Royal Society B, 364*, 1977–1984. https://doi.org/10.1098/rstb.2008.0304

APME. (2004). *An analysis of plastics consumption and recovery in Europe*. Association of Plastic Manufacturers Europe.

Arena, U., Mastellone, M., & Perugini, F. (2003). Life cycle assessment of a plastic packaging recycling system. *International Journal of Life Cycle Assessment, 8*, 92–98. https://doi.org/10.1007/BF02978432

Arvanitoyannis, I., & Bosnea, L. (2001). Recycling of polymeric materials used for food packaging: Current status and perspectives. *Food Review International, 17*, 291–346. https://doi.org/10.1081/FRI-100104703

ASTM Standard D5033. (2000). *Standard guide to development of ASTM standards relating to recycling and use of recycled plastics*. ASTM International. https://doi.org/10.1520/D5033-00

Barnes, D. K. A., Galgani, F., Thompson, R. C., & Barlaz, M. (2009). Accumulation and fragmentation of plastic debris in global environments. *Philosophical Transactions of the Royal Society B, 364*, 1985–1998. https://doi.org/10.1098/rstb.2008.0205

Campanelli, J. R., Kamal, M. R., & Cooper, D. G. (1993). *Journal of Applied Polymer Science, 48*, 443.

Datta, J., & Kopczynska, P. (2016). *Critical Reviews in Environment Science and Technology, 46*, 905.

DEFRA. (2007). Waste strategy factsheets. Retrieved November 26, 2008, from http://www.defra.gov.uk/environment/waste/strategy/factsheets/landfilltax.htm

Desai, S. D., Patel, J. V., Patel, M. R., & Sinha, V. K. (2005). *Indian Journal Chemical Technology, 12*, 82.

Drelich, J., Payne, J., Kim, T., & Miller, J. (1998). Selective froth floatation of PVC from PVC/PET mixtures for the plastics recycling industry. *Polymer Engineering & Science, 38*, 1378. https://doi.org/10.1002/pen.10308

Du, S., Valla, J. A., Parnas, R. S., Bollas, G. M., & Sustain, A. C. S. (2016). *Chemical Engineering, 4*, 2852.

EEA. (2008). *Better management of municipal waste will reduce greenhouse gas emissions*. European Environment Agency.

Fisher, M. (2003). Plastics recycling. In A. Andrady (Ed.), *Plastics and the environment* (pp. 563–627). Wiley Interscience.

Fletcher, B., & Mackay, M. (1996). A model of plastics recycling: Does recycling reduce the amount of waste? *Resources, Conservation & Recycling, 17*, 141–151. https://doi.org/10.1016/0921-3449(96)01068-3

Froelich, D., Maris, E., Haoues, N., Chemineau, L., Renard, H., Abraham, F., & Lassartesses, R. (2007). *Minerals Engineering, 20*, 902.

Frosch, R., & Gallopoulos, N. (1989). Strategies for manufacturing. *Scientific American, 261*, 144–152.

Garforth, A., Ali, S., Hernandez-Martinez, J., & Akah, A. (2004). Feedstock recycling of polymer wastes. *Current Opinion in Solid State and Materials Science, 8*, 419–425. https://doi.org/10.1016/j.cossms.2005.04.003

Genta, M., Iwaya, T., Sasaki, M., Goto, M., & Hirose, T. (2005). *Industrial and Engineering Chemistry Research, 44*, 3894.

George, N., & Kurian, T. (2014). *Industrial and Engineering Chemistry Research, 53*, 14185.

Geyer, B., Lorenz, G., & Kandelbauer, A. (2016). *Express Polymer Letters, 10*, 559.

Gilpin, R., Wagel, D., & Solch, J. (2003). Production, distribution, and fate of polycholorinated dibenzo-p-dioxins, dibenzofurans, and related organohalogens in the environment. In A. Schecter, T. Gasiewicz (Eds.), *Dioxins and health* (2nd edn.). Wiley.

Goto, M., Koyamoto, H., Kodama, A., Hirose, T., & Nagaoka, S. (2002). *Journal of Physics: Condensed Matter, 14*, 11427.

Gregory, M. R. (2009). Environmental implications of plastic debris in marine settings—entanglement, ingestion, smothering, hangers-on, hitch-hiking and alien invasions. *Philosophical Transactions of the Royal Society B, 364*, 2013–2025. https://doi.org/10.1098/rstb.2008.0265

Hoang, C. N., & Dang, Y. H. (2013). *Polymer Degradation and Stability, 98*, 697.

Ignatyev, I. A., Thielemans, W., Vander Beke, B. (2014). *ChemSusChem, 7*, 1579.

Institute for Local Self-Reliance. (2002). Western Europe's experience with refillable beverage containers. Retrieved November 26, 2008, from http://www.grrn.org/beverage/refillables/Europe.html

Karthikeyan, S., Sivakumar, N., Manimekalai, T. K., & Sathiskumar, C. (2012). *Elixir International Journal, 29*, 8291.

Kenny, S. T., Runic, J. N., Kaminsky, W., Woods, T., Babu, R. P., Keely, C. M., Blau, W., & O'Connor, K. E. (2008). *Environmental Science and Technology, 42*, 7696.

Langer, E., Waskiewicz, S., Lenartowicz-Klik, M., & Bortel, K. (2015). *Polymer Degradation and Stability, 119*, 105.

Leung, D. Y. C., Wu, X., & Leung, M. K. H. (2010). *Applied Energy, 87*, 1083.

Liu, B., Lu, X., Ju, Z., Sun, P., Xin, J., Yao, X., Zhou, Q., & Zhang, S. (2018). *Industrial and Engineering Chemistry Research, 57*, 16239.

Liu, B., Fu, W., Lu, X., Zhou, Q., Zhang, S., & Sustain, A. C. S. (2019). *Chemical Engineering, 7*, 3292.

López, M. D. M. C., Pernas, A. I. A., López, M. J. A., Latorre, A. L., Vilariño, J. M. L., & Rodríguez, M. V. G. (2014). *Materials Chemistry and Physics, 147*, 884.

Mecit, O., & Akar, A. (2001). Macromol. *Materials Engineering, 286*, 513.

Mohanraj, C., Senthilkumar, T., & Chandrasekar, M. (2017). *International Journal of Energy Research, 41*, 1534.

Paszun, D., & Spychaj, T. (1997). *Industrial and Engineering Chemistry Research, 36*, 1373.

Polk, M. B., Leboeuf, L. L., Shah, M., Won, C.-Y., Hu, X., & Ding, W. (1999). *Polymer-Plastics Technology and Engineering, 38*, 459.

Rahimi, A., & García, J. M. (2017). *Nature Reviews Chemistry, 1*, 0046.

Soni, R. K., Singh, S., & Dutt, K. (2010). *Journal of Applied Polymer Science, 115*, 3074.

Suh, D. J., Park, O. O., & Yoon, K. H. (2000). *Polymer, 41*, 461.

Tang, H., Li, N., Li, G., Wang, A., Cong, Y., Xu, G., Wang, X., & Zhang, T. (2019). *Green Chemistry, 21*, 2709.

Tawfik, M. E., & Eskander, S. B. (2010). *Polymer Degradation and Stability, 95*, 187.

Tawfik, M. E., Ahmed, N. M., & Eskander, S. B. (2011). *Journal of Applied Polymer Science, 120*, 2842.

Teotia, M., Tarannum, N., & Soni, R. K. (2017). *Journal of Applied Polymer Science, 134*, 45153.

Troev, K., Grancharov, G., Tsevi, R., & Gitsov, I. (2003). *Journal of Applied Polymer Science, 90*, 1148.

US EPA. EPA530-F-18-004. (2018).

US EPA. (2019). *Documentation for greenhouse gas emission and energy factors used in the waste reduction model (WARM)*.

Wang, H., Li, Z., Liu, Y., Zhang, X., & Zhang, S. (2009). *Green Chemistry, 11*, 1568.

Wang, Q., Geng, Y., Lu, X., Zhang, S., & Sustain, A. C. S. (2015). *Chemical Engineering, 3*, 340.

Wang, Q., Yao, X., Tang, S., Lu, X., Zhang, X., & Zhang, S. (2012). *Green Chemistry, 14*, 2559.

Wong, S. L., Ngadi, N., Abdullah, T. A. T., & Inuwa, I. M. (2015). *Renewable and Sustainable Energy Reviews, 50*, 1167.

Yoshioka, T., Okayama, N., & Okuwaki, A. (1998). *Industrial and Engineering Chemistry Research, 37*, 336.

Yoshioka, T., Motoki, T., & Okuwaki, A. (2001). *Industrial and Engineering Chemistry Research, 40*, 75.

Yue, Q. F., Wang, C. X., Zhang, L. N., Ni, Y., & Jin, Y. X. (2011). *Polymer Degradation and Stability, 96*, 399.

Yue, Q. F., Xiao, L. F., Zhang, M. L., Bai, X. F. (2013). *Polymers (Basel, Switz.), 5*, 1258.

Pyrolysis-Based Fuel Extraction from Household Waste Plastics for Diesel Generators

K. R. Deepa, Nayana shetty, N Shamala, P. Ashwini Kumari, R. Likitha, and K. A. Manohar

Abstract

Plastics are currently a serious hazard to our planet since they have become an essential component of our everyday life. Globally, more than a hundred million tons of plastic are manufactured each year, and its discarded plastics are now frequently found in waste facilities and dumpsters that are filling up. Although efforts are being made to create modern biodegradable polymers but haven't been much significant measures taken to address the issue at present. In this instance, a workable method for repurposing plastics is described for turning garbage plastic to value-added fuel. Researchers are looking for fuel alternatives for gasoline and diesel engines because of increasing energy requirements, tougher emission regulations, and a shortage of fossil fuel sources. However, due to the worldwide issues with disposal, plastic waste poses a very major environmental dilemma. Analyzing and comparing oil that's produced from using plastics against petroleum-based products revealed that this oil possesses characteristics that are comparable to diesel fuel. In this research, an effort was put into looking towards the pyrolysis method ability to turn residential garbage plastics to fuel, for different types of plastic trash, pyrolysis equipment has been developed, constructed, and tested. The qualities that the fuels produced are identified, then the resulting fuels are utilized for diesel-powered generators.

Keywords

Plastic • Pyrolysis • Oil • Alternative fuels • Recycling of plastics

K. R. Deepa · P. Ashwini Kumari (✉) · K. A. Manohar
REVA University, Bengaluru, 560064, India
e-mail: ashwinikumari.p@reva.edu.in

K. R. Deepa
e-mail: deepa.kr@reva.edu.in

N. shetty
Nitte (Deemed to be University) NMAMIT, Nitte, India

N. Shamala
VVIT Mysyru, Mysuru, India

R. Likitha
Nitte Meenakshi Institute of Technology Bengaluru, Bengaluru, 560064, India

1 Introduction

Plastics have recently become extensively utilized. Plastics are flexible, sanitary, lightweight, and long-lasting. Polymers, which are lengthy chains of molecules, are used to make plastics. While a readily available material, like crude oil or petroleum-based substances, is converted into another material with entirely distinct qualities, hence, polymers are produced. Plastic is being regularly utilized across the entire globe (Lam et al., 2019; Huchappa & Patil, 2022; Jha & Kannan, 2021; Jiao et al., 2021; Mangesh et al., 2020). Term "plastic" is frequently employed to refer to a wide variety of synthetic and partially synthetic substances. They additionally generate synthetic products including apparel, wrapping, vehicles, gadgets, aviation, healthcare equipment, and leisure goods. The list of advantages might go on forever, but in the absence of plastic, many things we take for granted nowadays wouldn't be feasible. The price of plastic is the way that altered the globe. Synthetic polymers are manufactured in enormous amounts with a wide range of characteristics and applications. In addition, some polymers are produced periodically. These comprise cellulose, protein molecules, and starch. When several monomer molecules are used to form polymer chains, they connect like a series of links. Regardless of how the molecules are incorporated, the polymer will alter. An additional polymer is one in which atoms from both monomers and polymers are joined. Condensation polymers are those that have had part of their

M. Sumesh et al. (eds.), *2nd International Conference on Smart Sustainable Materials and Technologies (ICSSMT 2023)*,
Advances in Science, Technology & Innovation, https://doi.org/10.1007/978-3-031-49826-8_23

monomer atoms liberated to smaller molecule sizes, such as those seen in fluid. Many of the carbon particles in added polymer have double bonds (Das et al., 2020; Jahirul et al., 2022; Jha et al., 2021; Kumari & Geethanjali, 2021; Puttaramaiah & Geethanjali, 2022). The polymers known as thermoplastic or thermo-softening plastics transform into a fluid at a certain temperature or become highly glassy while freezing. Higher molecule mass polymer tends to be what thermoplastic materials are. Such plastics are useful for recycling since they are capable of being melted and re-molded, like Bakelite, that has the property of thermosetting plastic that when solidified remains in its original form. According to how strong the inter-molecular forces are, its melting level will change. This melting point is going to increase in the presence of hydrogen bonding. Induction heating is a fascinating concept with a lot of possibilities for plastic reflow. This may be beneficial for connecting polymers like PVC or adhering plastic onto various substrates because of the quick regulated handling of heat. But since plastics don't conduct electricity, a charge must be created like when someone rubs a balloon against their head to generate an electrostatic charge. Similar considerations go into the concept of induction forces with plastic. It is presumed that attractive forces between surfaces with spherical curvature arise while discussing dispersion forces. The radius is formed by the degree of plastic deformation experienced during compression. Then we have hydrogen bonds, which are an attractive contact connecting a hydrogen atom and an atom that is electron negative, typically oxygen, fluorine, or nitrogen. The hydrogen now has a half-positive charge. To form the bond, the hydrogen must be covalently connected to an additional electronegative atom. The bonds may occur inside molecules or in several locations within an individual molecule Plastic characteristics can also be changed. This can happen by altering the original polymers or by adding additives, colorants, reinforcement, or fillers. Since using an addition always involves making some form of trade-off, additives must be compatible. To achieve the intended enhancement, such as resistance to burning, chemists strive to maintain many of the other material's qualities as high as they can. Other chemicals, such as fungicides or flame retardants, may be employed to enhance high-temperature stability. Plastics with reinforcements have greater dimensional stability and can withstand loads and higher temperatures. Due to this, the airline and aerospace industries may now design with more freedom and make important technological advancements (Yusuf et al. 2022; Alawa & Chakma, 2023; Deepa et al., 2022; Faraca & Astrup, 2019; Song et al., 2022). Industries may now design with more freedom and make important technological advancements. The only real worry about plastics now that they are a staple item in society is how they will impact the environment. The most recent news reports discuss plastics that emit chemicals, concerns over trash, and the impact on landfills and rivers. The price of the fuels required for its manufacture has steadily increased despite formerly cheap expenses. Although the solution is still mysterious, science continues to investigate the issue and has discovered several attractive alternatives, like fructose and oil shale. Only time can tell what plastics' future holds, although it is unclear whether we will ever live in a world without them. In today's world, plastics are becoming an essential component. These plastics are used in a wide range of industrial and residential domains due to their low size, durability, energy efficiency, rapid rate of manufacturing, and design flexibility. Hydrocarbons make up most plastics, which are derived from petroleum and contain stabilizers, antioxidants, and other additives. The ecology is greatly endangered by the disposal of waste plastics, yet no practical solution has yet been developed. The main ingredients found in plastics, which are non-biodegradable polymers, include carbon, hydrogen, and a little number of other elements like nitrogen. Plastic garbage considerably adds to the issue of managing trash because it is not biodegradable. According to a national survey carried out in 2000, India produced about 6000 tons of plastic, of which only 60% could be recycled and the remaining 40% could not be disposed of. Currently, 129 million tons of plastic are manufactured globally each year, with 77 million tons coming from petroleum. The demand for plastics is roughly 8 million tons a year in India alone. Every day, India generates more than 10,000 metric tons of plastic, and it imports nearly as much from other nations. In comparison to the 30–40 kg per capita usage of plastics in wealthy nations, India uses just approximately 3 kg per capita. These are primarily from the food and packaging sectors. While most plastics are recycled, there are occasionally exceptions because there is not enough market demand. Most polyethylene waste plastics, or 43% of it, are found in packaging and containers. Pyrolysis and plastic-to-fuel technologies only partially address the root causes of plastic pollution, such as overproduction and poor waste management. These technological advancements utilize a lot of energy and could increase greenhouse gas emissions. Because there is little demand for the produced fuels on the market, more processing is necessary. They are criticized by the public and take resources away from recycling. A comprehensive strategy that incorporates reduction, recycling, environmentally friendly options, and enhanced waste management systems is needed to address the plastics problem.

2 Types of Plastics

Waste plastics come in the following types: LDPE, HDPE, PP, PS, and PVC. Because safety deposits are expensive and incineration encourages the rising emission of dangerous greenhouse gases like CO_2, NO_2, SO_2, and others, the difficulties of waste plastics cannot be managed by land filling or burning. These methods of getting rid of waste plastics create harmful gas, which harms the environment. According to their sources, plastic waste may also be divided into two categories: industrial and municipal (Chandran et al., 2020; Damodharan et al., 2019; Kumari et al., 2022; Murthy et al., 2020; Parthasarathy et al., 2021). These groups have varied characteristics and properties and are managed using various techniques. Over 78% of the weight of the entire plastic waste is made up of thermoplastics, while the remaining portion is made up of thermosets. Polyolefins, such as polyethylene, polypropylene, polystyrene, and polyvinyl chloride, are the building blocks of thermoplastics, which are recyclable. However, thermosets, which mostly contain epoxy resins and polyurethanes, cannot be recycled. Types of plastic are shown in Table 1.

3 Methodology

Block diagram of plastic to energy converter plastics are increasingly considered essential materials in the modern world, and their use in industry is constantly expanding (Fig. 1). Waste plastic solid is presently gaining fresh interest in this aspect. In the first place, plastic is made from natural gas, namely from ethane, a gaseous component. As a result, it may be recreated from discarded plastic. A reactor was built to give the necessary temperature for the process as pyrolysis requires a high temperature of roughly 300 °C and operates without oxygen. Plastic trash may be turned into fuel, which has enormous potential for both the environment and the economy. Consequently, the process of turning plastics into fuel has now converted the issues into a chance to create income from the trash. There are two benefits to turning plastic into oil. The oil generated initially may be utilized as a fuel for homes as well as for automobiles and industry after additional refinement. Second, the many forms of pollution brought on by leftover plastics can be reduced. When plastic is heated, a chemical process occurs in which bigger molecules break down into smaller ones. At every given temperature, the molecule is in a state of molecular vibration, which increases as the temperature rises. High molecular vibration causes molecules to be stretched and disturbed to the point where they begin to break down into smaller molecules. Through this procedure, oil is produced that may be utilized as fuel for generators. To achieve the appropriate temperature, a temperature controller was integrated with the electric heater as part of the equipment design (Abomohra et al., 2021; Mishra et al., 2019; Palos et al., 2021; Sikdar et al., 2020; Sonil Nanda & Franco Berruti., 2021). It also contains the frame for the reactor tank, the tank for cleaning and condensing smoke, and the tank for the reactor. The design also includes how each machine component should be placed and how it should be arranged. The machine's design took the environment into account as well. To clean the smoke and condense the leftover vapor before it enters the atmosphere, an oil condensing and smoke cleaning tank were created. When all the required supplies were available, fabrication started. Angle bar, stainless pipe, copper tube, hose, hose clips, air chucks, stainless steel pots, thermocouples, temperature controllers, circuit breakers, contactors, compressors, condensing tanks, and wastewater collection tanks were the materials utilized. Plastic waste undergoes thermal degradation at high temperatures without oxygen during plastic pyrolysis for oil production. As a result, vapors and gases are produced. These vapors are sent to a condensation structure, wherein they're cooled and changed from a gaseous state to a liquid state. The solid leftovers and non-condensable gases are removed from the liquid oil. The quality of the oil can be improved by applying further treatment procedures to remove contaminants. The resulting oil can then go through additional processes like filtering, distillation, or refining to produce a higher-grade fuel. By burning non-condensable gases produced during pyrolysis, such as methane or hydrogen, to create heat for the pyrolysis procedure or other applications, these gases can be used for energy recovery. The temperature profile during this process is 350–500°C.

Table 1 Types of plastic

Plastic type	Moisture (wt %)	Fixed carbon (wt %)	Volatile (wt %)	Ash (wt %)
Polyethylene terephthalate (PET)	0.46	7.77	91.75	0.02
High-density polyethylene	0.0	0.01	99.81	0.18
Polyvinyl chloride (PVC)	0.84	6.3	93.70	0.0
Low-density polyethylene	0.30	1.22	99.70	0.0
Polypropylene	0.15	0.12	95.08	3.55
Polystyrene	0.25	0.04	99.63	0.0

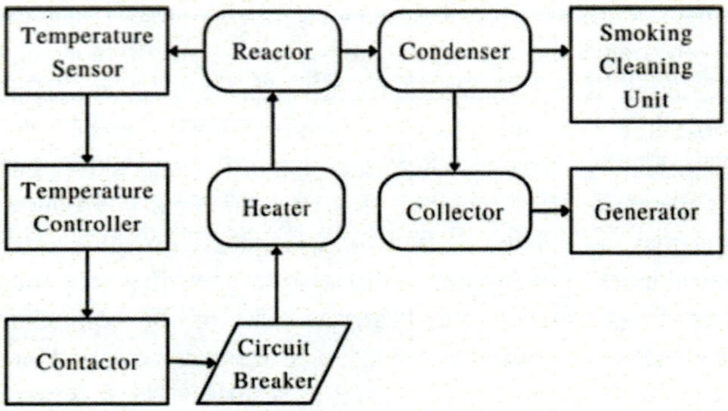

Fig. 1 Block diagram

A. Reactor Assembly and Condensing Chamber

Fig. 2 Reactor assembly

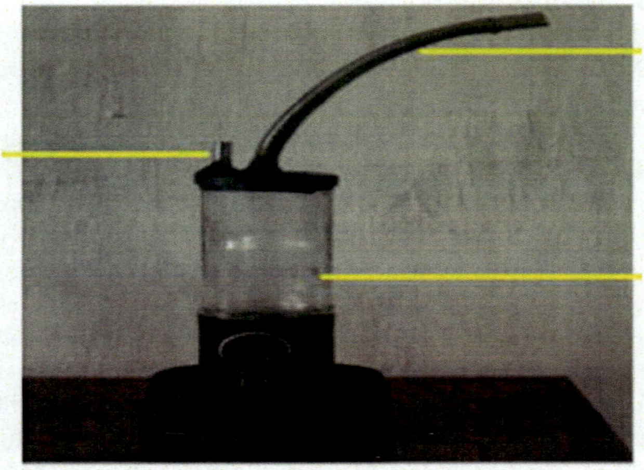

Fig. 3 Condensing chamber

4 Physical Setup

Reactor assembly and condensing chamber are shown in Figs. 2 and 3. The physical configuration of garbage plastic to energy conversion system is composed of a reactor in which plastic is heated to a controlled temperature and melted to produce vapors. Plastic is inserted into the opening provided. A thermocouple is used to measure its temperature inside the reactor and transmit the information to a temperature controller. The temperature controller constantly monitors the atmosphere's temperature using a sensor. The control circuit operates and can be changed as necessary whenever the temperature surpasses the predetermined point. An alarm function is activated if the temperature exceeds a certain level. A tripping mechanism stops the equipment and prevents damage in the event of poor control. When the reactor temperature rises over the specified temperature, the temperature controller, which is set at 380 degrees, opens the circuit, and shuts off the power to the heater. The reactor has a 9-inch diameter, and an 8-inch height, and is composed of stainless steel with a 0.5mm thickness. The vapors generated when plastic melts are condensed in a condensing chamber. It has a 2-inch diameter and a 6-inch height and is constructed of glass. There are two openings in it, one for the intake and one for the outflow. In the condenser, the oil will be collected. Since the uncondensed vapors are poisonous, they must be transported via a compressor rather than being released into the atmosphere (Ashwini Kumari & Geethanjali, 2018; Kumar et al., 2022).

Smoke cleansing unit and compressor inner assembly are shown in Fig. 4. Utilizing a smoke-cleaning device, waste plastic is converted into oil while producing smoke. It consists of the following components: (a) Compressor: It sucked up smoke or uncondensed vapor and sent it to the smoke cleaning tank. (b) Smoke cleansing tank: The smoke cleansing tank, made of PET plastic and with a water capacity of 4 gallons, submerges smoke to remove ash and particulates. The goal of emissions control systems, also known as smoke cleaning devices, is to reduce or eliminate contaminants in pyrolysis plant exhaust gases. To effectively clean smoke and regulate emissions, these devices employ a number of technologies, including as activated carbon filters, bag filters, wet or dry scrubbers, cyclone separators, and catalytic converters. Each technique focuses on a different type of pollutant, such as volatile organic compounds, organic compounds, nitrogen oxides, carbon monoxide, and solid particles, acidic gases, and gaseous pollutants. The smoke is dipped into the water in this section, which removes the ash and particles. Additionally, it turns the leftover vapor into oil. (c) Smoke service line: Acts as the path for smoke to travel from the compressor to the tank used for cleaning up smoke. It is constructed of a 1/4-inch copper tube and submerged in water inside the smoke-purging tank. (d) Connecting line: It is comprised of

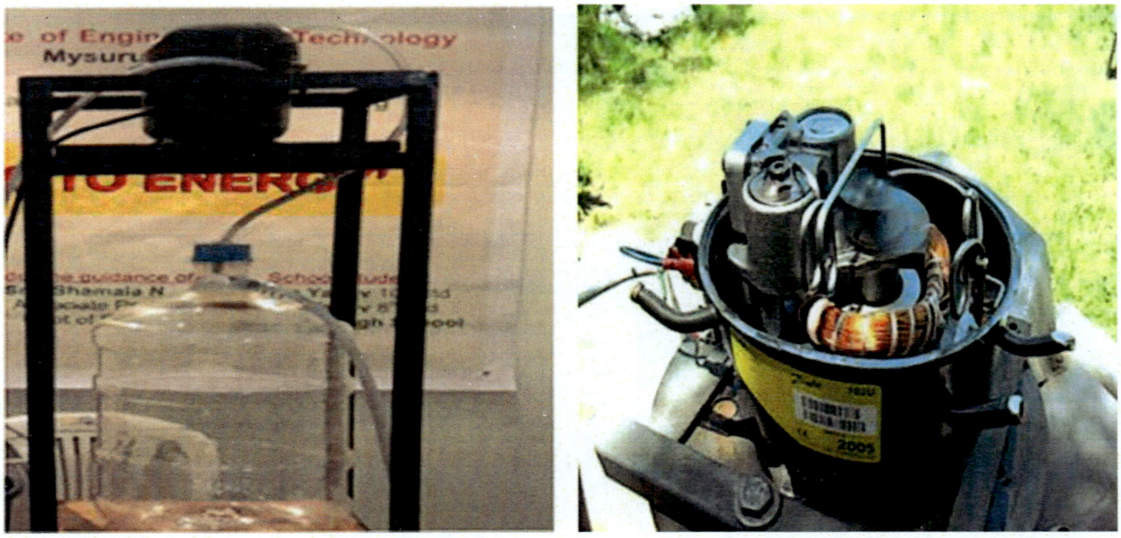

Fig. 4 Smoke cleansing unit and compressor inner assembly

a clear hose and joins the compressor discharge and smoke service flow.

5 Results and Discussions

The equipment's capacity to transform plastic waste directly into gasoline in terms of weight is known as conversion efficiency. This is calculated by multiplying the initial weight of the plastic by 100 and dividing the weight of the oil recovered. According to the study's findings, the waste plastic to oil converter has a 78.1% conversion efficiency [26–30]. The effectiveness of a piece of equipment in decreasing waste in terms of weight is known as waste reduction efficiency. The study's findings indicated that the equipment was 94.3% efficient in reducing waste.

$$\textit{waste reduction efficiency} = \frac{\textit{weight of char} - \textit{intial plastic waste}}{\textit{original weight}} * 100 \tag{1}$$

Measuring how efficiently the equipment eliminates waste allows for the calculation of waste reduction efficiency in terms of weight. Calculating this involves subtracting the weight of the char from the initial weight of the plastic, dividing the result by the original plastic weight, and multiplying the result by 100. The device showed a 94.3% efficiency in waste reduction, according to the study. Complete setup of the work with char collection is shown in Fig. 5.

As a result of this arrangement, for every kilogram of plastics, the pyrolysis process took 95 min to complete,

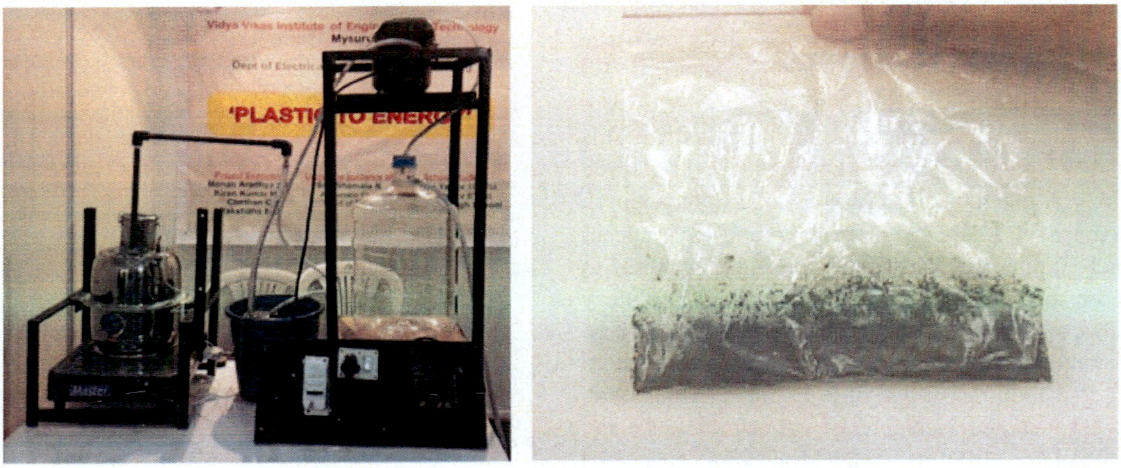

Fig. 5 Complete setup of the work with char collection

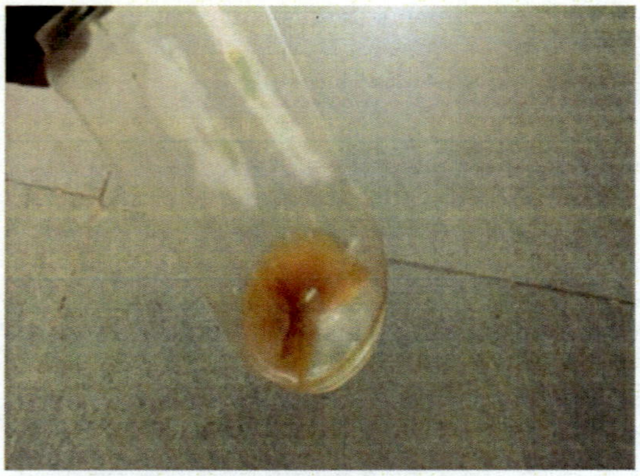

Fig. 6 Oil obtained

Fig. 7 Kinematic viscosity versus change in temperature

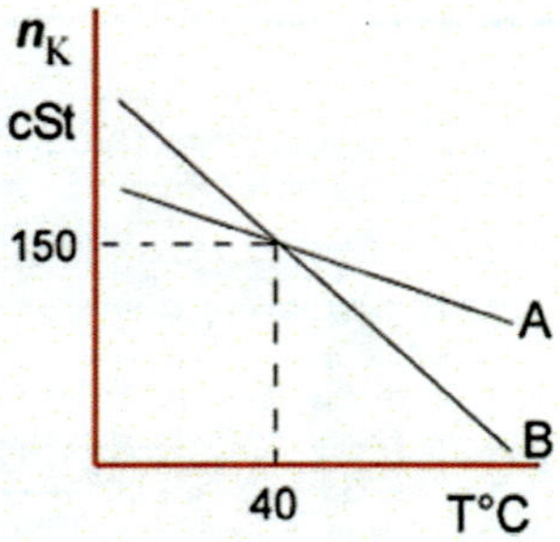

Table 2 Comparison of waste plastic oil, diesel, and petrol

Properties	WPO	Diesel	Petrol
Color	Pale yellow	Orange	Bluish green
Calorific value, Cal/g	10,222.0	10,000	1146.50
Kinematic viscosity cst@40°C	1.4	1.3–3.6	7.62
Sulfur content	0.01	0.001	0.1
Ash content (%)	NIL	0.01	0

yielding 800 ml of oil. This suggests that the reactor's plastics pyrolysis process can produce fuel at a temperature of 400 °C. Oil obtained is shown in Fig. 6, and kinematic viscosity versus change in temperature is shown in Fig. 7. Comparison of waste plastic oil, diesel, and petrol is shown in Table 2.

6 Conclusion

The International Energy Outlook 2008 shows that the global use of petroleum oil is 84 million barrels per day, indicating a steady growth in usage and, consequently, the cost of petroleum oil. In a thermal pyrolysis machine, waste plastics were converted into liquid hydrocarbon fuel. By transforming plastics into gasoline, we can solve two problems: the vast plastic oceans and the fuel deficit. This two-fold advantage, although it will only endure if the waste plastics do, will undoubtedly give us a solid foundation on which to construct a sustainable, healthy, and environmentally friendly future. It would greatly assist our economy to take into consideration the financial gains of such a project. This approach is preferable in every way, including economically and environmentally. Utilize this technique to effectively transform the weight of used liquid hydrocarbon fuels into 75% of the viable fuels without releasing any pollutants. Additionally, it would deal with dangerous plastic trash and lower crude oil imports. At this point, the depletion of non-renewable energy sources like fossil fuels necessitates upgrades to this method.

References

Abomohra, A. E. F., Sheikh, H. M. A., El-Naggar, A. H., & Wang, Q. (2021). Microwave vacuum co-pyrolysis of waste plastic and seaweeds for enhanced crude bio-oil recovery: Experimental and feasibility study towards industrialization. *Renewable and Sustainable Energy Reviews*, *149*, 111335.

Alawa, B., & Chakma, S. (2023). Experimental investigation in compression ratio on performance, combustion and emission of VCR engine using pyrolysis-oil produced from waste plastic materials. *Materials Today: Proceedings, 72*, 2593–2609.

Ashwini Kumari, P., & Geethanjali, P. (2020). Artificial neural network-based smart energy meter monitoring and control using global system for mobile communication module. In *Soft Computing for Problem Solving: SocProS 2018, Volume 2* (pp. 1–8). Springer Singapore. Nanda, S., Berruti, F. (2021). Thermochemical conversion of plastic waste to fuels: A review. *Environmental Chemistry Letters*, *19*, 123–148.

Chandran, M., Tamil kolundu, S., & Murugesan, C. (2020). Characterization studies: Waste plastic oil and its blends. *Energy Sources, Part A: Recovery, Utilization, and Environmental Effects*, *42*(3), 281–291.

Damodharan, D., Rajesh Kumar, B., Gopal, K., De Poures, M. V., & Sethuramasamyraja, B. (2019). Utilization of waste plastic oil in diesel engines: A review. *Reviews in Environmental Science and Bio/Technology*, *18*, 681–697.

Das, A. K., Hansdah, D., Mohapatra, A. K. & Panda, A. K. (2020). Energy, exergy and emission analysis on a DI single cylinder diesel engine using pyrolytic waste plastic oil diesel blend. *Journal of the Energy Institute*, *93*(4), 1624–1633.

Deepa, K. R., Chaitra, A. S., Jhansi, K., Anitha Kumari, R. D., & Kodabagi, M. M. (2022). Development of fire detection surveillance using machine learning & IoT. In *2022 IEEE 2nd Mysore Sub Section International Conference (MysuruCon)* (pp. 1–6). IEEE.

Faraca, G., & Astrup, T. (2019). Plastic waste from recycling centres: Characterisation and evaluation of plastic recyclability. *Waste Management, 95*, 388–398.

Huchappa, R., & Patil, K. K. (2022). Evolutionary model to guarantee quality of service for tactical worldwide interoperability for microwave access networks. *IAES International Journal of Artificial Intelligence*, *11*(2), 687.

Jahirul, M. I., Rasul, M. G., Schaller, D., Khan, M. M. K., Hasan, M. M., & Hazrat, M. A. (2022). Transport fuel from waste plastics pyrolysis–a review on technologies, challenges and opportunities. *Energy Conversion and Management, 258*, 115451.

Jha, K. K., & Kannan, T. T. M. (2021). Recycling of plastic waste into fuel by pyrolysis-a review. *Materials Today: Proceedings*, *37*, 3718–3720.

Jha, K.K., Kannan, T. T. M., Senthilvelan, N. (2021). Optimization of catalytic pyrolysis process for change of plastic waste into fuel. *Materials Today: Proceedings*, *39*, 708–711.

Jiao, X., Zheng, K., Zexun, H., Zhu, S., Sun, Y., & Xie, Y. (2021). Conversion of waste plastics into value-added carbonaceous fuels under mild conditions. *Advanced Materials, 33*(50), 2005192.

Kumar, P., Kumari, A., Lakkannavar, M., Chakraborty, S., Ramachandra, A. C., & Kongkham, D. (2022). Distributed wideband sensing on cognitive radio using hidden markov model. In *2022 IEEE 2nd Mysore Sub Section International Conference (MysuruCon)* (pp. 1–6). IEEE.

Kumari, P. A., & Geethanjali, P. (2022). Implementation of constraint handling techniques for photovoltaic parameter extraction using soft computing techniques. In *Mathematical Modeling, Computational Intelligence Techniques and Renewable Energy: Proceedings of the Second International Conference, MMCITRE 2021* (pp. 487–501). Springer.

Kumari, A., Geethanjali, P., Bairwa, B., & Ravishankar, H. (2022). Modelling and performance analysis of photovoltaic system for low-cost lawn cutting application. In *2022 IEEE 2nd Mysore Sub Section International Conference (MysuruCon)* (pp. 1–6). IEEE.

Lam, S. S., Mahari, W. A. W., Ok, Y. S., Peng, W., Chong, C. T., Ma, N.L., Chase, H. A. et al. (2019). Microwave vacuum pyrolysis of waste plastic and used cooking oil for simultaneous waste reduction and sustainable energy conversion: Recovery of cleaner liquid fuel and techno-economic analysis. *Renewable and Sustainable Energy Reviews*, *115*, 109359.

Mangesh, V. L., Padmanabhan, S., Tamizhdurai, P., & Ramesh, A. (2020). Experimental investigation to identify the type of waste plastic pyrolysis oil suitable for conversion to diesel engine fuel. *Journal of Cleaner Production, 246*, 119066.

Mishra, R. K., Iyer, J. S., & Mohanty, K. (2019). Conversion of waste biomass and waste nitrile gloves into renewable fuel. *Waste Management*, *89*, 397–407.

Murthy, K., Shetty, R. J., & Shiva, K. (2020). Plastic waste conversion to fuel: A review on pyrolysis process and influence of operating parameters. *Energy Sources, Part A: Recovery, Utilization, and Environmental Effects*, 1–21.

Nanda, S., & Berruti, F. (2021). Thermochemical conversion of plasticwaste to fuels: A review. *Environmental Chemistry Letters*, *19*, 123–148.

Palos, R., Gutierrez, A., Vela, F. J., Olazar, M., Arandes, J. M., & Bilbao, J. (2021). Waste refinery: The valorization of waste plastics and end-of-life tires in refinery units. A review. *Energy & Fuels, 35* (5), 3529–3557.

Parthasarathy, M., Ramkumar, S., Elumalai, P. V., Gupta, S. K., Krishnamoorthy, R., Iqbal, S. M., Dash, S. K., & Silambarasan, R. (2021). Experimental investigation of strategies to enhance the homogeneous charge compression ignition engine characteristics powered by waste plastic oil. *Energy Conversion and Management*, *236*, 114026.

Puttaramaiah, A. K., & Geethanjali, P. (2022). Ensemble of constraint handling techniques for PV parameter extraction using differential evolutionary algorithms. *International Journal of Power Electronics and Drive Systems*, *13*(3), 1645.

Sikdar, S., Siddaiah, A., & Menezes, P. L. (2020). Conversion of waste plastic to oils for tribological applications. *Lubricants, 8*(8), 78.

Song, Z., Xiu, F.-R., & Qi, Y. (2022). Degradation and partial oxidation of waste plastic express packaging bags in supercritical water: Resources transformation and pollutants removal. *Journal of Hazardous Materials, 423*, 127018.

Yusuf, A. A., Ampah, J. D., Soudagar, M. E. M., Veza, I., Kingsley, U., Afrane, S, Jin, C, Liu, H., Elfasakhany, A., & Buyondo, K. A. (2022). Effects of hybrid nanoparticle additives in n-butanol/waste plastic oil/diesel blends on combustion, particulate and gaseous emissions from diesel engine evaluated with entropy-weighted PROMETHEE II and TOPSIS: Environmental and health risks of plastic waste. *Energy Conversion and Management*, *264*, 115758.

Experimental and Performance Study of Diesel Engine Using Inlet Preheated Air by Heat Pipe Heat Exchanger

P. Ramkumar, M. Sivasubramanian, S. Ramasamy, C. M. Vivek, and P. Latha

Abstract

The utilization of waste exhaust heat energy into inlet preheated air is investigated using heat pipe heat exchanger. This study focuses on waste heat energy, which comes out as exhaust from standalone diesel engine is around 60–70%, this energy is converted into preheated air and supplied as inlet air for the engine using heat pipe with DI water as working fluid. This preheated air makes improvement in efficiency of the engine and is economically effective in all conditions. In this study, the effect of specific fuel consumption (SFC), brake thermal efficiency (BTE), smoke intensity, oxides of nitrogen (NOx), carbon monoxide (CO), and hydrocarbon (HC) emissions is reported for various loads ranging from no load to loaded (0–90%). The results are compared with and without heat pipe conditions and performance is analyzed. The study warranted that at 38 ° C as input temperature and 50% load condition, the better performance is achieved for SFC as 0.23 kg/kW h and 37% for BTE. The observation reports as influence of heat pipe for utilizing the waste exhaust heat makes improved in the performance of the engine.

P. Ramkumar (✉) · M. Sivasubramanian
Department of Mechanical Engineering, Kalasalingam Academy of Research and Education, Virudhunagar, Tamilnadu, India
e-mail: rkmailmech@gmail.com

S. Ramasamy
Department of Mechanical Engineering, Annamalai University, Chidambaram, Tamilnadu, India

C. M. Vivek
Department of Mechanical Engineering, Periyar Maniammai Institute of Science and Technology, Thanjavur, Tamilnadu, India

P. Latha
Department of Civil Engineering, Periyar Maniammai Institute of Science and Technology, Thanjavur, Tamilnadu, India

Keywords

Heat pipe • Diesel engine • Deionized water (DI) • Waste heat recovery • Efficiency

1 Introduction

Awareness towards Internal Combustion Engines (ICEs) is increased due to abrupt changes along fuel cost and the presence of standard emission norms over universe. The various methods are available to utilize the waste heat energy from the engines are integrated with unused heat recovery system had been proved best technique for lowering consumption of fuel and harmful emissions like CO_2, NOX, and increase engine power output. Out of various heat recovery methods, exhaust heat recovery using a heat pipe heat exchanger plays a vital role creates dropping pressure and also affects the performance of the engine by producing preheated air, so its design and usage in the engines are major importance. The various heat pipes used for specific applications according to the design concept for best heat recovery system emphasize the effective manner of operation and cost significant (Mochizuki et al., 2011; Srimuang & Amatachaya, 2012). The heat pipe was analyzed with 50, 65, and 75% of fill ratios with 5–15 W heat input using methanol. The 65% of filling ratio shows better performance (Chandrakant et al., 2023). The heat pipe used for different industrial applications for different geometrical structures used for cryogenic systems (Jouhara et al., 2017). The study focuses on composite wicks for ultra-thin flattened heat pipes. The results inferred that the pore size was determined for fluid pumping pressure. The permeability determined the frictional losses to the flow through the wick (Li et al., 2016). The micro heat pipes analysis concluded that heat transfer rates were maximized by increasing the working temperature at the heat input of 18 W (Tang et al., 2018). The grooved cylindrical heat pipe was analyzed for different

M. Sumesh et al. (eds.), *2nd International Conference on Smart Sustainable Materials and Technologies (ICSSMT 2023)*, Advances in Science, Technology & Innovation, https://doi.org/10.1007/978-3-031-49826-8_24

angles, the result proves that horizontal angle and internal grooves enhance the efficiency (Zeghari et al., 2023). The concentric tube heat pipe heat exchanger was investigated using methanol and acetone with different various mass flow rates, temperature inputs, and experimental results are compared with numerical using neuro-fuzzy system (Nair et al., 2022; Ramkumar et al., 2021a, 2021b). The investigation was done on diesel engine using waste heat recovery heat exchanger. The results state that preheated temperature of air at 150 °C is optimum, this decreases the smoke level of the engine and BTE is raised to 12.5% due to increased suction temperature, and brake-specific fuel consumption is reduced by 18.6% due to better combustion takes place in the engine cylinder. Emissions like CO, HC, and NOX were reduced by 6.9%, 24.66, and 15.5, respectively (Sabarish Kumar et al., 2019). In this work, industrial waste heat was recovered by using newly designed heat recovery heat exchanger such that lower part was shell and tube similarly the upper portion has bundle of finned tubes. The result stated that operating continuously had saved 15% of natural gas (Tian, 2017). The study focuses on exhaust gas recovery for IC engine. The operation shows that to obtain 90 °C for engine oil it requires 740 s, similarly with EHRS, this time decreased to 685 s, this leads to 7% enhancement in fuel and time duration (Vu et al., 2019). The experimental study focuses on implementation of heating vessel for preheating the inlet air of the engine, this shows at 55 °C the CO and NOx are reduced while comparing with air at environmental condition (Sharma et al., 2014). The influence of silver nanoparticles and composites makes improved in the performance of the result (Deepak et al., 2011; Sadanand et al., 2016).

The literatures focus on utilization of preheated air for inlet air using heating components. This paper is focused on utilization of exhaust waste gas from the IC engine using heat pipe heat exchanger technique and performance of the engine is analyzed. To reduce the generation of nitrogen oxide emission, many engines have exhaust gas recycling such that some of the exhaust flow ducted back into the intake system. With use of heat pipes can minimizes carbon and oxides of nitrogen emissions to the environment through by recovering part of the heat energy from the exhaust gas and used for preheating of inlet air to the engine. The hot exhaust gas is prevented to mix directly with the environment to minimize global warming.

2 Experimentation and Working Procedure

2.1 Experimental Fabrication

In this investigation, the copper cylindrical heat pipes of 150 mm are fabricated with steel wire mesh of 37 holes/inch as wick. The fabricated heat pipe heat exchanger is attached with the state-alone diesel engine. The heat pipes are filled with de-ionized water (DI) with 50% fill ratio (Washburn, 1930). The five numbers of heat pipe with 6.25 mm diameter and segmented into 80, 40, and 30 mm for evaporator, condenser, and adiabatic sections, respectively, are given in Fig. 1. The fabricated heat pipe is placed vertically inside the heat exchanger channel of length 300 mm, height 160 mm, and inlet and outlet vent diameter of 20 mm, shown in Fig. 2. Inside the heat exchanger channel, the upper portion is the evaporator region, the 80 mm of heat pipe is kept in

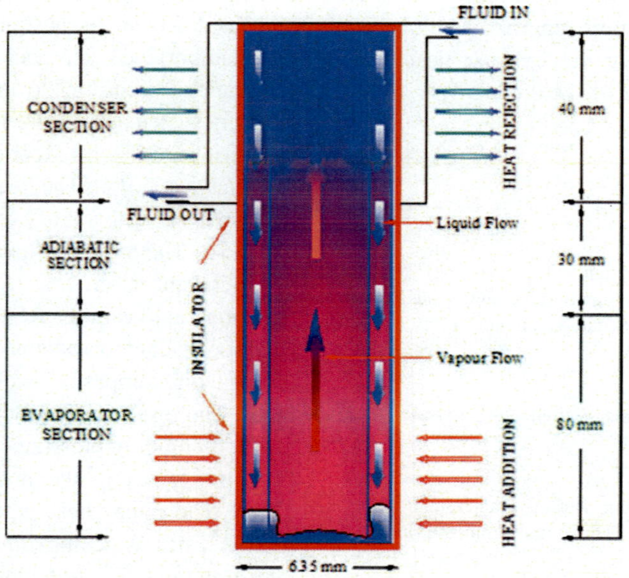

Fig. 1 Schematic diagram of heat pipe

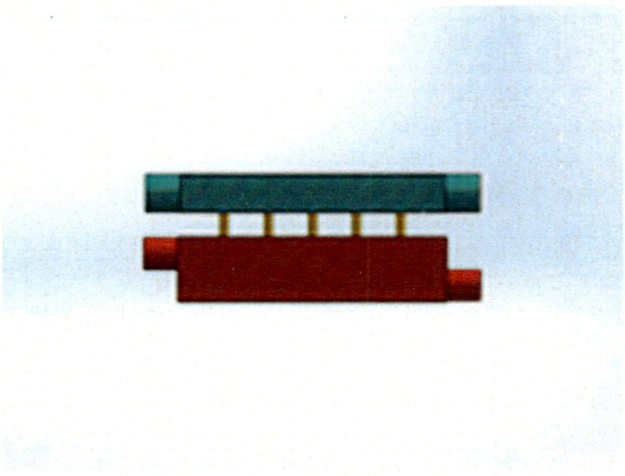

Fig. 2 Schematic view of heat pipe heat exchanger

Table 1 Engine specifications

Parameters	Description
Engine	4-stroke stand-alone oil engine
Make	KIRLOSKAR
Cylinder	1
Bore diameter	80 mm
Stroke length	110 mm
Power	4.4 kW (rated)
Cooling	Water
Orifice diameter	20 mm
Speed	650 rpm (rated)
Fuel	Diesel

this zone, the exhaust gas from the engine comes out through the channel, similarly, the lower portion is the condenser region, the 40 mm of heat pipe is placed at this section, in which inlet air enters into the engine.

In the adiabatic section, 30 mm is fully insulated with black foam material to minimize the heat loss with the environment. All the node temperatures are monitored using K-type thermocouple.

2.2 Experimental Working Procedure

The stand-alone diesel engine is fitted with the fabricated heat pipe heat exchanger as given in Table 1. The exhaut of the engine is connected with the evaporator section of HPHE, in which heat pipes with DI water absorbs the heat and converted into vapor and moves towards the adibatic section, at this juncture no heat interaction takes place and this reaches the condenser region given in Fig. 3. This zone is connected with inlet air of the engine, the fresh air passes

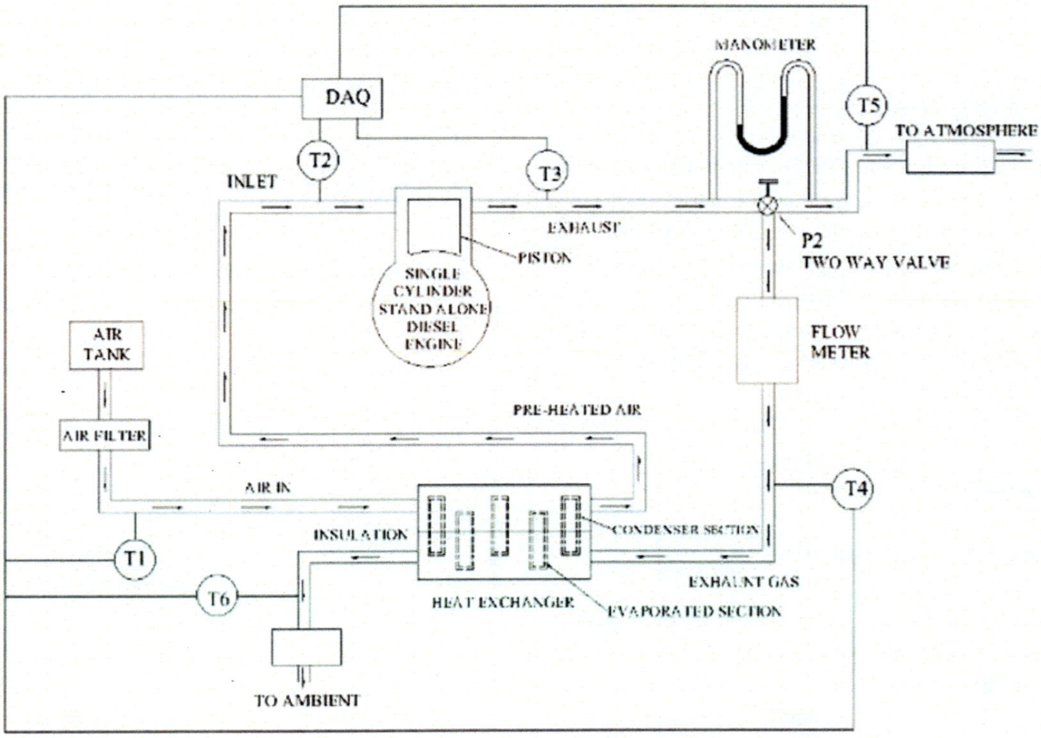

Fig. 3 Layout of experimental setup

Fig. 4 Photographic view of experimental setup

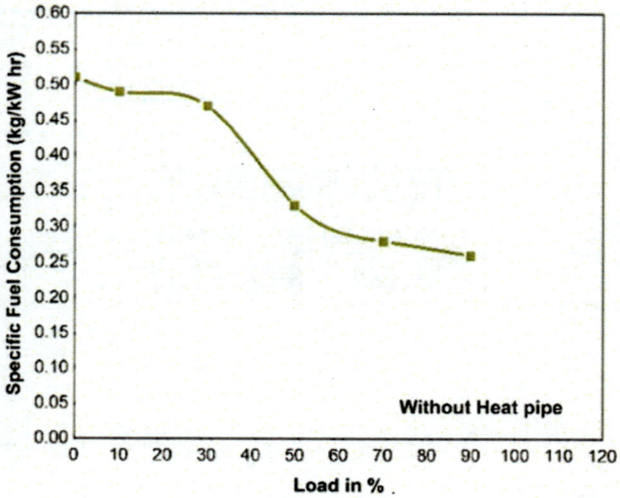

Fig. 5 SFC for without heat pipe diesel engine

through heat exchanger at condenser zone, thus fresh air absorbs the heat and converts to preheated air and enters into the engine, similarly, the vapor working fluid after rejecting the heat converts to liquid and reaches the evaporator. This preheated air makes full combustion in the engine and fuel consumption also gets reduced, meanwhile thermal efficiency of the engine gets improved and exhaust gas of the engine reduces, thus performance gets improved. The flow meter is used to regulate the flow rate. The initial test is done for no load condition (0%) and further with loaded conditions as 10, 30, 50, 70, and 90% as shown in Fig. 4.

2.3 Uncertainty Study

The experimentation uncertainty is calculated using the Kline and McClintock's method (Holman, 2007). The specific fuel consumption possesses 1.25%, and brake thermal efficiency shows 0.86%. The uncertainties occurred in the instruments are calibrated to adhere the given standards of flow meters, thermocouples, pumps, heaters, temperature indicators. The periodical calibrations are done frequently for reducing the instrumentation errors.

3 Result and Discussion

3.1 Specific Fuel Consumption

In this study, the Kirloskar stand-alone diesel engine is investigated for without and with heat pipe conditions. The investigation is carried out in no-load and load conditions, and performance of the engine is studied and reported.

In without heat pipe, the preliminary test is conducted for no-load condition (0%) and with loaded conditions such as 10, 30, 50, 70, and 90%. The Specific Fuel Consumption (SFC) is calculated for the above parameters, and results show that at without heat pipe condition minimum results are observed as 0.26 kg/kW h for 90% load condition as given in Fig. 5. In case of with heat pipe condition for different heat input temperatures such as 34, 36, 38, and 40 °C, the observation shows the minimum SFC at 38 °C for 50% load condition as 0.23 kg/kW h, depicted in Fig. 6. This shows that SFC reduces considerably by allowing preheated air by the heat pipe, while comparing with the conventional engine without heat pipe. This indicates a decrease in the flow rate of fuel consumption per kW of power generation. The time required to reach engine oil temperature is reduced in the with heat pipe compared with conventional diesel engine. The effect of preheated air makes the reduction in the quantity of fuel consumption and heat required for the engine.

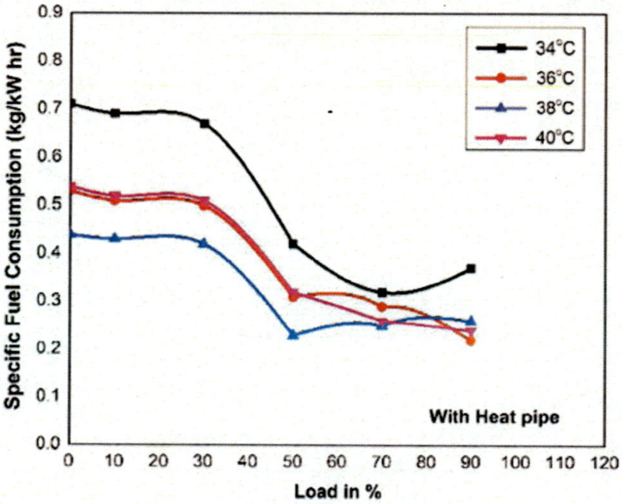

Fig. 6 SFC for with heat pipe diesel engine

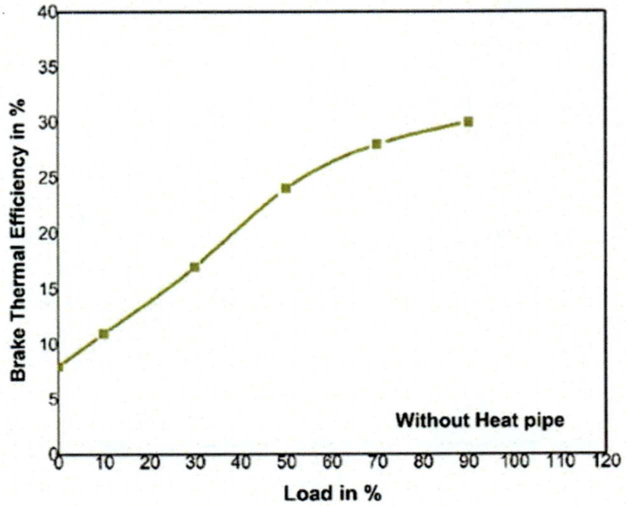

Fig. 7 BTE for without heat pipe diesel engine

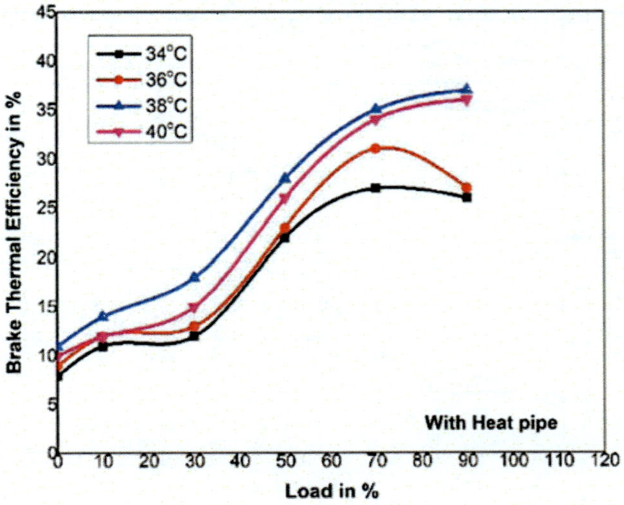

Fig. 8 BTE for with heat pipe diesel engine

3.2 Brake Thermal Efficiency

Figure 7 shows the Brake Thermal Efficiency (BTE) of the without heat pipe condition. The results are plotted for above above-stated conditions, the higher efficiency is achieved for 90% load condition, it produces 30% efficiency. The investigation is done for the same parameters for various input temperatures of 34, 36, 38, and 40 °C for with heat pipe conditions, the results are denoted as for 38 °C at 90% load condition, the maximum efficiency achieved as 37%, as given in Fig. 8. Influence of preheated air entering into the engine increases the brake thermal efficiency. These results indicate that complete combustion took place and produced more output at maximum load condition in the test engine with preheated air produced by heat pipe than conventional system. This preheated air induces the combustion of fuel in the chamber while cranking the engine and enhances the fuel vaporization, which leads to the cold start ability. By the presence of less dense warm air due to the preheating, throttle opening losses are reduced due to small opening of throttle valve to get the same volume of air. This improves the efficiency of the engine by the influence of heat pipe.

3.3 Oxides of Nitrogen

Figure 9, depicts the effect of oxides of nitrogen for load conditions for without heat pipe conditions. The results show that NO_x emission is observed for no load condition (0%) as 90 ppm and 120 to 980 ppm for 10, 30, 50, and 90% load conditions, respectively. Similarly, in case of with heat pipe conditions for above-stated parameters for 34, 36, 38, and 40 °C input temperatures, the results show that minimum NO_x emission is observed for 34 °C heat input conditions, the emission ranges from 40 to 250 ppm, while comparing with other conditions of the system as in Fig. 10. The results indicate that due to the presence preheated air, complete combustion took place and more amount of COs is converted to CO_2. The results proved that optimum air preheated temperature is less than 40 °C, otherwise NO_x emissions will be more due to the presence of higher temperature in the combustion chamber. The maximum temperature of preheated air might lead to less ignition delay, which causes the NO_X emission in the engine. Meanwhile, the preheated air leads to uniform combustion, this minimizes the noise in the engine. The preheated air induces the uniform mixing of air and fuel due to easy vaporization, this leads to decrease in CO emissions and fuel consumption.

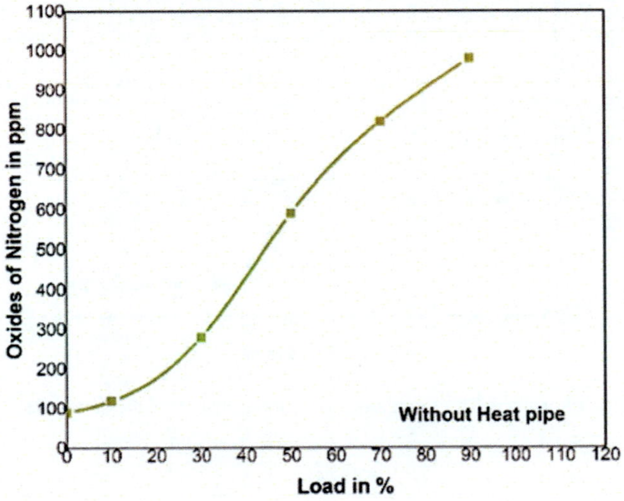

Fig. 9 Effect of NO_x emission for without heat pipe diesel engine

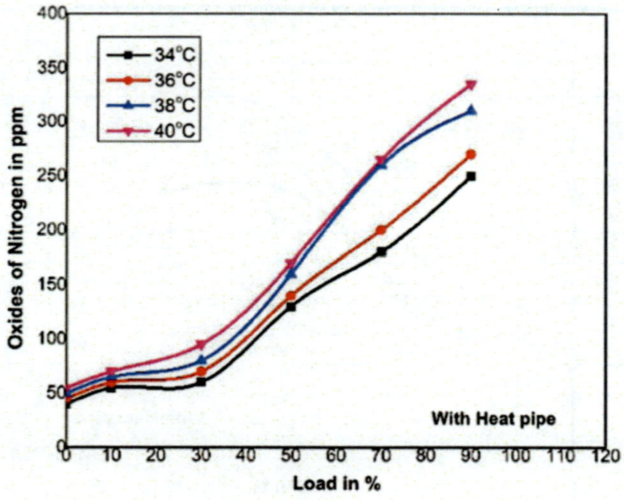

Fig. 10 Effect of NOx emission for with heat pipe diesel engine

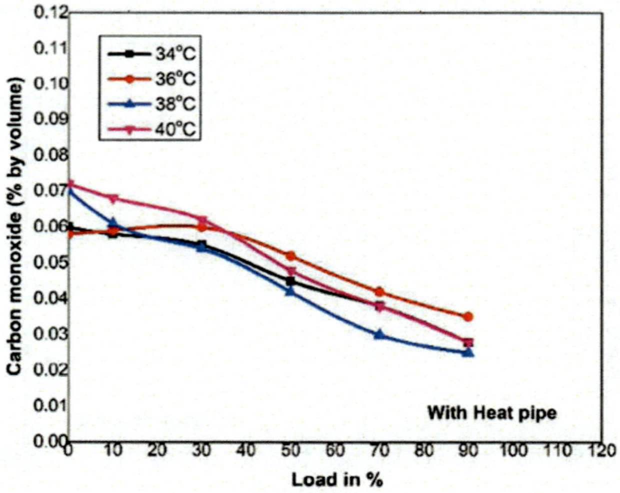

Fig. 12 Effect of CO emission for with heat pipe diesel engine

3.4 Carbon Monoxide Emission

The effect of carbon monoxide emission for no load and load conditions for without heat pipe is reported in Fig. 11. The results show that minimum CO emission is observed at 70% load condition as 0.04%. The effect of with heat pipe for above-stated parameters for various heat input temperatures of 34, 36, 38, and 40 °C, reveals that at 38 °C conditions, the minimum CO emission is 0.025% as given in Fig. 12. This evaluation shows that preheat air makes uniform combustion and air–fuel mixing ratio, thus decreasing the CO emissions in the engine.

3.5 Hydrocarbon Emission

The effect of hydrocarbon emission is stated for without heat pipe condition for no load and load conditions and depicted in Fig. 13. The results reveal that minimum HC emission is observed for 70% load condition as 37 ppm. In case of with heat pipe conditions for different temperatures of 34, 36, 38, and 40 °C, the observation shows that minimum HC emissions are observed at 38 °C and values range from 7 to 12 ppm are shown in Fig. 14. The plotted graph infers that HC emissions are also reduced because of the influence of the heat pipe in the engine. It is fact that, consumption of fuel

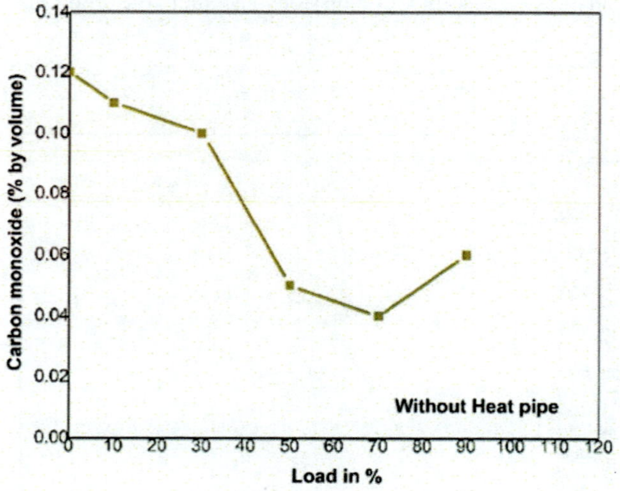

Fig. 11 Effect of CO emission for without heat pipe diesel engine

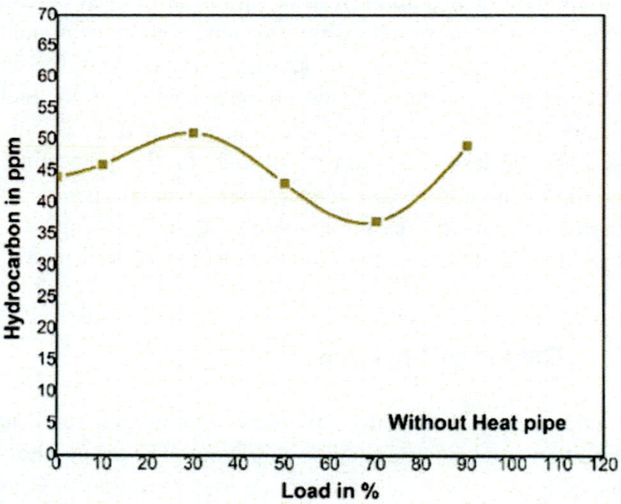

Fig. 13 Effect of HC emission for without heat pipe diesel engine

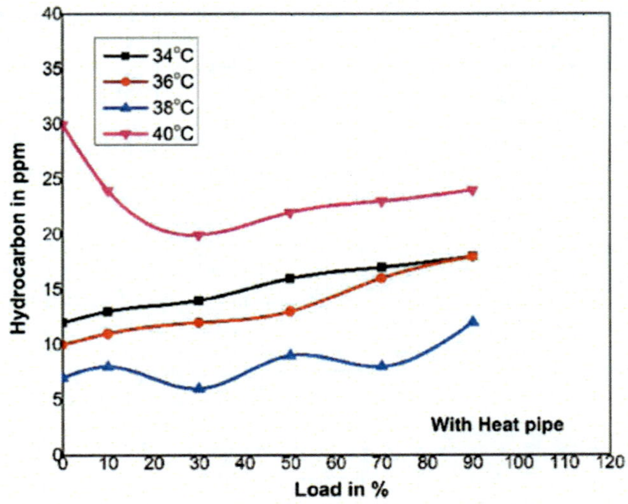

Fig. 14 Effect of HC emission for with heat pipe diesel engine

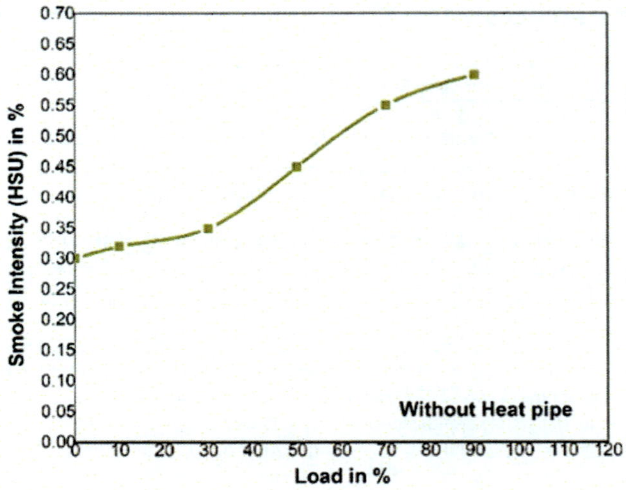

Fig. 15 Effect of smoke intensity for without heat pipe diesel engine

rate is slightly reduced and also preheated temperature of air which aids better combustion inside the engine. In case of conventional engine, HC emissions are comparatively high due to imperfect combustion and high fuel intake.

3.6 Smoke Intensity

The effect of smoke intensity for no load and loaded conditions that are stated for without heat pipe is given in Fig. 15. The results shows that the minimum smoke density ranges from 0.3 to 0.6% from 0 to 90% load conditions. The effect of with heat pipe for different input temperatures of 34, 36, 38, and 40 °C, the results show that minimum smoke density is observed at 38 °C and it ranges as 0.31 to 0.36%, from 0 to 90% load conditions shown in Fig. 16. The results clearly show that minimum smoke intensity is obtained for with heat pipe than without heat pipe, which are similar parameters and conditions. The use of heat pipes in the heat exchanger can recover adequate heat energy from the exhaust gas under ideal conditions and is used for maintaining air preheating temperature nearly optimum value as 38 °C. This denotes higher energy saving, and similarly heat pipe is a year around energy saving device, and this can be employed in the utilization of waste heat into useful energy.

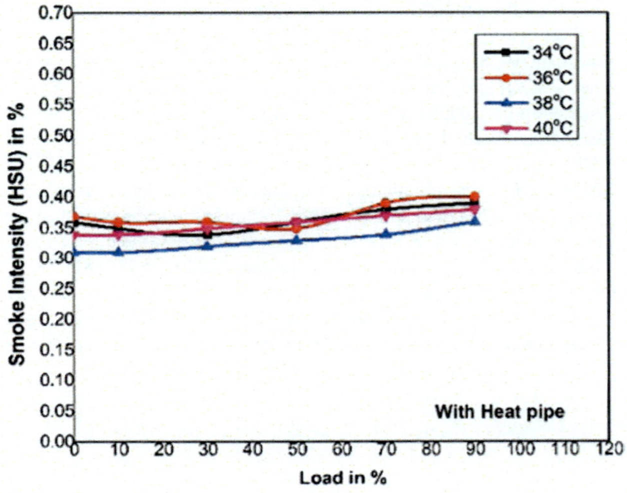

Fig. 16 Effect of smoke intensity for with heat pipe diesel engine

4 Conclusion

The IC engine exhaust gas is dissipated to the environment as waste gas, and this gas is utilized for preheating the inlet air using the heat pipe techniques. The analysis is done for with and without heat pipe for no load and loaded conditions. The DI water is used as working fluid, and air to air heat exchange takes place in the system. The results warranted that the influence of heat pipe for preheating the air makes uniform combustion and increases in performance of the engine. Heating of inlet air helps to decrease engine warm-up time, which improves fuel economy and emissions. Heat pipe heat exchanger is well acted as an air preheater without consuming electrical energy as that of conventional system, which reduces the cost of energy consumption. The observations show that there is an increase in brake thermal efficiency, a decrease in specific fuel consumption, and NO_x, HC, CO emissions are reduced while comparing for without heat pipe condition. The heat pipe with 38 °C as input temperature shows good performance in the engine and observation of 50% load condition has minimum emissions for this proposed experimentation.

References

Deepak, V., Umamaheshwaran, P. S., Guhan, K., Nanthini, R. A., Krithiga, B., Jaithoon, N. M. H., & Gurunathan, S. (2011). Synthesis of gold and silver nanoparticles using purified URAK. *Colloids and Surfaces B: Biointerfaces, 86*(2), 353–358.

Holman, J. P. (2007). *Experimental Methods for Engineers* (7th edn.). McGraw-Hill.

Jouhara, H., Chauhan, A., Nannou, T., Almahmoud, S., Delpech, B., & Wrobel, L. C. (2017). Heat pipe based systems—Advances and applications. *Energy, 128*, 729–754. https://doi.org/10.1016/j.energy.2017.04.028

Li, Y., Zhou, W., He, J., Yan, Y., Li, B., & Zeng, Z. (2016). Thermal performance of ultra-thin flattened heat pipes with composite wick structure. *Applied Thermal Engineering, 102*, 487–499. https://doi.org/10.1016/j.applthermaleng.2016.03.097

Mochizuki, M., Nguyen, T., Mashiko, K., Saito, Y., Nguyen, T., & Wuttijumnong, V. (2011). A review of heat pipe application including new opportunities. *Frontiers in Heat Pipes, 2*(1). https://doi.org/10.5098/fhp.v2.1.3001

Nair, A., Ramkumar, P., Mahadevan, S., Prakash, C., Dixit, S., Murali, G., Vatin, N. I., Epifantsev, K., & Kumar, K. (2022) Machine learning for prediction of heat pipe effectiveness. *Energy, 15*(9), 3276.

Ramkumar, P., Sivasubramanian, M., Rajesh Kanna, P., & Raveendiran, P. (2021a). An experimental inquisition of waste heat recovery in electronic component system using concentric tube heat pipe heat exchanger with different working fluids under gravity assistance. *Microprocessors and Microsystems, 83*, 104033.

Ramkumar, P., Sivasubramanian, M., Rajesh Kanna, P., & Raveendiran, P. (2021b). Heat transfer behaviour on influence of an adiabatic section on concentric tube shell assisted heat pipe heat exchanger. *International Journal of Ambient Energy, 42*(06), 672–681.

Sabarish Kumar, P., Santhosh, D., Arun Vasantha Geethan, K., & Arun Raja, A. K. (2019). Experimentation of a diesel engine with waste heat recovery heat exchanger. *International Journal of Engineering and Advanced Technology, 9*(2), 949–952

Sadanand, V., Rajini, N., Satyanarayana, B., & Varada Rajulu, A. (2016). Preparation and properties of cellulose/silver nanoparticle composites with in situ-generated silver nanoparticles using Ocimum Sanctum leaf extract. *International Journal of Polymer Analysis and Characterization, 21*(5), 408–416.

Sharma, V., Babu, J. M., Naresh, R., Gowthaman, S., & Mariappan, R. (2014). *Design and fabrication of air preheater for diesel engine*. Lecture notes in mechanical engineering (pp. 261–267). https://doi.org/10.1007/978-81-322-1871-5_32

Sonawane, C. R., Tolia, K., Pandey, A., Kulkarni, A., Punchal, H., Sadasivuni, K.K., Kumar, A., & Khalid, M. (2023). Experimental and numerical analysis of heat transfer and fluid flow characteristics inside pulsating heat pipe. *Chemical Engineering Communications, 210*(4), 549–565. https://doi.org/10.1080/00986445.2021.1974413

Srimuang, W., & Amatachaya, P. (2012). A review of the applications of heat pipe heat exchangers for heat recovery. *Renewable and Sustainable Energy Reviews, 16*(6), 4303–4315.

Tang, H., Tang, Y., Wan, Z., Li, J., Yuan, W., Lu, L., Li, Y., & Tang, K. (2018). Review of applications and developments of ultra-thin micro heat pipes for electronic cooling. *Applied Energy, 223*, 383–400. https://doi.org/10.1016/j.apenergy.2018.04.072

Tian, E., He, Y.-L., & Tao, W.-Q. (2017). Research on a new type waste heat recovery gravity heat pipe exchanger. *Applied Energy, 188*, 586–594. https://doi.org/10.1016/j.apenergy.2016.12.029

Vu, Q. K., Minh, D. V., Duy, T. N., Minh, T. P., & The, L. N. (2019). A study of exhaust waste heat recovery in internal combustion engines. *IOP Conference Series: Materials Science and Engineering, 507*, 012028. https://doi.org/10.1088/1757-899X/507/1/012028

Washburn, E. W. (1930). *International critical tables of numerical data, physics, chemistry and technology*. National Research Council, The National Academies Press. https://doi.org/10.17226/20230

Zeghari, K., Louahlia, H., Tiffonnet, A.-L., & Schaetzel, P. (2023). Micro-grooved circular miniature heat pipe for thermal management: experimental and analytical investigations. *Thermal Science and Engineering Progress, 40*, 101714. https://doi.org/10.1016/j.tsep.2023.101714

Metal Fused Filament Fabrication Process—A Review

S. Jasper and M. Ravichandran

Abstract

Rapid prototyping to expensive items like sophisticated spare parts are just a few of the high-end products that can now be made using additive manufacturing (AM). The future of the manufacturing industry has been influenced by AM, which has advantages of lower material utilisation, geometric freedom, and production automation imaginable. With the rapid expansion of additive manufacturing (AM) applications, feedstock materials have undergone a significant transformation. These materials now include metals, composites, polymers, and ceramics. The development in metal feedstock material discoveries has made it possible to investigate the use of new AM methods. The very popular and economical material extrusion AM technique is fused filament fabrication (FFF). The article outlines the key concepts for FFF-based 3D printing of metal items. The process involves stacking filament that contains molten metal. Steel, ceramic, carbide, aluminium, and copper powders are utilised in the Metal Fused Filament Fabrication (MFFF) process. The powder particles need to be a particular size and form. The combination is then used to create filament. The preparation of the filament and the printing of the object are both covered in the paper. Analyses are done on the potential applications for MFFF technology. It has been shown that the technique holds great promise for various domains. The potential for further study in this field is also highlighted.

Keywords

Additive manufacturing • Metal fused filament fabrication • 3D printing

1 Introduction

Over the past few decades, rapid prototyping and the fabrication of spare parts have given way to larger-scale uses of additive manufacturing (AM), such as the construction of bridges and skyscrapers Because of AM's numerous benefits, a reduced amount of material waste, geometric freedom, and automation compatibility are a few examples, Industry 4.0 technologies are transforming. The foundations for the transition from production-focused design to function-focused design were laid by AM (Duda & Raghavan, 2016; Ngo et al., 2018). Depending on the application and the raw material utilised, each of the aforementioned groupings can be further divided into sub-categories (Calderaro et al., 2020). Advanced composites, metal powders, polymers, and ceramics are all examples of feedstock materials utilised in additive manufacturing (AM) processes (Garcia et al., 2020; Kukla et al., 2017; Ligon, 2017; Medellin et al., 2019; Schmid & Wegenerb, 2016). Leading AM techniques including Powder Bed Fusion (PBF) and Direct Energy Deposition (DED) are successful at incorporating metallic feedstock materials (Chohan et al., 2017). What distinguishes one feedstock material from another is whether it is in powder or wire form. DED's energy source is an electron beam or electric arc, whereas PBF's is a laser beam. Bound Powder Extrusion (BPE), a relatively new method for metal 3D printers was used in 2017 when Markforged TM introduced Metal X. A nozzle is used, in both BPE and Fused Filament Fabrication (FFF), [to take material feedstock out of a filament (Kukla et al., 2018; Kumar et al., 2019a; Thavanayagam et al., 2015). One of the material extrusion techniques used in additive manufacturing

S. Jasper (✉)
Department of Mechanical Engineering, Jeppiaar Institute of Technology, Chennai, Tamil Nadu 631604, India
e-mail: joeljasper7@gmail.com

M. Ravichandran
Department of Mechanical Engineering, K.Ramakrishnan College of Engineering, Tiruchirappalli, Tamil Nadu 621112, India

M. Sumesh et al. (eds.), *2nd International Conference on Smart Sustainable Materials and Technologies (ICSSMT 2023)*,
Advances in Science, Technology & Innovation, https://doi.org/10.1007/978-3-031-49826-8_25

is FFF technology. In contrast to other AM techniques, FFF has the advantages of being cheaper and producing parts more quickly, but the two biggest disadvantages are its poor surface finish and weak mechanical qualities. It is critical to distinguish between Metal X's classification under BPE technology and FFF, which is widely utilised as a feedstock for plastics and polymer materials. Utilising BPE technology offers the added advantage of reducing the infrastructure requirements and safety issues often connected with conventional metal AM methods (Lee et al., 2023; Wang et al., 2023; Zhang et al., 2020). The raw material is fed into a nozzle, heated to the required temperature, and then layered down in the FFF process. while being continuously streamed at a steady pressure. After each layer is deposited, the platform travels vertically along the Z axis while the nozzle moves horizontally along the X and Y axes. The layers are joined while still being semi-molten. Metal X uses bound powder filament as its printing medium, with the polymer matrix serving as a binder and holding the metal powder in place (Andersen et al., 2018). Metal X can be made from a variety of metals, including copper, Inconel 625, stainless steel, and other alloys. The advancement of feedstock materials for additive manufacturing leads to an increase in the uses of the technology (Haleem & Javaid, 2020; Kumar et al., 2019b; Leary & Binder jetting., 2020; Lee et al., 2017; Lores et al., 2019; Momeni et al., 2017; Rayatea & Jaina, 2018), particularly in the production of spare parts. It has been demonstrated that fabricating metal parts using FFF technology can produce premium items with impressive dimensional accuracy and complicated geometries. Atzeni and Salmi showed that metal additive manufacturing (AM) has the potential to produce small- to medium-sized batches of goods at a significant cost-savings compared to traditional production techniques (Atzeni & Salmi, 2012). And the traditional manufacturing approach for die casting was compared in the analysis. Particularly with material extrusion methods, the design and fabrication parameters might affect the final properties of items made via additive manufacturing. These qualities include dimensional accuracy, surface roughness, and resolution as well as mechanical traits like strength and toughness (Gibson et al., 2015; Khaliq et al., 2017; Masood & Song, 2004; Turner & Gold, 2015). The infill design or pattern is a crucial design and production parameter (Cano et al., 2020; Suteja & Soesanti, 1569). Using FFF technology, a variety of infill patterns are feasible, including honeycomb, rectangular, grid, and rectilinear. Numerous mechanical parameters including tensile strength, yield strength, precision and repeatability, stiffness, and ductility are greatly impacted by the infill pattern. Mohammed Hikmat and colleagues (2021) used the Taguchi method to study the tensile properties of PLA pieces printed using the FFF technology. The outcome demonstrates that the tensile strength was statistically significantly improved by construct orientation, nozzle diameter, and infill density (Aguirre-Cortés et al., 2023). With ABS as the feedstock material, made by additive manufacturing was examined by Fernandez-Vicente et al. (2016). The results reveal a considerable difference in the test specimens' tensile mechanical properties as the infill pattern is changed, with the honeycomb infill pattern performing best among the three types at various infill densities. Khan et al.'s (2018) study used PLA test specimens in the form of honeycomb, concentric, rectilinear, and Hilbert curve patterns to assess the effects of four different infill types. The findings reveal that the tensile strength was highest for the rectilinear infill pattern, then the honeycomb infill pattern. The study on PLA components by Khaliq et al. (2017) with different infill densities: grid, concentric, and trihexagonal. The findings indicate that, in comparison to the other two infill patterns examined, the concentric infill pattern performed best in terms of all tensile mechanical parameters at all infill densities. Al Rashid and Koç (2023) investigated the impact of items additively built using the FFF process and reinforced with carbon fibre. The mechanical properties, price, and lead time of metal items made using additive manufacturing (Ahn, 2016; Li et al., 2017; Revilla-Leon et al., 2019) may also be impacted by the infill patterns when bonded powder metal filaments are used as the raw material. Aerospace and medical technology are only a couple of the interesting industries where metal additive manufacturing has applications (Bandyopadhyay et al., 2020; Huang et al., 2013; Wong, 2012).

2 Materials and Methods

Metal X can be made from a variety of metals, including copper, Inconel 625, 17-4 PH stainless steel, and other metal alloys (Cano et al., 2019; Gonzalez-Gutierrez et al., 2017; Mirzababaei & Pasebani, 2019; Strano et al., 2019a; Thompson et al., 2019a, 2019b). The three steps to make up the production process are: 3D printing, washing, and sintering. A CAD model is the first step in the metal FFF production process. The STL model is processed by the integrated online platform Eiger, before being effectively transferred to the printer. The STL file for this model is then saved, and it is submitted to the specialised slicing programme Eiger. The raft usage once the item has been uploaded to Eiger. The user has the option of choosing a solid, triangular, or most recently, gyroid infill. Eiger automatically calculates the washing and sintering parameters, which Mark forged has tuned for each metal in their filament products. Classifications of additive manufacturing processes are shown in Fig. 1. Additive manufacturing technologies are shown in Fig. 2.

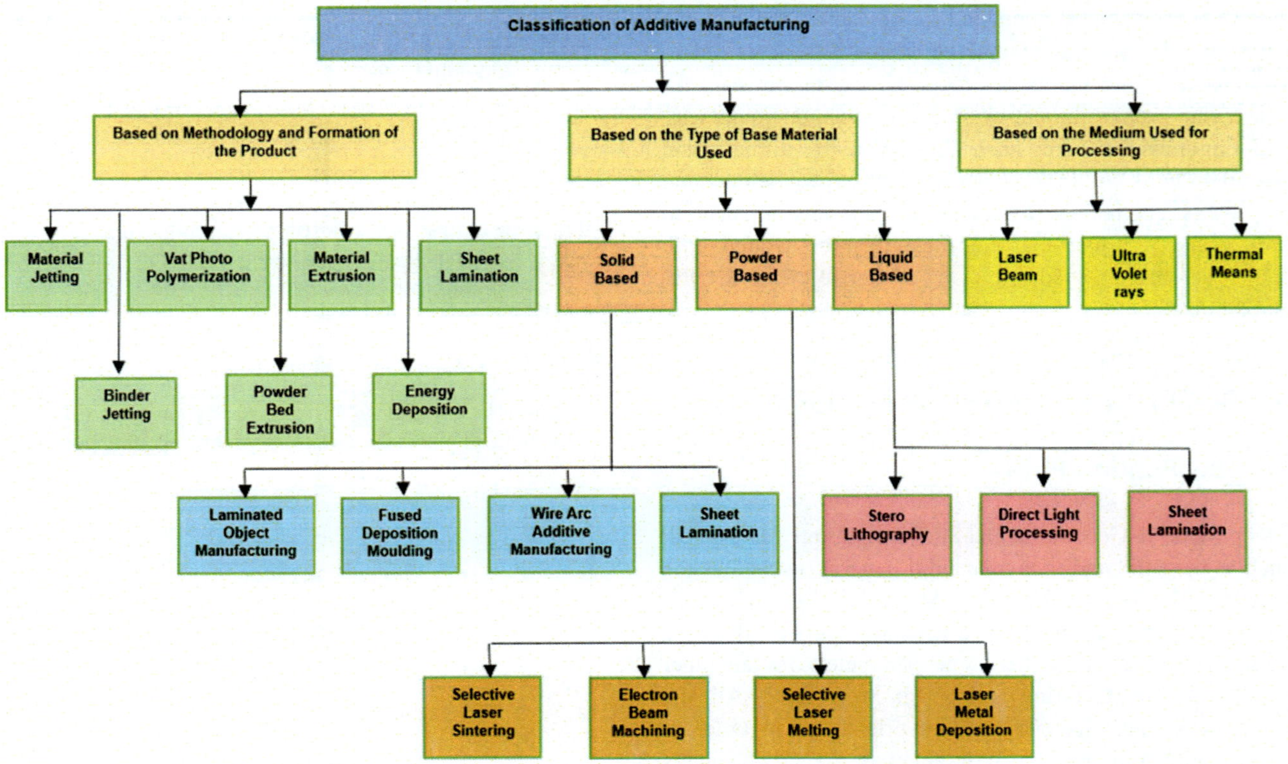

Fig. 1 Classifications of additive manufacturing processes

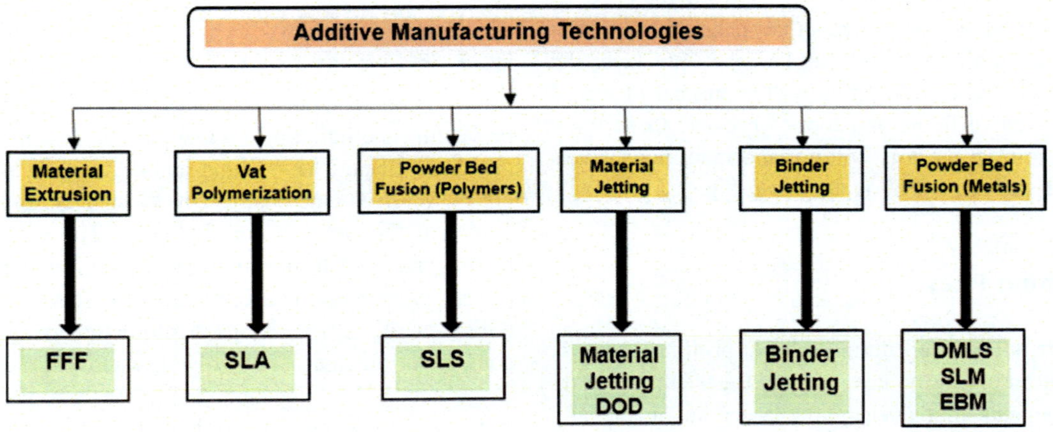

Fig. 2 Additive manufacturing technologies

3 Filament Preparation

It takes more than just making the right composition choice for the future filament to produce a successful finished item. The component must be mixed properly to eliminate the risk of insufficient filament quality and deformation of product during printing. Both tiny batches and continuous mixing are possible (Cañadilla et al., 2022; Gonzalez-Gutierrez et al., 2018; Hertle et al., 2020; Lengauer et al., 2019; Strano et al., 2019b). Different mixers and grinders might be used in the initial arrangement. Extruders that use screws were used in the second setup. The most often used mixers include blades that are shaped like S and Z. FFF twin-screw extruder diagram Twin-screw extruders are advised for better mixing when the powder particles are prone to agglomeration (Gonzalez-Gutierrez et al., 2017; Singh et al., 2020). The powder must first be sieved in order to increase homogeneity. To further avoid agglomeration, some powders require stearic acid treatment. The metal powder and binder system must then be combined before being fed into the extruder. The process of making filament is described in various ways in a number of texts. Giving the finished filament, the proper amount of stiffness, strength, roundness, and toughness is the key goal. High viscosity, stiffness, and poor roundness will make it challenging to feed the filament, but low strength and stiffness will cause the filament to break. The resulting filament must then be wound onto spools before being fed into a 3D printer to create the finished object.

4 Metal Fused Filament Fabrication Process

4.1 3D Printing Phase

Using Metal X-Generation One, the samples are 3D printed, such as the solid and triangular infill patterns, as well as the toolpath (Riecker et al., 2017; Xia et al., 2020). For all specimens, the construction orientation was fixed to be horizontal on the print bed. The parameters for all other aspects of manufacturing and design were fixed. All test components must have a constant orientation in order to prevent anisotropy's effects (Alkindi et al., 2021). The layers encircling the sides of the main body are represented by the "number of wall layers." The user must also determine "the number of floor and roof layers." This word refers to the number of completely solid layers that make up the bottom and top of the infill in triangle infills. The only layer that may be changed with solid infills is "the number of walls," not "the number of floors and roofs," because these layers are already solid at the top, bottom, and main body levels. Sequence of steps in MFFF is shown in Fig. 3.

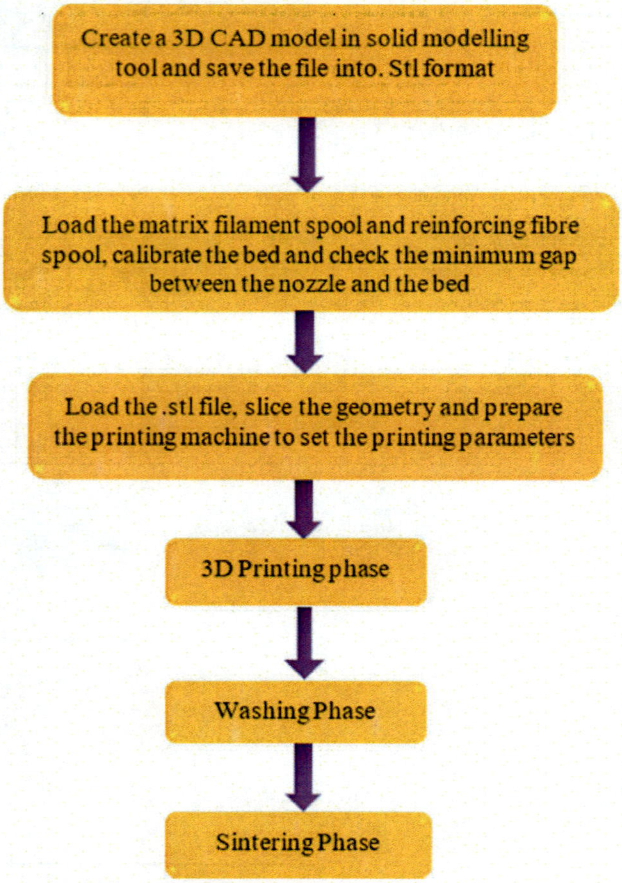

Fig. 3 Sequence of steps in MFFF

4.2 Washing Phase

A green part is produced during the 3D printing stage. To adjust for shrinking in the next phases, Eiger created brittle green components that are slightly larger and made of a blend of the metal powder and the waxy binding agent. The raft will be separated from the part during the sintering process. For this separation, a ceramic release layer was positioned between the raft and the part during the printing process; a little tap with a hammer in a vice may be necessary at this stage. A precise scale is used to measure the part's mass before it is transferred to the washing phase and again after it has been cleaned and dried. The mass loss must be calculated in this step, and a predetermined threshold must be met before the part can be sintered. It needs to be rewashed and then redried, or both if a particular component failed to satisfy the mass loss requirement. This stage aids in preventing the buildup of waxy binding material and keeps the sinter clean. The washing and drying process is carried out by Wash 1. One chamber in Wash 1 is for washing the part, and the other is for drying it. The majority of the binding material in the mixture is removed from the component during this phase, which also serves to behind it (Gong et al., 2019). Trans-dichloroethylene is the active component of the HFO solvent chemical family-related Tergo Metal Cleaning fluid. The pieces are placed into the sinter for step (3) once they have been cleaned, dried, and the mass loss criterion has been met. MFFF process setup is shown in Fig. 4.

4.3 Sintering Phase

Phase 3 of the procedure entails sintering the component, which in this experiment was the Sinter 1. The item goes through a noticeable heat ramp during the sintering process, going from ambient temperature to as much as 85% of the metal's melting point. During the sintering process, a predetermined argon is used. Additionally, during the sintering

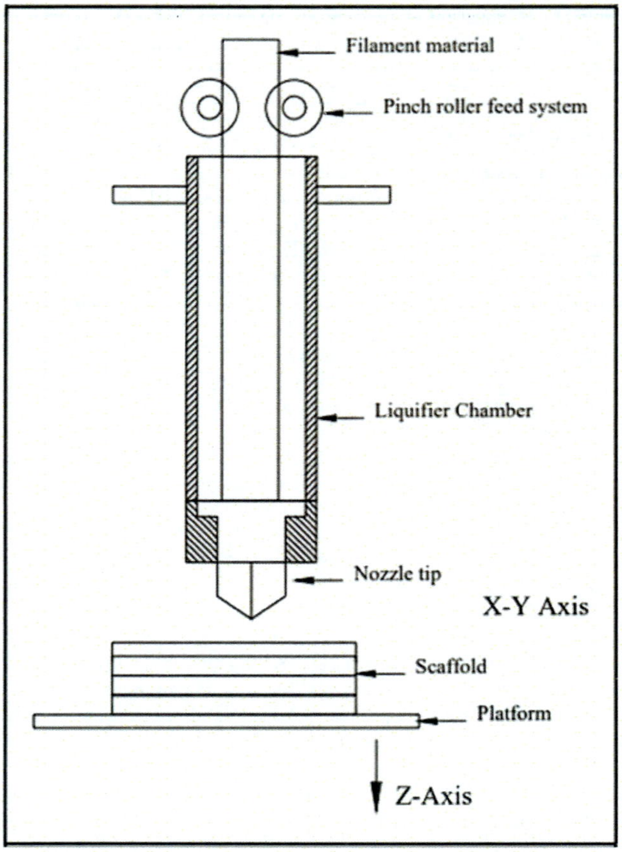

Fig. 4 MFFF process setup

process, heat is used to fuse the layers together, remove any remaining binder, and seal any pores. As a result, the part later experiences substantial shrinkage (Sharif et al., 2022). The ramp-up and ramp-down portions of the procedure for each batch total about 27 h. The structural characteristics of 3D printed products used in industrial applications, such as conductivity and mechanical tensile strength, are significantly influenced by the sintering process parameters. According to earlier studies, copper can sinter at 1030 °C and 17-4 PH stainless steel at temperatures between 1300 and 1380 °C (Gonzalez-Gutierrez et al., 2019; Shuppe & Griffiths, 2022). A metal item produced in phase 3 is in its final state and is prepared for use in the specified application or for additional processing (Sames et al., 2016), such as post-machining or heat treatment.

5 Conclusion

Since FFF makes printing metal items more economical than with other technologies, the ability to manufacture metal objects on traditional 3D printers has created new opportunities for the professional and amateur additive industries. The complete manufacturing process for a certain product consists of the steps of filament production, product printing, binder system removal, and product burning. There are many subtleties in each stage, yet some elements have an impact from beginning to end. They must be given priority. The following points can be quickly emphasised: the size, shape, and quantity of the metal powder; the make-up of the binder system; the procedure used to join the components; and the final shape of the filament. Techniques for 3D printing, technologies for removing binder systems, and final heat treatment of the product are used at the manufacturing step. The process flow diagram for producing a metal object using FFF technology. Future research should focus on developing 3D printing technology in all aspects and developing generic models of the production processes. The shrinkage following heat treatment is depicted and a set of broad guidelines for material selection, powder properties, binder system composition, and product manufacturing method. FFF 3D printers have enormous potential for usage in a variety of industries, including instrument manufacturing, medicine, mechanical engineering, and education.

References

Aguirre-Cortés, J. M., Moral-Rodríguez, A. I., Bailón-García, E., Davó-Quiñonero, A., Pérez-Cadenas, A. F., Carrasco-Marín, F. (2020). 3D printing in photocatalysis: Methods and capabilities for the improved performance. *Applied Materials Today, 32*, 101831.

Ahn, D. G. (2016). Direct metal additive manufacturing processes and their sustainable applications for green technology: A review. *International Journal of Precision Engineering and Manufacturing-Green Technology, 3*, 381–395.

Al Rashid, A., & Koç, M. (2023). Experimental validation of numerical model for thermomechanical performance of material extrusion additive manufacturing process: Effect of infill design & density. *Results in Engineering, 17*, 100860.

Alkindi, T., Alyammahi, M., Susantyoko, R. A., & Atatreh, S. (2021). The effect of varying specimens' printing angles to the bed surface on the tensile strength of 3D-printed 17–4PH stainless-steels via metal FFF additive manufacturing. *MRS Communications, 11*, 310–316.

Andersen, O., Riecker, S., Studnitzky, T., Hein, S., Lohse, U., & Kie-, B. (2018). Manufacturing and properties of metal parts made by fused filament fabrication. *Powder Metallurgy, 18*, 2–7.

Atzeni, E., & Salmi, A. (2012). Economics of additive manufacturing for end-useable metal parts. *International Journal of Advanced Manufacturing Technology, 62*, 1147–1155.

Bandyopadhyay, A., Zhang, Y., & Bose, S. (2020). Recent developments in metal additive manufacturing. *Current Opinion in Chemical Engineering, 28*, 96–104.

Calderaro, D. R., Lacerda, D. P., & Veit, D. R. (2020). Selection of additive manufacturing technologies in productive systems: A decision support model. *Gestão & Produção, 27*.

Cañadilla, A., Romero, A., Rodríguez, G. P., Caminero, M. Á., & Dura, Ó. J. (2022). Mechanical, electrical, and thermal characterization of pure copper parts manufactured via material extrusion additive manufacturing. *Materials (Basel), 15*.

Cano, S., Gonzalez-Gutierrez, J., Sapkota, J., Spoerk, M., Arbeiter, F., Schuschnigg, S., Holzer, C., & Kukla, C. (2019). Additive manufacturing of zirconia parts by fused filament fabrication and solvent debinding: Selection of binder formulation. *Additive Manufacturing, 26*, 117–128.

Cano, S., Lube, T., Huber, P., Gallego, A., Naranjo, J. A., Berges, C., et al. (2020). Influence of the infill orientation on the properties of zirconia parts produced by fused filament fabrication. *Materials (Basel), 13*.

Chohan, J. S., Singh, R., Boparai, K. S., Penna, R., & Fraternali, F. (2017). Dimensional accuracy analysis of coupled fused deposition modeling and vapour smoothing operations for biomedical applications. *Composites Part B Engineering, 117*, 138–149.

Duda, T., & Raghavan, L. V. (2016). 3D metal printing technology. *IFAC-PapersOnLine, 49*, 103–110.

Fernandez-Vicente, M., Calle, W., Ferrandiz, S., & Conejero, A. (2016). Effect of infill parameters on tensile mechanical behavior in desktop 3D printing, 3D print. *Additive Manufacturing, 3*, 183–192.

Garcia, E. A., Ayranci, C., & Qureshi, A. J. (2020). Material property-manufacturing process optimization for Vat-photo polymerization 3D Printers. *Journal of Manufacturing and Materials Processing, 4*, 12.

Gibson, I., Rosen, D., & Stucker, B. (2015). *Additive manufacturing technologies: 3D printing, rapid prototyping and direct digital manufacturing* (498 pp.). Springer (2015)

Gong, H., Snelling, D., Kardel, K., & Carrano, A. (2019). Comparison of stainless steel 316L parts made by FDM- and SLM-based additive manufacturing processes. *JOM (Journal of Occupational Medicine), 71*, 880–885.

Gonzalez-Gutierrez, J., Godec, D., Kukla, C., Schlauf, T., Burkhardt, C., & Holzer, C. (2017). *Shaping, debinding and sintering of steel components via fused filament fabrication* (pp. 99–104). Croatian Association of Production Engineering.

Gonzalez-Gutierrez, J., Arbeiter, F., Schlauf, T., Kukla, C., & Holzer, C. (2019). Tensile properties of sintered 17–4PH stainless steel fabricated by material extrusion additive manufacturing. *Materials Letters, 248*, 165–168.

Gonzalez-Gutierrez, J., Cano, S., Schuschnigg, S., Kukla, C., Sapkota, J., & Holzer, C. (2018). Additive manufacturing of metallic and ceramic components by the material extrusion of highly-filled polymers: A review and future perspectives. *Materials, 11*(5), 840. https://doi.org/10.3390/ma11050840

Haleem, A., & Javaid, M. (2020). 3D printed medical parts with different materials using additive manufacturing. *Clinical Epidemiology and Global Health, 8*(1), 215–223.

Hertle, S., Kleffel, T., Wörz, A., & Drummer, D. (2020). Production of polymer-metal hybrids using extrusion-based additive manufacturing and electrochemically treated aluminum. *Additive Manufacturing, 33*, 101135.

Hikmat, M., Rostam, S., & Ahmed, Y. M. (2021). Investigation of tensile property-based Taguchi method of PLA parts fabricated by FDM 3D printing technology. *Results in Engineering, 11*, 100264.

Huang, S. H., Liu, P., Mokasdar, A., & Hou, L. (2013). Additive manufacturing and its societal impact: A literature review. *International Journal of Advanced Manufacturing Technology, 67*, 1191–1203.

Khaliq, H., Nobrega, J. M., & Carneiro, O. S. (2017). On the use of high viscosity polymers in the fused filament fabrication process. *Rapid Prototyping Journal, 23*, 727–735.

Khan, S. F., Zakaria, H., Chong, Y. L., Saad, M. A. M., & Basaruddin, K. (2018). Effect of infill on tensile and flexural strength of 3D printed PLA parts. *IOP Conference Series: Materials Science and Engineering, 429*.

Kukla, C., Gonzalez-Gutierrez, J., Duretek, I., Schuschnigg, S., & Holzer, C. (2017). *Effect of particle size on the properties of highly-filled polymers for fused filament fabrication*. AIP Publish.

Kukla, C., Gonzalez-gutierrez, J., Felfer, P., & Holzer, C. (2018). Material extrusion with filaments for the production of metal parts and feedstock therefore. *Metal Additive Manufacturing, 4*, 1–9.

Kumar, S., Singh, R., Singh, T. P., et al. (2019a). Investigations of polylactic acid reinforced composite feedstock filaments for multimaterial three-dimensional printing applications. *Proceedings of the Institution of Mechanical Engineers, Part c: Journal of Mechanical Engineering Science, 233*, 5953–5965.

Kumar, S., Singh, R., Batish, A., et al. (2019). Additive manufacturing of smart materials exhibiting 4-D properties: A state of art review. *Journal of Thermoplastic Composite Materials*.

Leary, M. (2020). Binder jetting. In *Design for additive manufacturing: Additive manufacturing materials and technologies* (pp. 335–339). Elsevier.

Lee, A. Y., An, J., & Chua, C. K. (2017). Two-way 4D printing: A review on the reversibility of 3D-printed shape memory materials. *Engineering, 3*, 663–674.

Lee, J. J., Mohammed, A. A., Pullen, A., Myant, C. W., & Proud, W. G. (2023). Mechanical characterisation of 3D printed lightweight lattice structures with varying internal design alterations. *Materials Today Communications, 36*, 106456.

Lengauer, W., Duretek, I., Fürst, M., Schwarz, V., Gonzalez-Gutierrez, J., Schuschnigg, S., Kukla, C., Kitzmantel, M., Neubauer, E., Lieberwirth, C., & Morrison, V. (2019). Fabrication and properties of extrusion-based 3D-printed hardmetal and cermet components. *International Journal of Refractory Metals & Hard Materials, 82*, 141–149.

Li, Y., Jia, G., Cheng, Y., & Hu, Y. (2017). Additive manufacturing technology in spare parts supply chain: A comparative study. *International Journal of Production Research, 55*, 1498–1515.

Ligon, S. C., Liska, R., Stampfl, J., Gurr, M., & Mülhaupt, R. (2017). Polymers for 3D printing and customized additive manufacturing. *Chemical Reviews, 117*, 10212–10290.

Lores, A., Azurmendi, N., Agote, I., & Zuza, E. (2019). A review on recent developments in binder jetting metal additive manufacturing: Materials and process characteristics. *Powder Metallurgy, 62*, 267–296.

Masood, S. H., & Song, W. Q. (2004). Development of new metal/polymer materials for rapid tooling using fused deposition modelling. *Materials and Design, 25*(7), 587–594.

Medellin, A., Du, W., Miao, G., Zou, J., & Pei, Z. (2019). Vat photopolymerization 3D printing of nanocomposites: A literature review. *Journal of Micro and Nano-Manufacturing, 7*, 031006.

Mirzababaei, S., & Pasebani, S. (2019). A review on binder jet additive manufacturing of 316L stainless steel. *Journal of Manufacturing and Materials Processing, 3*, 82.

Momeni, F., Seyed, M., Hassani, M. N., Liu, X., & Ni, J. (2017). A review of 4D printing. *Materials and Design, 122*, 42–79.

Ngo, T. D., Kashani, A., Imbalzano, G., Nguyen, K. T. Q., & Hui, D. (2018). Additive manufacturing (3D printing): A review of materials, methods, applications and challenges. *Composites Part B Engineering, 143*, 172–196.

Rayatea, A., & Jaina, P. K. (2018). A review on 4D printing material composites and their applications. *Materials Today: Proceedings, 5*, 20474–20484.

Revilla-Leon, M., Meyer, M. J., & Ozcan, M. (2019). Metal additive manufacturing technologies: Literature review of current status and prosthodontic applications. *International Journal of Computerized Dentistry, 22*, 55–67.

Riecker, S., Hein, S., & Studnitzky, T. (2017). 3D printing of metal parts by means of fused filament fabrication-A non beam-based approach. Europe 2017–AM ALTEN Technology.

Sames, W. J., List, F. A., Pannala, S., Dehoff, R. R., & Babu, S. S. (2016). The metallurgy and processing science of metal additive manufacturing. *International Materials Reviews, 61*, 315–360.

Schmid, M., & Wegenerb, K. (2016). Additive manufacturing: Polymers applicable for laser sintering (LS). *Procedia Engineering, 149*, 457–464.

Sharif, A., Farid, N., & O'Connor, G. M. (2022). Ultrashort laser sintering of metal nanoparticles: A review. *Results in Engineering, 16*, 10073.

Shuppe, A., & Griffiths, J. (2022). *Sintering of 3-D printed metals*. Williams Honors College, Honors Research Projects.

Singh, P., Balla, V. K., Tofangchi, A., Atre, S. V., & Kate, K. H. (2020). Printability studies of Ti6Al-4V by metal fused filament fabrication (MF3). *International Journal of Refractory Metals & Hard Materials, 91*, 105249.

Strano, M., Rane, K., Briatico Vangosa, F., & Di Landro, L. (2019). Extrusion of metal powder-polymer mixtures: Melt rheology and process stability. *Journal of Materials Processing Technology, 273*, 11625.

Strano, M., Rane, K., Briatico Vangosa, F., & Di Landro, L. (2019). Extrusion of metal powder-polymer mixtures: Melt rheology and process stability. *Journal of Materials Processing Technology, 273*, 116250.

Suteja, T. J., & Soesanti, A. (2020). Mechanical properties of 3D printed polylactic acid product for various infill design parameters: A review. *Journal of Physics: Conference Series, 1569*.

Thavanayagam, G., Pickering, K. L., Swan, J. E., & Cao, P. (2015). Analysis of rheological behaviour of titanium feedstocks formulated with a water-soluble binder system for powder injection moulding. *Powder Technology, 269*, 227–232.

Thompson, Y., Gonzalez-Gutierrez, J., Kukla, C., & Felfer, P. (2019a). Fused filament fabrication, debinding and sintering as a low cost additive manufacturing method of 316L stainless steel. *Additive Manufacturing, 30*, 100861.

Thompson, Y., Gonzalez-gutierrez, J., Kukla, C., & Felfer, P. (2019b). Fused filament fabrication, debinding and sintering as a low cost additive manufacturing method of 316L stainless steel. *Additive Manufacturing, 30*, 220–228.

Turner, B.N., & Gold, S.A. (2015). A review of melt extrusion additive manufacturing processes: II. Materials, dimensional accuracy, and surface roughness. *Rapid Prototyping Journal, 21*, 250–261.

Wang, K., Engelbrecht, K., & Bahl, C. R. H. (2023). Additive manufactured thermoplastic elastomers for low-stress driven elastocaloric cooling. *Applied Materials Today, 30*, 101711.

Wong, K. V., & Hernandez, A. (2012). A review of additive manufacturing. *ISRN Mechanical Engineering*, 1–10.

Xia, L., Lin, S., & Ma, G. (2020). Stress-based tool-path planning methodology for fused filament fabrication. *Additive Manufacturing, 32*, 101020.

Zhang, Y., Bai, S., Riede, M., Garratt, E., & Roch, A. (2020). A comprehensive study on fused filament fabrication of Ti-6Al-4V structures. *Additive Manufacturing, 34*, 101256.

Mineralogical Identification of Clays Using K-Mean Clustering

Siva Kumar Prasad Chebiyyam and Mallikarjuna Rao Kattamuri

Abstract

The degree of expansivity of clay depends on the mineral (s) present; therefore, identifying mineral(s) in clay is essential to assess its swelling and shrinkage characteristics. Fine-grained soils could be classified as Kaolinitic or Montmorillonitic where Montmorillonitic soils are relatively more expansive in nature. The expansive soils are very problematic as they affect the stability of structures found on them. X-ray diffraction (XRD), differential thermal analysis scanning electron microscopy (SEM) etc. techniques could be able to predict the mineral (s) in clay with high accuracy; however, employing such techniques in soil investigation is not possible due to their sophistication and handling of bulk heterogeneous soil mass. Many researchers and codes suggested the expansivity of soils based on index properties such as liquid limit, plastic limit, shrinkage limit, etc. This study aims to identify the Kaolinitic, Montmorillonitic and Mixture of both soils by applying an unsupervised learning clustering technique namely K-Mean clustering.

Keywords

Kaolinitic • Montmorillonitic • Mixture • K-mean clustering

1 Introduction

Kaolinite, Illite and Montmorillonite are the popular minerals in most clays. Identification of mineral composition during the soil investigation is essential to assess expansive behaviour of clay. Moisture-sensitive soils are regarded as expansive and problematic as they cause instability for the foundation (Asuri & Keshavamurthy, 2016). Based on the physicochemical studies, the fine-grained soils could be Kaolinitic that are nonexpanding lattice-type soils or montmorillonitic, which are expanding lattice type (Prakash & Sridharan, 2004). The mineralogical composition of clays will be established through XRD, SEM and differential thermal analysis. Such methods from a soil investigation viewpoint are rather impractical due to a wide variety of reasons, such as the volume and heterogeneity of soil to be handled and the need for more facility and expertise in regular soil mechanics labs. By studying index properties, namely free swelling index/ratio, liquid limit and plastic limit/plasticity index, many researchers and codes of practice suggested the expansivity of fine-grained soils. For different liquid limit ranges, swelling potentials were recommended by Chen (1965), Snethan et al. (1977). Based on the plasticity index, Holtz and Gibs and Chen categorized swelling potentials under four heads: low, medium, high and very high (Holtz & Gibbs, 1956). Holtz and Gibbs published shrinkage limit and the corresponding volume changes for expansive soils (Holtz & Gibbs, 1956). Chen and Holtz and Gibbs have suggested the degree of expansivity of clays into four based on their percentage clay fraction (Chen, 1965; Holtz & Gibbs, 1956). With the help of the free swell ratio, Sridharan and Prakash have tried to predict the dominant clay mineral and its swelling potential (Sridharan & Prakash, 2000).

Previously clustering was widely used in statistical data analysis; however, over time, its applications are provided to be more success in many fields, such as data mining, machine learning, image analysis, bio information and, importantly, pattern recognition (Soni, 2012). Application of clustering algorithm is gaining popularity in civil engineering also.

Traffic flow analysis and forecasting, identification of traffic blackspots and identification of traffic patterns are common applications of clustering in traffic engineering and transport management.

S. K. P. Chebiyyam (✉) · M. R. Kattamuri
Department of Civil Engineering, S. V. University College of Engineering, Tirupati, India
e-mail: sivakumarprasad.c@gmail.com

M. Sumesh et al. (eds.), *2nd International Conference on Smart Sustainable Materials and Technologies (ICSSMT 2023)*,
Advances in Science, Technology & Innovation, https://doi.org/10.1007/978-3-031-49826-8_26

Preparation of contractor prequalification needs a study of several contracts, and every arrangement has many dimensions. Clustering techniques are instrumental in preparing contractor prequalification and evaluating and interpreting contractors based on the professionality, adequacy of resources and overall ability of the contractor to award a contract (Holt, 1996).

Literature is available related to application of clustering techniques for soil classification and analysis of penetration test data.

Good amount of literature exists that gives expansivity of soils based on liquid limit, shrinkage limit, plasticity index, particle size, activity and free swelling index/free swelling ratio. Mishra et al., Shah et al., Chavali et al., Sato et al., Sruthi et al., Zhang and Frederick, Prakash and Sridharan and Prakash et al. in their research publications have presented 98 soil, which belongs to either Kaolinitic, Montmorillonitic or Mixture with their index properties chiefly liquid limit and plastic limit (Chavali et al., 2017; Lakshmi Sruthi & Hari Prasad Reddy, 2019; Mishra et al., 2011; Prakash & Sridharan, 2004; Qiong et al., 2011; Saha et al., 2019; Sato et al., 2016; Shah et al., 2017; Zhang & Frederick, 2017). All the mentioned 98 soils are considered as data for analysis in the present work.

The aim of current work is to group the data with the help of an unsupervised clustering technique namely K-Mean clustering based on *1. Liquid Limit (LL) and Plastic Limit (PL)* and *2. Liquid Limit (LL) and Plasticity Index (PI)* to analyse if the groups thus formed are based on the clay minerals and to investigate for a possible pattern based on the type of clay in the groups thus formed. The K-Mean clustering is implemented in Python programming language supported with the help of SCIKIT-LEARN, a library.

2 Clustering and K-Mean Clustering

The categorization of data of a dataset into groups based on similarity is termed clustering. Thus, a formed group is a cluster; records of dataset in the cluster are similar to each other and dissimilar to records of dataset in different clusters. According to data modelling, clustering is an offshoot of mathematical sciences. When it comes to machine learning, clustering is searching for hidden patterns that come under unsupervised learning (Rai & Shubha, 2010).

The K-mean clustering algorithm divides the dataset into a given K number of clusters interactively. From the K randomly chosen points, the Euclidean distances are calculated from a point in the dataset to decide which cluster it belongs based on the shortest distance. The centroid is located for each cluster, and the previously chosen random points are shifted to this initial centroid. The process is repeated until a cluster's centroidal distance becomes constant during iterations (Hot Dervic & Vesna, 2016).

$$d = \sum_{k=1}^{k} \sum_{i=1}^{n} ||(x_i - c_k)||^2$$

where k represents K cluster centres, c_k represents the kth centre, and x_i represents the ith record in the data set (Yuan & Yang, 2019).

The solution to the centroid, $c_k = \frac{1}{n}\sum_{i=1}^{n} x_i$.

In order to determine the optimum number of clusters, *Elbow Method* is adopted wherein a simple graph is drawn between number of clusters and WCSS (within-cluster sum of squares) (Kuraria et al., 2018). The slope of Elbow Method graph changes with number of clusters, where the slope will be steep at the initial stage and becomes more and more flat as the number of clusters increases. The optimum number of clusters is considered as that point where there is no significant change between the slopes of that point and the succeeding point (Goktepe et al., 2015).

3 Data for K-Mean Clustering

The data used for the clustering analysis are extracted from the peer-reviewed journals published by different researchers as mentioned in *Introduction* of this paper. Figure 1 shows the number of soils having particular dominant mineral(s). Out of 98 clays considered for this analysis 20 of Kaolinitic, 27 are have both Kaolinitic and Montmorillonitic designated as Mixture and remaining 51 of Montmorillonitic.

All the 98 clays are divided into three groups, namely, 1. Kaolinitic (Kaolinite, Kaolinite with non-expansive minerals namely Chlorite, Muscovite and Mica), 2. Mixture (combination of Kaolinite, Montmorillonite or Kaolinite, Montmorillonite with Muscovite), and 3. Montmorillonitic (all types of Montmorillonite clays).

4 Strategy for K-Mean Clustering Implementation

As discussed in *Introduction*, previous researchers have made attempts to predict the soil expansivity based on the index properties. However, none of these studies directly spell out the presence of possible dominant clay mineral(s) in the soil based on the index properties. Clustering and its examination of mineralogically known clays reveals the possible dependence of clay minerals on the index properties. The consistency limits namely Liquid Limit, Plastic Limit and Plasticity Index are considered as variables for clustering. The clustering is done by considering two variables at a time viz. (1) LL and PL and (2) LL and PI.

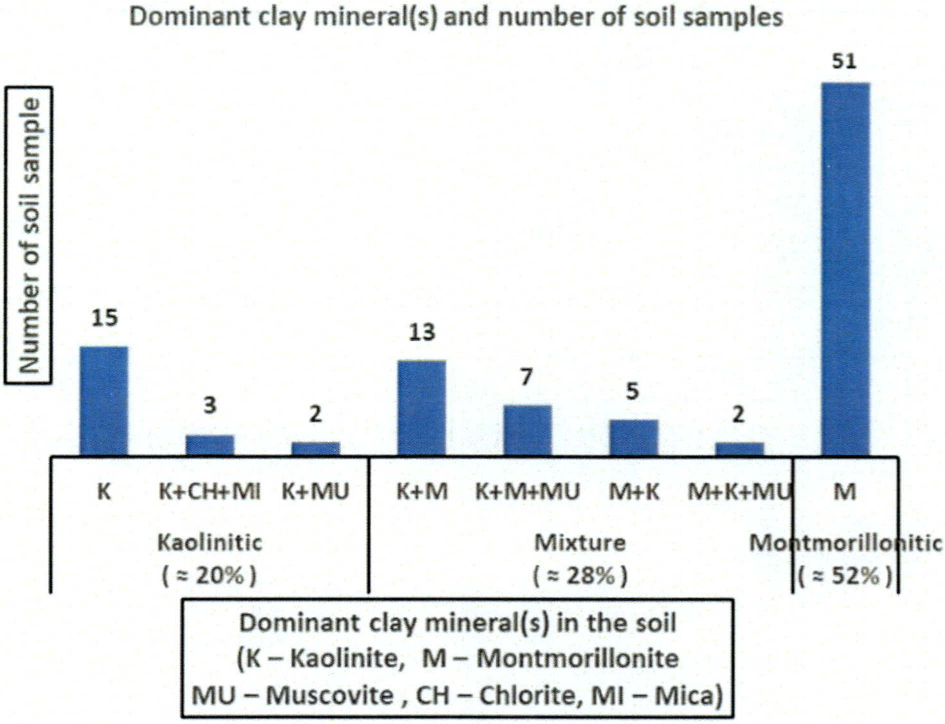

Fig. 1 Soils and their dominant clay minerals

5 Results and Analysis of K-Mean Clustering

K-Mean clustering algorithm is applied on clay sample by considering two variables at time as discussed below.

5.1 LL and PL

For the two variables, namely, liquid limit and plastic limit used in the clustering analysis, a graph is drawn between the number of clusters and WCSS. The graph, thus, obtained, namely, *Elbow Method* graph that is used to identify optimum number of clusters is shown in Fig. 2.

Clustering analysis is performed by considering three, four, six and eight8 clusters. Better cluster groups are identified when number of clusters is 8. Figure 3 shows eight clusters in which six clusters are having 100% Montmorillonitic and remaining two clusters have both Kaolinitic and Montmorillonitic soils.

The results obtained for eight group cluster in tabulated are shown in Table 1. Beyond eight cluster groups, there is no significant change in the proportions of soils under mixed clusters namely I and II, and hence further clustering is of no big use.

In order to get better demarking between Kaolinitic, Mixture and Montmorillonitic soils, further cluster analysis is performed on cluster groups I and II obtained from eight group clusters on the whole data. Under the cluster groups I and II, there are 65 soils and again cluster analysis is performed on these 65 soils. Elbow Method graph is drawn as shown in Fig. 4.

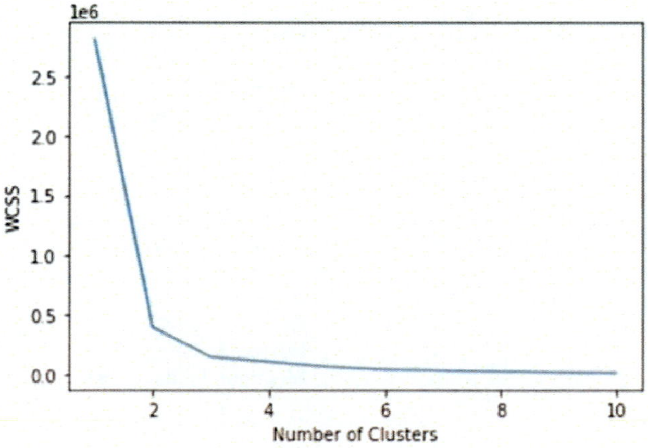

Fig. 2 Elbow method graph for LL and PL

Though Elbow Method graph is showing slope variation till 10 number of clusters, number of clusters is limited to 3 as there is no significant variation in proportion of soils beyond three cluster groups. Figure 5 shows the three sub groups that are formed on the cluster I and II data. The results of the three sub groups are shown in Table 2.

Based on the clustering analysis performed, a conceptual drawing is prepared to better represent the results obtained. While developing the conceptual diagram shown in Fig. 6,

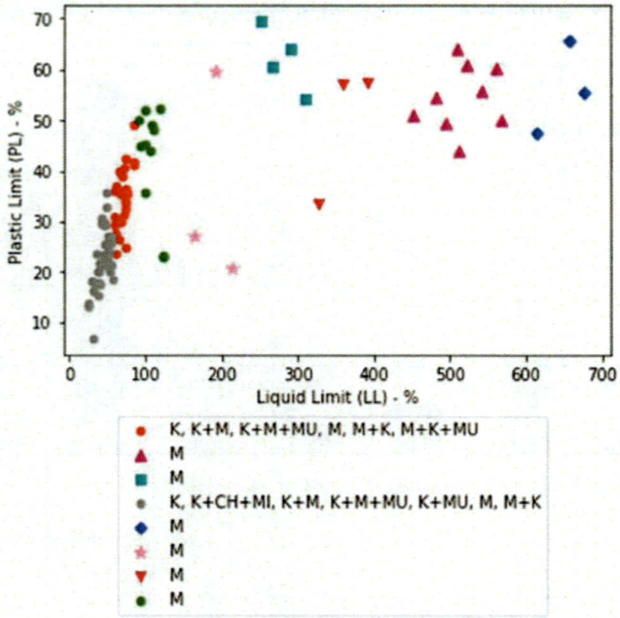

Fig. 3 LL and PL for eight clusters using K-mean

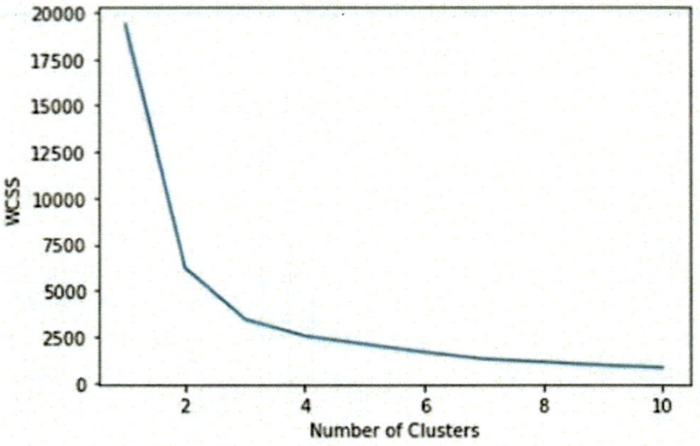

Fig. 4 Elbow method graph for LL and PL sub-group

the authors have taken liberty to round the liquid limit boundaries and to set plastic limit boundaries keeping aside the certain amount of overlapping. The conceptual drawing is divided into four zones, namely, Z0, Z1, Z2, Z3 and Z4. Z0 is free zone where no soil sample falls inside. There is an

Table 1 8 Group cluster data for variables LL and PL using K-mean

Eight clusters	LL (%)		PL (%)		Composition	Total
	Min	Max	Min	Max		
I Kaolinitic = 46% Mixture = 33% Montmorillonitic = 21%	24.8	56.2	6.7	35.6	K	13
					K + CH + MI	3
					K + M	8
					K + M + MU	2
					K + MU	2
					M	8
					M + K	3
II Kaolinitic = 8% Mixture = 54% Montmorillonitic = 38%	58	84	24	49	K	2
					K + M	5
					K + M + MU	5
					M	10
					M + K	2
					M + K + MU	2
III Montmorillonitic = 100%	92	124	23	52.3	M	11
IV Montmorillonitic = 100%	165	215	21	59.7	M	3
V Montmorillonitic = 100%	251	311	54	69.6	M	4
VI Montmorillonitic = 100%	328	391	34	57.2	M	3
VII Montmorillonitic = 100%	452	567	44	63.9	M	9
VIII Montmorillonitic = 100%	616	678	48	65.5	M	3

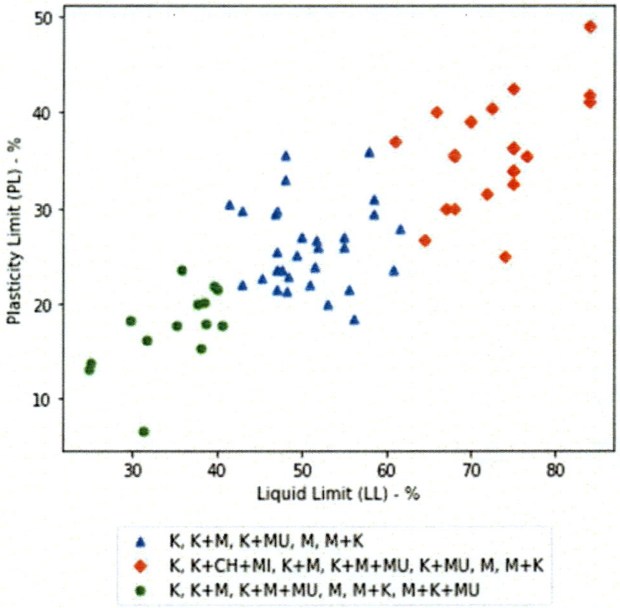

Fig. 5 LL and PL sub-group of three clusters

overlapping of plastic limit values that come under Z2 and Z3. This overlap is depicted in the conceptual drawing.

5.2 LL and PI

Elbow Method graph was drawn as shown in Fig. 7 when clustering analysis was done on LL and PI to find optimum number of clusters.

From the Elbow Method graph shown in Fig. 7, it is understood that from the number of clusters 6 to the number of clusters 10, the graph is gradually becoming asymptotic.

Better segregation of Kaolinitic, Mixture and Montmorillonitic is occurring when number of clusters is 10. Table 3 shows the 10-cluster data, wherein clusters I and II are having both Kaolinitic, Mixture and Montmorillonitic soil whereas all other clusters have only Montmorillonitic soil. Thus formed 10 clusters are visualized in the form of a graph drawn between the variables liquid limit and plasticity index as shown in Fig. 8.

Table 2 Data of LL and PL 3 sub-group clusters

Sub-group 3 clusters	LL (%)		PL (%)		Composition	Total
	Min	Max	Min	Max		
I Kaolinitic = 57% Mixture = 36% Montmorillonitic = 7%	24.8	40.7	6.7	23.6	K	8
					K + M	3
					K + MU	1
					M	1
					M + K	1
II Kaolinitic = 20% Mixture = 50% Montmorillonitic = 30%	41.4	61.6	18.5	36	K	6
					K + CH + MI	3
					K + M	6
					K + M + MU	2
					K + MU	1
					M	9
					M + K	3
III Kaolinitic = 5% Mixture = 57% Montmorillonitic = 38%	61	84	25	49	K	1
					K + M	4
					K + M + MU	5
					M	8
					M + K	1
					M + K + MU	2

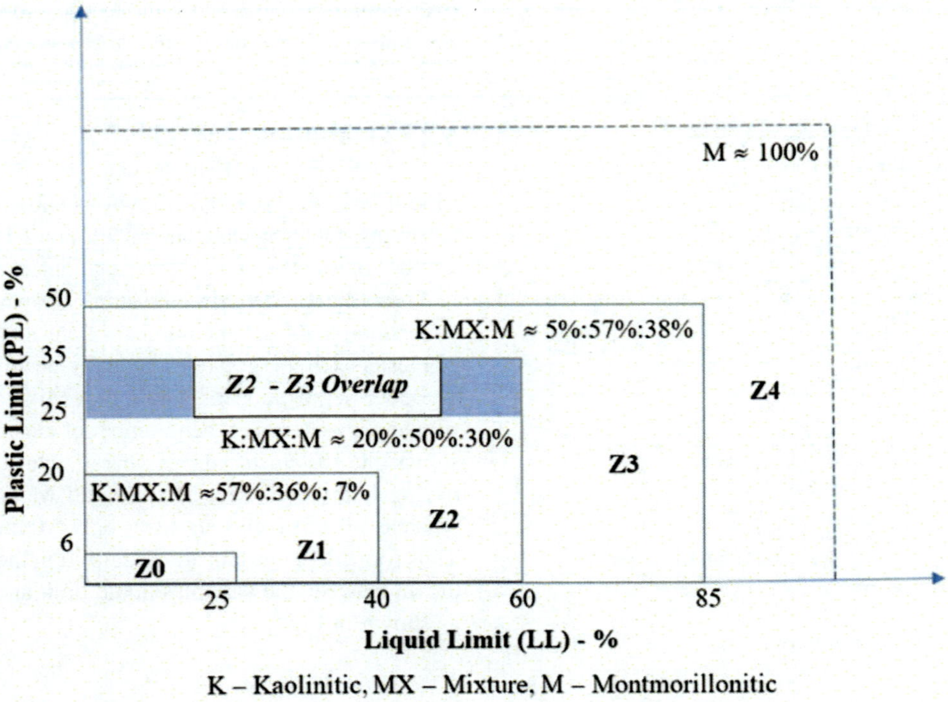

Fig. 6 Zones for clay types based on cluster analysis of LL and PL

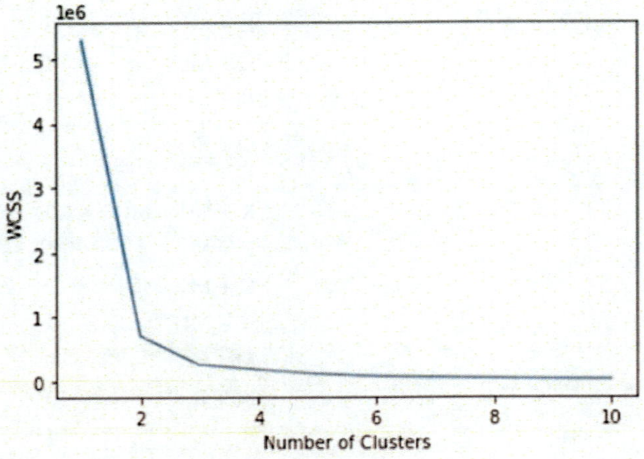

Fig. 7 Elbow method graph for LL and PI

To get better segregation, both clusters I and II are combined to get 66 soils having both Kaolinitic and Montmorillonitic and cluster analysis is performed on these 66 soils. In order to find the optimum number of clusters, the Elbow Method graph is drawn as shown in Fig. 9. Three clusters formed within the sub-group are visualized as shown in Fig. 10, and the relevant data are presented in Table 4.

A conceptual drawing is developed to represent the segregation of soils depending on its type as shown in Fig. 11. The boundary values are little bit rounded to get more general conceptual drawing.

The purpose of conceptual drawings shown in Figs. 6 and 11 is to develop a chart that gives an idea of mineralogical composition of clay based on the index properties.

6 Conclusions

Following are observations when K-Mean cluster algorithm is applied on Liquid Limit and Plastic Limit:

- As the number of clusters is increased from 3 to 8, better segregation of Kaolinitic, Mixture and Montmorillonitic clays has taken place.
- When the number of cluster groups is 8, except clusters I and II all the remaining clusters are having 100% Montmorillonitic soil.
- Cluster I and II groups are having 65 soils, which are both Kaolinitic, Mixture and Montmorillonitic type, when again clustered into three sub-groups showed 57%, 20% and 5% of Kaolinitic soils; 36%, 50% and 57% of

Table 3 10 Group cluster data for variables LL and PL using K-mean

10 clusters	LL (%)		PI (%)		Composition	Total
	Min	Max	Min	Max		
I Kaolinitic = 47% Mixture = 37% Montmorillonitic = 16%	24.8	58	10.9	33	K	13
					K + CH + MI	3
					K + M	9
					K + M + MU	2
					K + MU	2
					M	6
					M + K	3
II Kaolinitic = 7% Mixture = 46.5% Montmorillonitic = 46.5%	55.6	92	24	49	K	2
					K + M	4
					K + M + MU	5
					M	13
					M + K	2
					M + K + MU	2
III Montmorillonitic = 100%	94	124.2	48.9	101	M	10
IV Montmorillonitic = 100%	165	192.2	132.5	138	M	2
V Montmorillonitic = 100%	214.5	266.8	181.72	206.4	M	3
VI Montmorillonitic = 100%	290.2	310.5	226.1	256.4	M	2
VII Montmorillonitic = 100%	328.2	391.4	294.7	334.2	M	3
VIII Montmorillonitic = 100%	452.2	522.6	401.3	467.3	M	6
IX Montmorillonitic = 100%	542	567	486.2	516.8	M	3
X Montmorillonitic = 100%	615.5	678	567.9	622.7	M	3

Mixture soil and; 7%, 30% and 38% of Montmorillonitic in the three sub-groups, respectively, which are shown as Z1, Z2 and Z3 in conceptual drawing. Also, an overlapping of plastic limit values is noticed between Z2 and Z3.

Following conclusions can be drawn when K-Mean clustering algorithm is applied by considering two variables, namely, liquid limit and plasticity index at a time:

- Better segregation of clay type is noticed when number of clusters is 10.
- Under the 10 clusters, the presence of both Kaolinitic, Mixture and Montmorillonitic clay is observed in clusters I and II whereas the remaining clusters are having Montmorillonitic clay
- Cluster I and II groups are having 66 soils, which are both Kaolinitic, Mixture and Montmorillonitic type, when again clustered into three sub-groups showed 64%, 17% and 5% of Kaolinitic soils; 27%, 50% and 50% of Mixtures and; 9%, 33% and 45% of Montmorillonitic clay in the three sub-group, respectively, that are shown as Z1, Z2 and Z3 in conceptual drawing.

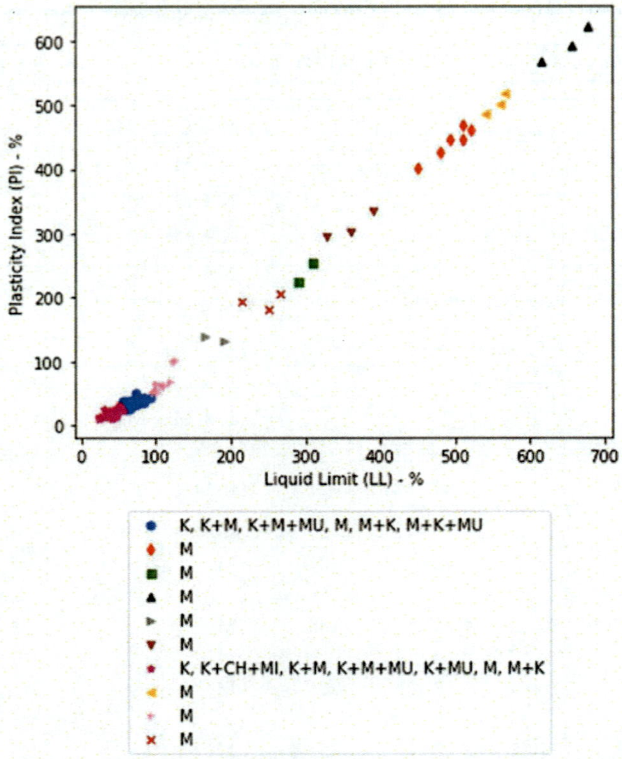

Fig. 8 LL and PL for eight clusters using K-mean

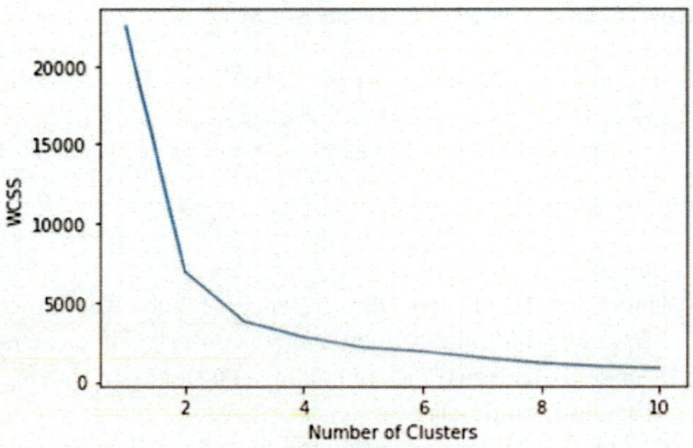

Fig. 9 Elbow method graph for LL and PI sub-group

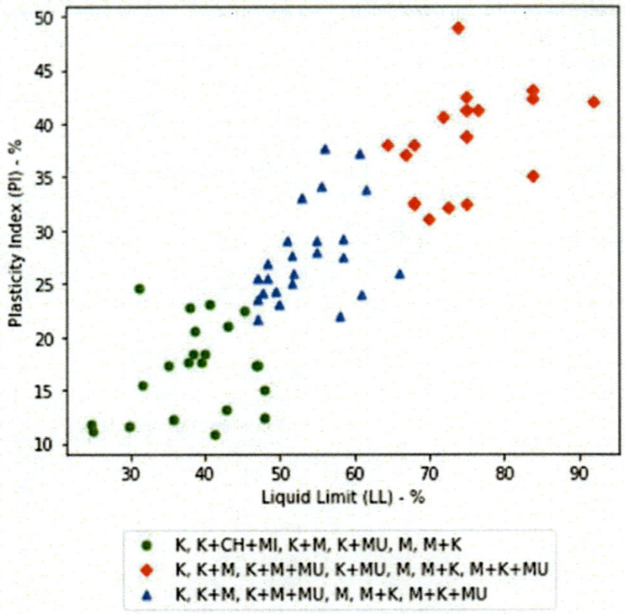

Fig. 10 LL and PI sub-group of three clusters

7 Scope for Further Study

The researchers have collected 98 clay samples that are identified mineralogically as Kaolinitic, Mixture and Montmorillonitic along with liquid limit and plastic limit test data from peer-reviewed journals. On the available data, K-Mean clustering algorithm is applied to establish boundary values for index properties to recognize the clay mineral. The current work can be extended by incorporating more soils and thereby improving the accuracy of limits. Apart from liquid limit and plastic limit, other properties such as shrinking limit, cation exchange capacity etc. may also be examined for identification of clay mineral.

Table 4 Data of LL and PI sub-group 3 clusters

Sub-group 3 clusters	LL (%)		PI (%)		Composition	Total
	Min	Max	Min	Max		
I Kaolinitic = 64% Mixture = 27% Montmorillonitic = 9%	24.8	48	10.9	24.5	K	11
					K + CH + MI	3
					K + M	4
					K + MU	1
					M	2
					M + K	1
II Kaolinitic = 17% Mixture = 50% Montmorillonitic = 33%	47	66	21.6	37.7	K	3
					K + M	6
					K + M + MU	2
					K + MU	1
					M	8
					M + K	3
					M + K + MU	1
III Kaolinitic = 5% Mixture = 50% Montmorillonitic = 45%	64.6	92	31	49	K	1
					K + M	3
					K + M + MU	5
					M	9
					M + K	1
					M + K + MU	1

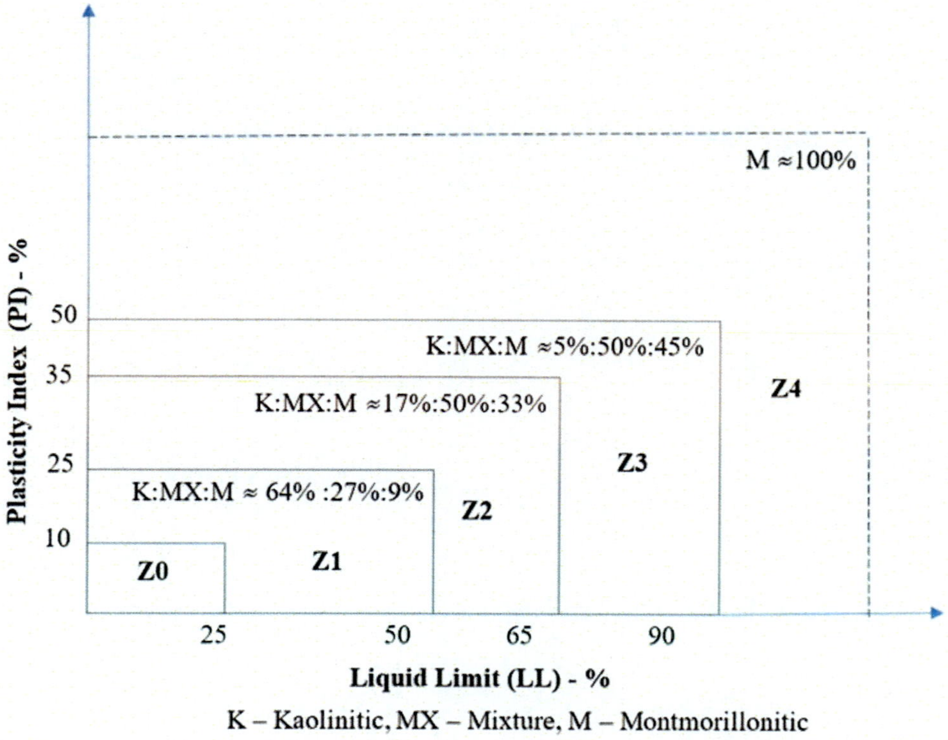

Fig. 11 Zones for clay types based on cluster analysis of LL and PI

References

Asuri, S., & Keshavamurthy, P. (2016). Expansive soil characterisation: An appraisal. *INAEL, 1*, 29–33. https://doi.org/10.1007/s41403-016-0001-9

Chavali, R. V. P., Vindula, S. K., Hari Prasad Reddy, P., Babu, A., & Pillai. R. J. (2017). Swelling behavior of kaolinitic clays contaminated with alkali solutions: A micro-level study. *Applied Clay Science Journal, 135*, 575–582. https://doi.org/10.1016/j.clay.2016.10.045

Chen. F. H. (1965). The use of piers to prevent the uplifting of lightly loaded structure founded on expansive soil. In *Concluding Proceedings Engineering Effects of Moisture Change in Soils, International Research and Engineering Conference on Expansive Clay Soils, A&M, Texas* (pp. 152–171).

Goktepe, A., Altun, S., & Sezer, A. (2015). Evaluation and use of clustering algorithms for standard penetration test data classification. *Artificial Intelligence for Engineering Design, Analysis and Manufacturing, 29*(1), 55–64. https://doi.org/10.1017/S089006041400033X

Holt, G. D. (1996). Applying cluster analysis to construction contractor classification. *Building and Environment, 31*(6), 557–568. https://doi.org/10.1016/0360-1323(96)00028-5

Holtz, W. G., & Gibbs, H. J. (1956). Engineering properties of expansive clays. *Transactions of the ASCE, 121*, 641–663.

Hot Dervic, E., & Vesna, P. (2016). Soil data clustering by using K-means and fuzzy K-means algorithm. *Telfor Journal, 8*, 56–61.

Kuraria, A., Jharbade, N., & Soni, M. (2018). Centroid selection process using WCSS and elbow method for K-mean clustering algorithm in data mining. *International Journal of Scientific Research in Science, Engineering and Technology*, 190–195. https://doi.org/10.32628/IJSRSET21841122

Lakshmi Sruthi, P., & Hari Prasad Reddy, P. (2019). Swelling and mineralogical characteristics of alkali-transformed kaolinitic clays. *Applied Clay Science Journal, 183*, 105353. https://doi.org/10.1016/j.clay.2019.105353

Mishra, A. K., Ohtsubo, M., Li, L., & Higashi, T. (2011). Controlling factors of the swelling of various bentonites and their correlations with the hydraulic conductivity of soil-bentonite mixtures. *Applied Clay Science Journal, 52*, 78–84. https://doi.org/10.1016/j.clay.2011.01.033

Prakash, K., & Sridharan, A. (2004). Free swell ratio and clay mineralogy of fine-grained soils. *Geotechnical Testing Journal, 27* (2), 220–225. https://doi.org/10.1520/GTJ10860

Qiong, L., Jie, Y., & Jinfang, Z. (2011). Application of clustering algorithm in intelligent transportation data analysis. In M. Zhu (Eds.), *Information and Management Engineering. ICCIC 2011. Communications in Computer and Information Science* (Vol. 236, pp. 467–473). Springer. https://doi.org/10.1007/978-3-642-24097-3_70

Rai, P., & Shubha, S. (2010). A survey of clustering techniques. *International Journal of Computer Applications, 7*. https://doi.org/10.5120/1326-1808

Saha, R., Tariq, M. T., Hadi, M., & Xiao, Y. (2019). Pattern recognition using clustering analysis to support transportation system management, operations, and modeling. *Journal of Advanced Transportation*. https://doi.org/10.1155/2019/1628417

Sato, K., Barast, G., Razakamanantsoa, A. R., Djeran-Maigre, I., Katsumi, T., & Levacher, D. (2016). Comparison of prehydration and polymer adding effects on Na activated Ca-bentonite by free swell index test. *Applied Science Journal, 142*, 69–80. https://doi.org/10.1016/j.clay.2016.10.009

Shah, M. V., Pandya, H. J., & Shukla, A. D. (2017). Influence of chemical additives on shrinkage and swelling characteristics of bentonite clay. *Procedia Engineering Journal, 189*, 932–937. https://doi.org/10.1016/j.proeng.2017.05.144

Snethan, D. R., Johnson, L. D., & Patrick, D. M. (1977). *An evaluation of expedient of methodology for identification of potentially expansive soils*. Soils and Pavements Laboratory. U.S. Army Engineer Waterways Experiment Station, Vicksburg, MS, Report No. FHWARE-77-94, NTIS PB-289-164.

Soni, M. T. (2012). An overview on clustering methods. *IOSR Journal of Engineering, 2*(4), 719–725.

Sridharan, A., & Prakash, K. (2000). Classification procedures for expansive soils. In *Proceedings of the Institution of Civil Engineers-geotechnical Engineering* - PROC INST CIVIL ENG-GEOTECH E (pp. 143, 235–240). https://doi.org/10.1680/geng.2000.143.4.235

Yuan, C., & Yang, H. (2019). Research on K-value selection method of K-means clustering algorithm. *J—Multidisciplinary Scientific Journal, 2*, 226–235. https://doi.org/10.3390/j2020016

Zhang, K., & Frederick, C. N. (2017). Experimental investigation on compaction and Atterberg limits characteristics of soils: Aspects of clay content using artificial mixtures. *KSCE Journal of Civil Engineering, 21*, 546–553. https://doi.org/10.1007/s12205-017-1580-z